Lumped and Distributed Passive Networks

A Generalized and Advanced Viewpoint

ELECTRICAL SCIENCE

A Series of Monographs and Texts

Edited by

Henry G. Booker
UNIVERSITY OF CALIFORNIA AT SAN DIEGO
LA JOLLA, CALIFORNIA

Nicholas DeClaris
UNIVERSITY OF MARYLAND
COLLEGE PARK, MARYLAND

JOSEPH E. ROWE. Nonlinear Electron-Wave Interaction Phenomena. 1965

MAX J. O. STRUTT. Semiconductor Devices: Volume I. Semiconductors and Semiconductor Diodes. 1966

AUSTIN BLAQUIERE. Nonlinear System Analysis. 1966

VICTOR RUMSEY. Frequency Independent Antennas. 1966

CHARLES K. BIRDSALL AND WILLIAM B. BRIDGES. Electron Dynamics of Diode Regions. 1966

A. D. KUZ'MIN AND A. E. SALOMONOVICH. Radioastronomical Methods of Antenna Measurements. 1966

CHARLES COOK AND MARVIN BERNFELD. Radar Signals: An Introduction to Theory and Application. 1967

J. W. CRISPIN, JR., AND K. M. SIEGEL (eds.). Methods of Radar Cross Section Analysis. 1968

GIUSEPPE BIORCI (ed.). Network and Switching Theory. 1968

ERNEST C. OKRESS (ed.). Microwave Power Engineering:
Volume 1. Generation, Transmission, Rectification. 1968
Volume 2. Applications. 1968

T. R. BASHKOW (ed.). Engineering Applications of Digital Computers. 1968

R. LYON-CAEN. Diodes, Transistors, and Integrated Circuits for Switching Systems. 1968

JULIUS T. TOU (ed). Applied Automata Theory. 1968

M. RONALD WOHLERS. Lumped and Distributed Passive Networks. 1969

MICHEL CUENOD AND ALLEN E. DURLING. A Discrete-Time Approach for System Analysis. 1969

In Preparation

GEORGE TYRAS. Radiation and Propagation of Electromagnetic Waves.

GEORGE METZGER AND JEAN PAUL VABRE. Transmission Lines with Pulse Excitation.

Lumped and Distributed Passive Networks

A Generalized and Advanced Viewpoint

M. Ronald Wohlers

RESEARCH DEPARTMENT
GRUMMAN AIRCRAFT ENGINEERING CORPORATION
BETHPAGE, LONG ISLAND, NEW YORK

1969

ACADEMIC PRESS New York and London

ACADEMIC PRESS, INC.
111 Fifth Avenue, New York, New York 10003

United Kingdom Edition published by
ACADEMIC PRESS, INC. (LONDON) LTD.
Berkeley Square House, London W.1

Library of Congress Catalog Card Number: 68-8426

PRINTED IN THE UNITED STATES OF AMERICA

This book is dedicated to my children
Rebecca, Jennifer, and Jessica

May they and their children always have the opportunity to experience the joys and hardships of man's ultimate concern—the pursuit of knowledge.

Preface

The initial motivation for a systematic study of passive networks came, during the 1920's, from problems that arose in the design of electric filters. Subsequent investigations focused on the limitations imposed on the performance of such filters by the components used for their construction, and on the development of systematic design procedures for obtaining filters whose terminal performance was specified a priori. It was soon realized that the basic property governing the performance of such devices was their inability to generate electrical energy. This property, which we will define later more precisely, is referred to as passivity. The fact that similar properties have been noted in many physical phenomena is the justification for our study of generalized passive networks. We use the term "generalized" to indicate the fact that the networks to be considered need not be composed of the usual collection of lumped electrical components such as resistors, capacitors, etc. In fact we will be concerned with the mathematical study of a subset of linear operators, specifically those which we will define as passive. Our primary attention will be devoted to questions of analysis and representation of such operators, but we will illustrate the results of these analyses by obtaining some of the limitations that are imposed on the performance of those systems that are passive. The inverse problem of network theory, that of finding the internal structure of a system given its terminal performance, will not receive much attention in this book with the exception of certain relatively new results in Chapter 4 concerning the realization of passive operators in the form of nonuniform transmission lines of finite length. This decision was made at the outset of the preparation of this book and is based in large part on the fact that many excellent books are now available which deal with these questions at least in the special case of lumped or rational networks. Moreover, it is the contention of this book that the systematic study of passive operators can be justified by the results obtainable from analysis alone. In fact, these results are pertinent to many areas of physics in which the inversion problem is of little or no interest. However, because the results of these studies of the inversion problem for rational networks are of considerable interest in some areas we will take the liberty of quoting some of the key results, without proof but with proper reference.

The first two chapters deal with the structure of general linear passive operators. Because we use the theory of distributions (generalized functions) in these discussions, we have collected some of the basic results of distribution theory and placed them in an appendix. The third and fourth chapters illustrate the application of passive operator theory to rational (lumped) and irrational (distributed) systems, and the fifth chapter discusses some applications of optimization theory to the study of networks.

The author's first and most lasting impressions of network theory were gained while he was a student of Dr. H. Carlin and Professor D. Youla. Many of the basic tools in network theory, including the use of bounded real matrices and the complex normalization of scattering matrices, were developed and applied to network problems by these gentlemen. Moreover, the idea that network theoretic concepts are pertinent to many areas of applied physics has its greatest proponent in Dr. Carlin. To both of these gentlemen, network theorists and the author in particular owe a great debt of gratitude.

The author would like to acknowledge the cooperation of the Research Department of the Grumman Aircraft Engineering Corporation in the preparation of this book and particularly the splendid services of Mildred Sudwischer who typed the initial manuscript. He also acknowledges the support obtained from the Air Force Office of Scientific Research under Contract AF49(638)-1512, for those investigations on which Chapter 4 of this book are based.

The encouragment received from Professor N. DeClaris during all stages of the preparation of this book and the many suggestions made by Professor R. Newcomb who read a first draft of the manuscript are both gratefully acknowledged.

Finally, the author wishes to express his thanks to his wife Jane for her unfailing moral support during the preparation of this book.

October, 1968 M. RONALD WOHLERS

Contents

I

Linear Systems

INTRODUCTION

Before embarking on the study of linear passive systems we will review some of the basic properties of the more general class of linear systems. Our purpose is twofold: first we would like to have precise definitions of certain properties that further delineate the class of linear systems, such as causality, time-invariance, etc.; and second, since we will be using the theory of generalized functions (distributions) in some of our future discussion, we want to acquaint the reader both with this theory and with the very general results than can be obtained by its use.

Throughout this book we will be using an essentially deductive approach to develop a mathematical theory for passive systems. In particular, we will be dealing with mathematical models that are intended to be accurate representations of physical systems. If along the way we are driven to a conclusion which we judge to be nonphysical then the fault will lie with the model and will only be rectifiable through an alteration of the model. This possibility is not unlikely, since in order to proceed in a reasonable fashion, one tries to make the simplest possible model for the physical situation. Moreover, it is not usually possible to determine initially whether the model will yield conclusions that are consistent with other physical requirements. Although we assume that this distinction between a physical system and its mathematical model is appreciated by the reader, we feel the necessity of reiterating it, since we will be using the theory of generalized functions in our studies and at certain points in our discussion we may find that a parameter, such as current or voltage, is a generalized function. As we will see, it will prove quite convenient to deal with these functions even though some may question their physical relevance. Moreover, we might note that unless one is willing to impose some restrictive assumptions on the types of systems being considered, and it is not at all clear what these assumptions should be or how to impose them in a reasonable manner, then one is forced to deal with generalized functions at certain points in the theory.

We have another purpose in this chapter, and that is to illustrate what we will call the network viewpoint. It is hoped that the utility of this viewpoint and an appreciation of what one could expect to gain from its application will become clear as we proceed. In lieu of trying to present an abstract discussion of what we mean by this viewpoint we will in Section 1.3 describe some specific physical situations in which we will make direct use of it.

We assume that the reader is acquainted with some of the basic facts concerning the use of transform techniques in the study of linear systems. However, we will rephrase some of these results in the more general context of distribution theory. We do not want to encumber the discussion needlessly with mathematical detail, but at the same time we want to be able to state some of the more basic conclusions in a rigorous fashion, specifically indicating their limitations.

In keeping with our desire to avoid needless detail, we will adopt a notation in which many of the better known results will still be familiar. For example, in lieu of the notation $\langle f, \varphi \rangle$ for the inner product of a distribution f and a testing function φ we will write the integral $\int f(t)\varphi(t)\,dt$. A complete discussion, at least for our purposes, of those results of distribution theory that we will make use of is given in Appendix I. We summarize here only certain facts that will be used repeatedly. Thus a distribution f is a member of the set $\mathscr{D}'$, i.e., $f \in \mathscr{D}'$, if (a) $\int f\varphi\,dt$ exists for all those testing functions $\varphi(t)$ that are infinitely differentiable and vanish outside some finite interval of t, i.e., for $\varphi \in C_0^{\infty}$, and (b) $\lim_{n\to\infty} \int f\varphi_n\,dt = 0$ for any sequence of φ_n in which each element is zero outside some fixed interval and such that the sequence (φ_n) converges to zero uniformly on that interval together with their derivatives of any order. The subset of $\mathscr{D}'$, consisting of all distributions that are zero outside some finite interval, the interval in general being different for each distribution, is designated as $\mathscr{E}'$. In a similar fashion one defines the set $\mathscr{S}'$ as all those distributions for which $\int f\varphi\,dt$ exists when $\varphi \in \mathscr{S}$ (the collection of all infinitely differentiable functions such that they and all their derivatives vanish faster than any polynomial as $t \to \pm\infty$), and such that $\lim_{n\to\infty} \int f\varphi_n\,dt = 0$ for every sequence $\varphi_n \in \mathscr{S}$ such that

$$\sup_{-\infty<t<\infty} \left| t^m \frac{d^l\varphi_n}{dt^l} \right| \to 0 \qquad \text{as} \quad n \to \infty,$$

for all m and l. This last set of distributions can be given a very concrete representation: if $f \in \mathscr{S}'$, then

$$f(t) = \frac{d^n}{dt^n}\left[(1 + t^2)^m f_0(t)\right]$$

for some m and n and some bounded continuous function $f_0(t)$. Here, however, the derivative operation is to be interpreted in a distributional sense, i.e.,

$$\int \frac{d^n g}{dt^n} \varphi \, dt = (-1)^n \int g \frac{d^n \varphi}{dt^n} \, dt$$

for all $\varphi \in \mathcal{S}$. Note that the second integral is well defined since φ and all its derivatives are members of $\mathcal{S}$. Finally, a sequence of distributions is said to converge to zero if $\lim_{n\to\infty} \int f_n \varphi \, dt = 0$ for all testing functions, e.g., if $f_n \in \mathcal{D}'$ then this must hold for all $\varphi \in C_0^{\infty}$. This convergence is referred to as weak convergence and is sometimes written as $f_n \xrightarrow{w} 0$. The other mode of convergence discussed has been defined on the set of testing functions, e.g.,

$$\lim_{n\to\infty} \int f\varphi_n \, dt \qquad \text{for} \qquad \varphi_n \in C_0^{\infty}.$$

For convenience we will refer to the testing functions and their topology or mode of convergence by a single letter. Thus by $\mathcal{D}$ we will mean the set of testing functions for the distributions $\mathcal{D}'$ including the mode of convergence to zero of such testing functions, and if we say that a sequence of testing functions converges to zero in the topology of $\mathcal{D}$ we will mean that this sequence of testing functions satisfies part (b) of the definition given above for distributions contained in $\mathcal{D}'$. On the other hand, we will say that a sequence of distributions converges to zero in the topology of, say, $\mathcal{D}'$ if $\lim_{n\to\infty} \int f_n \varphi \, dt = 0$ for all $\varphi \in C_0^{\infty}$. With this brief introduction we will now proceed with our discussion of linear systems.

1.1. THE AXIOMS OF LINEAR SYSTEM THEORY

Let us assume that in a physical phenomenon under investigation we can identify a certain number of variables—to be specific, we assume them to be functions of time—as independent parameters and another group of variables as dependent parameters. It is to be understood that the independent parameters, which we will designate as inputs, may be varied at our discretion, but that the dependent parameters, which we designate as outputs, are determined by the physical phenomenon once the inputs are specified. If we use the notation

$$\mathbf{f} = \begin{bmatrix} f_1(t) \\ \vdots \\ f_n(t) \end{bmatrix}$$

to specify a collection of n parameters, then we model the physical phenomenon by an operator T which maps inputs $\mathbf{f}$ into outputs $\mathbf{g}$ and we may state the following definition of a system:

Definition 1. A system is a unique (single-valued) mapping between a collection of inputs and the corresponding collection of outputs. Thus it is assumed that there exists some operator T such that for all inputs $\mathbf{f}$ belonging to the set D_T (the so-called domain of the operator) there exists a single $\mathbf{g}$ contained in R_T (the range of the operator) where $\mathbf{g} = T[\mathbf{f}]$.

We will have occasion to consider operator whose domain D_T consists of all $\mathbf{f}$ such that each $f_i(t) \in C_0^\infty$, and the corresponding outputs $\mathbf{g}$ (there may in general be more or less outputs than inputs) are such that each $g_i \in \mathscr{D}'$. However, depending upon our immediate purpose these domains and ranges may vary. In any event the physical content of Definition 1 is clear: we will always assume that we have identified all the possible independent parameters of the system (the inputs) so that our outputs are uniquely determined.

The definition of a linear system may now be phrased as follows:

Definition 2. A system is linear if for all $\mathbf{f} \in D_T$,

$$T[\alpha_1 \mathbf{f}_1 + \alpha_2 \mathbf{f}_2] = \alpha_1 T[\mathbf{f}_1] + \alpha_2 T[\mathbf{f}_2],$$

where α_1 and α_2 are arbitrary constants.

Unfortunately, this definition is still too broad to allow us to determine much of the structure of such operators. However, with the addition of the following definition of continuity we will be in a position to make some very definitive statements. It will be noted that this definition is meaningful only in the context of distribution theory; moreover, it is far less restrictive than any of the similar continuity requirements that may be imposed in the classical framework. The latter conclusion may be demonstrated directly by examples of specific linear operators.

Definition 3. A system is continuous on some domain D_T if for every sequence of inputs $\mathbf{f}_m \in D_T$ that converge to zero in the topology of D_T, the corresponding sequence of outputs $\mathbf{g}_m$ converge to zero in the topology of $\mathscr{D}'$, i.e., every element of $\mathbf{g}_m$ is such that $\lim_{m\to\infty} \int g_{im} \varphi \, dt = 0$ for all $\varphi \in C_0^\infty$. It is to be understood that convergence to zero of the input sequence is in the specific topology of the domain, e.g., if the domain contains all vectors $\mathbf{f}$ such that $f_i \in \mathscr{D}$, then convergence is in the topology of $\mathscr{D}$, whereas if $f_i \in \mathscr{E}'$ (the inputs are distributions), then weak convergence is implied, since that is the equivalent topology of $\mathscr{E}'$.

We are now in a position to obtain our first major result, although, as we will note, the crux of the argument rests upon a result in distribution theory which we will have to accept since its proof would take us too far afield.

Theorem 1.1. The output of a linear continuous system whose domain D_T contains $\mathscr{D}$ and whose range is $\mathscr{D}'$ may be expressed for all inputs $\mathbf{f} \in C_0^{\infty}$ as

$$\mathbf{g} \equiv T[\mathbf{f}] = \int [h(t, \tau)]\mathbf{f}(\tau)\, d\tau,$$

where $[h(t, \tau)]$ is a matrix of two-dimensional distributions in $\mathscr{D}'$. If $\mathbf{f}$ contains n elements and $\mathbf{g}$ contains m elements, then this matrix is $m \times n$.

Proof. We note first that we may reduce the problem to a consideration of a scalar input, since any input $\mathbf{f}$ may be written as

$$\mathbf{f} = \begin{bmatrix} f_1 \\ 0 \\ \vdots \\ 0 \end{bmatrix} + \begin{bmatrix} 0 \\ f_2 \\ 0 \\ \vdots \\ 0 \end{bmatrix} + \cdots + \begin{bmatrix} 0 \\ \vdots \\ 0 \\ f_n \end{bmatrix} = \mathbf{f}_1 + \cdots + \mathbf{f}_n,$$

and by the linearity of the system we obtain

$$\mathbf{g} = \begin{bmatrix} g_{11} + g_{12} + \cdots + g_{1n} \\ \vdots \\ g_{m1} + g_{m2} + \cdots + g_{mn} \end{bmatrix} = \begin{bmatrix} g_{11} \\ \vdots \\ g_{m1} \end{bmatrix} + \cdots + \begin{bmatrix} g_{1n} \\ \vdots \\ g_{mn} \end{bmatrix},$$

where g_{jk} is that portion of the jth element of the output that is produced by the kth input. Now, if we show that

$$g_{jk} = \int K(t, \tau) f_k(\tau)\, d\tau$$

for some two-dimensional distribution $K(t, \tau)$, then we may identify $K = h_{jk}$ and then form the matrix $[h(t, \tau)]$ to complete the proof. Now, consider

$$N(f_k, \beta) \equiv \int g_{jk} \beta\, dt \equiv \int T_j[\mathbf{f}_k]\beta(t)\, dt,$$

where $\beta(t) \in C_0^{\infty}$,

$$\mathbf{f}_k = \begin{bmatrix} 0 \\ \vdots \\ f_k \\ \vdots \\ 0 \end{bmatrix}$$

and $T_j[\mathbf{f}_k]$ denotes the jth element of the output due to $\mathbf{f}_k$. Since the system is linear, then for a fixed β, $N(f_k, \beta)$ is a linear operator on f_k. Moreover, if we fix f_k, then we see by its definition that N is linear in β. In addition, N is continuous with respect to β (for a fixed f_k) in the topology of $\mathscr{D}$, since by assumption the operator maps the inputs into outputs contained in $\mathscr{D}'$, i.e., $T_j[f_k] \in \mathscr{D}'$. If we consider a sequence $(f_k)_n$ that converges to zero in the topology of $\mathscr{D}$, then, since the system is assumed to be continuous in $\mathscr{D}$, the corresponding sequence of outputs $(g_{jk})_n$ converge weakly to zero in the sense that $\lim_{n\to\infty} \int (g_{jk})_n \beta \, dt = 0$ for all $\beta \in C_0^\infty$. We have thus shown that the assumptions of the theorem imply that $N(f_k, \beta)$ is a bilinear continuous functional on $\mathscr{D}$, i.e., it is linear in each parameter separately, and it is continuous in the topology of $\mathscr{D}$, again in each variable separately. We now appeal to one of the more powerful results of distribution theory, namely the kernel theorem of Schwartz (see [Sc2] or [Ge3]), which states that there exists a unique two-dimensional distribution $K(x, y)$ contained in $\mathscr{D}'$ which represents such bilinear functionals in the form

$$\begin{aligned} N(f_k, \beta) &= \iint K(y, x) f_k(x) \beta(y) \, dx \, dy \\ &= \int \beta(y) \, dy \int K(y, x) f_k(x) \, dx \\ &= \int f_k(x) \, dy \int K(y, x) \beta(y) \, dy. \end{aligned}$$

Thus we may finally identify

$$g_{jk}(t) = \int K(t, \tau) f_k(\tau) \, d\tau$$

or $K(t, \tau) = h_{jk}(t, \tau)$. QED

We observe that the matrix $[h]$ obtained in this representation has a simple physical significance if delta functions are among the allowed inputs to the system. Thus if we let

$$\mathbf{f} = \begin{bmatrix} 0 \\ \cdot \\ \delta(t-a) \\ \cdot \\ 0 \end{bmatrix},$$

where $f_k = \delta(t - a)$, then

$$\mathbf{g} = \int [h(t, \tau)] \begin{bmatrix} 0 \\ \cdot \\ \delta(\tau - a) \\ \cdot \\ 0 \end{bmatrix} d\tau = \begin{bmatrix} h_{1k}(t, a) \\ \vdots \\ h_{mk}(t, a) \end{bmatrix},$$

or $h_{jk}(t, a)$ is the jth element of the output when the input consists of a single nonzero element—namely, $f_k = \delta(t - a)$.

To this point the linear continuous systems under consideration may have time as one of the independent parameters necessary to specify it. For example, if $T[f] = \alpha(t)\mathbf{f}$, where $\alpha(t)$ is an arbitrary continuous function of time, then one may show directly that T is both linear and continuous, but obviously the output $\mathbf{g}$ due to the input $\mathbf{f}(t + a)$ will be entirely different than the output due to the input $\mathbf{f}(t)$, which is simply a translation in time of the input $\mathbf{f}(t + a)$. In this case we say that the system is a time-variable one, and we would like to define what is then meant by a time-invariant system.

Definition 4. A time-invariant system is one for which

$$T[\mathbf{f}(t + a)] = \mathbf{g}(t + a)$$

for all possible real constants a.

Physically, such systems are translation-invariant, since the output due to a translated (in time) input is simply obtained as a translation by the same amount of the original output. For such systems we obtain a convolution representation which we state as a corollary to Theorem 1.1.

Corollary 1.1. Let the system considered in Theorem 1.1 be time-invariant. Then

$$\mathbf{g} = \int [h(t - \tau)]\mathbf{f}(\tau)\, d\tau = \int [h(\tau)]\mathbf{f}(t - \tau)\, d\tau = [h] * \mathbf{f}$$

for all $\mathbf{f}$ such that $f_i \in C_0^{\infty}$. Specifically,

$$\mathbf{g}_j(t) = \sum_{k=1}^{n} h_{jk} * f_k,$$

where the scalar convolution $h_{jk} * f_k$ is well defined for all $f_k \in C_0^{\infty}$.

Proof. Follows from Theorem 1.1, since $h_{jk}(t, \tau)$ is the response to $f_k = \delta(t - \tau)$ and, since the system is time-invariant, this output must be only a function of the difference $t - \tau$. Moreover, since $h_{jk} \in \mathscr{D}'$, we note from Appendix I that the convolution operation is well defined for all $f_k \in C_0^{\infty}$. QED.

Since the convolution representation of linear time-invariant systems is central to our subsequent discussions and the proof leading to Theorem 1.1 made use of a deep result which we did not establish here, we will prove separately the following alternative to Corollary 1.1.

Theorem 1.2. The output of a linear time-invariant system that is continuous on $\mathscr{E}'$ and whose range is contained in $\mathscr{D}'$ may be represented as

$$\mathbf{g} = [h] * \mathbf{f}$$

for all inputs $\mathbf{f} \in \mathscr{E}'$. Moreover, the system will be continuous on $\mathscr{D}$.

Proof. As in the proof of Theorem 1.1 we may reduce the argument to a scalar one, i.e., we consider $g_{jk} = T_j[f_k]$, and note that T_j is a linear time-invariant operator that is continuous on $\mathscr{E}'$. First we assume that $f_k \in C_0^\infty$, and without loss of generality we may further assume that $f_k(t)$ is zero outside the interval $0 < t < 1$. Consider the distributional input

$$r_n(t) = \frac{1}{N} \sum_{n=0}^{N} f_k\left(\frac{n}{N}\right) \delta\left(t - \frac{n}{N}\right)$$

and its corresponding output

$$g_n(t) = T_j[r_n(t)] = \frac{1}{N} \sum_{n=0}^{N} f_k\left(\frac{n}{N}\right) h\left(t - \frac{n}{N}\right),$$

where we define $T_j[\delta(t)] = h(t)$ and we have used both the linearity and time-invariance of the operator to arrive at this conclusion. Thus we see that

$$g_n = h * \left[\frac{1}{N} \sum_{n=0}^{N} f_k\left(\frac{n}{N}\right) \delta\left(t - \frac{n}{N}\right)\right] = h * r_n .$$

But

$$\lim_{N \to \infty} \int r_n(t)\varphi(t)\, dt = \lim_{N \to \infty} \frac{1}{N} \sum_{n=0}^{N} f_k\left(\frac{n}{N}\right) \varphi\left(\frac{n}{N}\right) = \int f_k(t)\varphi(t)\, dt,$$

since by assumption $f_k \in C_0^\infty$, or we observe that the sequence r_n converges weakly to f_k, i.e.,

$$r_n(t) \xrightarrow{w} f_k(t).$$

We may then conclude that

$$g_n(t) \xrightarrow{w} g_{jk} = T_j[f_k],$$

since by assumption the operator is continuous in the topology of $\mathscr{E}'$, which is the weak topology. However, as noted in Appendix I, the convolution operation is also continuous with respect to either one of its components, and we may conclude that

$$g_n = h * r_n \xrightarrow{w} h * f_k ,$$

or that

$$g_{jk} = h * f_k,$$

for some distribution $h \in \mathscr{D}'$ that is defined by $h = T_j[\delta(t)]$. We thus have shown that the convolution representation holds for any input contained in C_0^∞, and we may now extend it to the case where the input is in $\mathscr{E}'$. First, if $f_k \in \mathscr{E}'$, we form the sequence $f_k * \beta_n = u_n$, where β_n is a sequence of functions contained in C_0^∞ such that $\beta_n \xrightarrow{w} \delta(t)$. By the continuity of the convolution, $f_k * \beta_n = u_n \xrightarrow{w} f_k$. Thus we find

$$g_n = h * [f_k * \beta_n] \xrightarrow{w} h * f_k,$$

but by the assumed continuity of the linear operator T_j we also have

$$g_n = T_j[f_k * \beta_n] \xrightarrow{w} T_j[f_k] = g_{jk},$$

and we have finally that $g_{jk} = h * f_k$ for an arbitrary $f_k \in \mathscr{E}'$. The theorem follows with one final observation—namely, that since convolution is also continuous in the topology of $\mathscr{D}$,

$$\int (g_{jk})_n \varphi(t)\, dt = \int [h * (f_k)_n]\varphi\, dt \to 0$$

if the sequence $(f_k)_n \in C_0^\infty$ converges to zero in the topology of $\mathscr{D}$. QED

We should point out that the representation in Theorem 1.2 was obtained by assuming that the system was continuous on $\mathscr{E}'$, and we were then able to conclude that it also must be continuous on $\mathscr{D}$. In Corollary 1.1 we obtained the same representation assuming initially that the system was continuous on $\mathscr{D}$. But we have already noted that the resultant representation, being a convolution, is continuous on $\mathscr{E}'$. Thus both of the two initial assumptions concerning continuity lead to the same conclusion. This fact is a direct consequence of the time-invariant nature of the system and need not hold for time-variable systems. In fact, if we consider the scalar time-invariant system with an input $f \in C_0^\infty$, then the corresponding output $g = h * f$ is well defined for all t as an ordinary function. Moreover, $d^n g/dt^n = h * d^n f/dt^n$ is also defined for all t in the classic sense, since, with $f \in C_0^\infty$, $d^n f/dt^n \in C_0^\infty$. Thus the output of a linear, continuous, time-invariant system is an infinitely differentiable (in the classic sense) function. The simple example noted previously of a linear continuous system, $g = \alpha(t) f$, where $f \in C_0^\infty$ and $\alpha(t)$ is a continuous function, shows that this is not the case for time-variable systems, since g is obviously not infinitely differentiable. Moreover, this system is continuous on $\mathscr{D}$ but is not even defined for inputs $f \in \mathscr{E}'$.

Another property which we may ascribe to a particular system is that of causality or lack of anticipation. Specifically:

Definition 5. A causal system is one in which, given two arbitrary inputs that are equal for $t < t_0$, the corresponding outputs are also equal for $t < t_0$. Note that equality here is in the distributional sense, i.e., $f_1 = f_2$ for $t < t_0$ if $\int (f_1 - f_2)\varphi \, dt = 0$ for all $\varphi \in C_0^{\infty}$ that are zero for $t \geq t_0$.†

In the case of linear time-invariant systems we obtain a very simple criterion for causality, namely:

Theorem 1.3. A linear, continuous, time-invariant system whose domain contains $\mathscr{E}'$ is causal if and only if the impulse response $h(t)$ is zero for $t < 0$.

Proof. Considering the scalar case for simplicity, we note that the necessity follows immediately, since $h(t) = T[\delta(t)]$ and $\delta(t) = 0$ for $t < 0$, but $0 = T[0]$, so that $h(t) = 0$ for $t < 0$.

Now as to the sufficiency, we form the function $g_1 - g_2 = h * (f_1 - f_2)$ and obtain

$$\begin{aligned}\int (g_1 - g_2)\varphi \, dt &= \int [h * (f_1 - f_2)]\varphi \, dt \\ &= \int h(t) \, dt \int [f_1 - f_2]\varphi(t + x) \, dx.\end{aligned}$$

But if $f_1 = f_2$ for $t < t_0$, then $\int [f_1 - f_2]\varphi(t + x) \, dx = 0$ for $t \geq 0$ if $\varphi(t)$ is assumed to be zero for $t \geq t_0$. Thus if $h(t) = 0$ for $t < 0$, then $\int (g_1 - g_2)\varphi \, dt$ is zero for all such φ, i.e., $g_1 = g_2$ for $t < t_0$. QED

The assumption of causality also allows us to extend the convolution representation as follows:

Theorem 1.4. The output of a linear, continuous, time-invariant, causal system whose domain contains $\mathscr{E}'$ may be written as

$$\mathbf{g} = [h] * \mathbf{f}$$

for all $\mathbf{f}$ such that $f_i \in \mathscr{D}'_{t_0}$ (distributions in $\mathscr{D}'$ that are zero for $t < t_0$). Moreover, the corresponding outputs are contained in $\mathscr{D}'_{t_0}$.

Proof. Again after reducing the problem to a scalar one we note that if $f \in \mathscr{D}'_{t_0}$, then it may always be written as $f = \alpha f + (1 - \alpha)f$, where α is an infinitely differentiable function that is one for $t < b$, with b an arbitrary real constant,

† However, if f_1 and f_2 are ordinary functions, this is equivalent to requiring that $f_1 = f_2$ pointwise for $t < t_0$.

and zero for $t > b + \varepsilon$, with ε an arbitararily small real constant (see Appendix I for an example of such a function). Now since $f \in \mathscr{D}'_{t_0}$, then $\alpha f \in \mathscr{E}'$ and is nonzero for $t_0 < t < b + \varepsilon$; moreover, $\alpha f = f$ for $t < b$. Since we assume the system to be causal, we see that for $t < b$

$$g = T[f] = T[\alpha f] = h * \alpha f,$$

where we have correctly used Theorem 1.2, since $\alpha f \in \mathscr{E}'$. However, since b is arbitrary, we may always compute $h * f$ for any $f \in \mathscr{D}'_{t_0}$.

Finally, using the causality of the system, it follows immediately that the corresponding output $g \in \mathscr{D}'_{t_0}$. QED

Before ending this section we will introduce one final definition, that of reciprocity:

Definition 6. A system T is said to be reciprocal if with $\mathbf{g}_1 = T[\mathbf{f}_1]$ and $\mathbf{g}_2 = T[\mathbf{f}_2]$

$$\mathbf{f}_1{}^T * \mathbf{g}_2 = \mathbf{f}_2{}^T * \mathbf{g}_1$$

for all $\mathbf{f}_1$ and $\mathbf{f}_2$ whose elements are contained in $C_0{}^\infty$. Here $\mathbf{f}^T$ denotes the transpose of the vector $\mathbf{f}$, so that explicitly

$$\mathbf{f}_1{}^T * \mathbf{g}_2 = \sum_{k=1}^{n} f_{1k} * g_{2k},$$

and we note that reciprocity is only defined for those systems with an equal number of inputs and outputs.

The physical significance of this axiom is essentially one of symmetry, as we may illustrate with linear time-invariant systems in which $\mathbf{g} = [h] * \mathbf{f}$. When such a system is reciprocal we obtain

$$\mathbf{f}_1{}^T * [h] * \mathbf{f}_2 = \mathbf{f}_2{}^T * [h] * \mathbf{f}_1 = \mathbf{f}_1{}^T * [h]^T * \mathbf{f}_2$$

or

$$\mathbf{f}_1{}^T * \{[h] - [h]^T\} * \mathbf{f}_2 = 0,$$

and, since $\mathbf{f}_1$ and $\mathbf{f}_2$ are arbitrary, we have

$$h_{jk}(t) = h_{kj}(t).$$

Thus the jth output when all inputs are zero except for $f_k = \delta(t)$ is identical to the kth output when all inputs are zero except for $f_j = \delta(t)$.

1.2. TRANSFORM TECHNIQUES AND THE SYSTEM FUNCTION

It was demonstrated during the development of linear system theory that the use of Fourier and Laplace transform techniques has brought many of the properties of these systems into sharp focus. In this section we will summarize many of these results whose classical counterparts are well known. It should be noted, however, that in the context of distribution theory many of these results can be stated in both a more general and a more satisfying manner.

Assume that the class of linear systems under consideration is not only continuous, time-invariant, and causal, but that in addition their impulse responses $h(t)$ are such that $h(t)e^{-at} \in \mathscr{S}'$ for some a. In this case the convolution representation may be extended so that the output of such systems are so represented for all inputs contained in $\mathscr{S}$; in particular, if $f = e^{pt}$, where $p = \sigma + j\omega$, we find that

$$g(t) = \int h(\tau)e^{p(t-\tau)}\, d\tau = e^{pt}\int h(\tau)e^{-p\tau}\, d\tau$$

is well defined for all $\operatorname{Re} p = \sigma > a$, since

$$\int h(\tau)e^{-p\tau}\, d\tau = \int h(\tau)e^{-a\tau}e^{-(p-a)\tau}\, d\tau$$

is defined for $\operatorname{Re}(p - a) > 0$ since $h(t) = 0$, $t < 0$, and $he^{-a\tau} \in \mathscr{S}'$. Thus we have shown that e^{pt} is an eigenfunction for such operators and $\lambda = \int he^{-p\tau}\, d\tau$ are the corresponding eigenvalues, i.e., e^{pt} for $\operatorname{Re} p > a$ are solutions to $T[f] = \lambda f$ when T is linear, continuous, time-invariant, causal, and continuous. We note that the eigenvalues are the Laplace transforms of the impulse response of the system. It is this fact that makes the Laplace or, more generally, the complex Fourier transform a natural tool in the study of such operators.

The basic result of the transform technique is the following representation theorem which we may easily establish using the fact that $\mathscr{L}(u * v) = \mathscr{L}(u)\mathscr{L}(v)$ for $\operatorname{Re} p > \sigma_0$, where $\mathscr{L}$ denotes the Laplace transform, $u \exp(-\sigma_0 t) \in \mathscr{S}'$ (u is zero for $t < 0$) and $v \in \mathscr{E}'$.

Theorem 1.5. The Laplace transform $\mathbf{G}(p)$ of the output of a linear, continuous, time-invariant, causal system whose impulse response matrix $[h(t)]$ is such that $h_{jk}(t)e^{-at} \in \mathscr{S}'$ may be represented in terms of the Laplace transform $\mathbf{F}(p)$ of its input as

$$\mathbf{G}(p) = [H(p)]\mathbf{F}(p) \qquad \text{for} \quad \operatorname{Re} p > a$$

where $[H(p)] = \mathscr{L}([h(t)])$ and the inputs f_i are all assumed to be in $\mathscr{E}'$.

One could extend this theorem to more general inputs; however, since the system is assumed to be causal and any input contained in $\mathscr{D}'_{t_0}$ may be represented for $t < b$ by a distribution in $\mathscr{E}'$ (see the proof of Theorem 1.4), we see that it is essentially in its most general form.

If in lieu of the Laplace transform one had considered the Fourier transform, $\mathscr{F}$, then the above result will still follow if $a = 0$, i.e., $h_{jk} \in \mathscr{S}'$. We see this directly since $\mathscr{F}(h_{jk}) \in \mathscr{S}'$ and with $f_k \in \mathscr{E}'$, $\mathscr{F}(f_k) = F_k(\omega)$ is an infinitely differentiable function of ω. Thus

$$\mathbf{G}(\omega) = [H(\omega)]\mathbf{F}(\omega)$$

is well defined. However, if the input is not in $\mathscr{E}'$ but was, for example, in $\mathscr{S}'$, then one could not expect the result to hold, since $F(\omega) \in \mathscr{S}'$ and, as noted in Appendix I, we cannot always define the product of two distributions.

The distribution $H_{jk}(\omega)$ or the function $H_{jk}(p)$ (for $\operatorname{Re} p > a$) are generally referred to as system functions, since they uniquely characterize the systems performance. In particular, the Laplace transform $H(p)$ finds wide application in the case of causal systems, since it will be an analytic function of p for $\operatorname{Re} p > a$. In particular, we may establish (for a proof see [Be1]):

Theorem 1.6. If the impulse response matrix $[h(t)]$ of a linear, time-invariant, causal system is such that $h_{jk}(t)e^{-at} \in \mathscr{S}'$ for $a > a_0$, then $\mathscr{L}(h_{jk}) = H_{jk}(p)$ is an analytic function of p in $\operatorname{Re} p > a_0$ and there exist constants K and A depending on σ_0 such that $|H_{jk}(p)| \leq A\,|p|^k$ for all $\operatorname{Re} p \geq \sigma_0 > a_0$. Conversely, if each element of the matrix $[H(p)]$ is analytic in $\operatorname{Re} p > a_0$ and satisfies such a boundedness criterion, then it is the Laplace transform of a matrix of distributions $[h(t)]$ that are zero for $t < 0$ and such that $h_{jk}(t)e^{-at} \in \mathscr{S}'$ for $a > a_0$. Therefore $[h(t)]$ is the impulse response matrix of some causal, linear, time-invariant, continuous system.

Thus it is the analyticity of $H(p)$ that reflects the causality of the system.

We should also note that in the case of multidimensional inputs and outputs the system matrix $[H(p)]$ also directly reflects the reciprocity of a system. Thus from Definition 6 and the subsequent discussion we observe that reciprocal systems have symmetric impulse response matrices, i.e., $h_{jk}(t) = h_{kj}(t)$. Therefore the system function, which in this case is a matrix $[H(p)]$, is also symmetric.

Now if we consider the Laplace transform as a complex Fourier transform, then we might expect that in some sense $H(p)$ should become $H(j\omega)$ as $\operatorname{Re} = \sigma \to 0$. First we must require that the Fourier transform $H(\omega)$ exist, or, since $H(\omega) = \mathscr{F}(h)$, then we must require that $h \in \mathscr{S}'$. In this case we see from Theorem 1.6 that $H(p) = \mathscr{L}(h)$ will be an analytic function of p for all $\operatorname{Re} p > 0$, and we might then expect that $H(\omega)$ is the boundary value of this analytic function, the boundary being $p = j\omega$ or $\operatorname{Re} p = 0$. These questions have been investigated (see, for example, [Be1] or [Br1]), and although we will not make direct use of these results, we will summarize some of those that are related to our present studies. The basic result states that when $h(t) = 0$ for $t < 0$ and

$h \in \mathscr{S}'$ then $H(p)$ takes on its boundary value, $H(\omega)$, in a distributional sense—namely,

$$\int H(\omega)\varphi(\omega)\,d\omega = \lim_{\sigma\to 0} \int H(\sigma + i\omega)\varphi(\omega)\,d\omega,$$

where $\varphi(\omega)$ is an arbitrary member of $\mathscr{S}$, i.e., $H(p) \xrightarrow{w} H(\omega)$ in the $\mathscr{S}'$ topology. Moreover, one may show the converse; if $H(p) \xrightarrow{w} H(\omega)$ and $H(p)$ satisfies the conditions of Theorem 1.6 with $a_0 = 0$, then $H(p)$ is the Laplace transform of some $h(t)$ that is not only zero for $t < 0$ but such that $h \in \mathscr{S}'$. This bond between $H(p)$ and $H(\omega)$ allows many useful and interesting conclusions to be drawn, some of which are listed below for the case when $H(\omega)$ is contained in a subset of $\mathscr{S}'$—namely, $\mathscr{D}'_{L_2}$, which consists of those distributions that can be expressed as a finite sum of generalized derivatives of L_2 functions. We make this restriction so that the results may be given in their simplest form. Thus, if $H(\omega) \in \mathscr{D}'_{L_2}$ and $h(t)$ is zero for $t < 0$, then

$$\mathscr{L}(h) = H(p) = \frac{1}{2\pi}\int \frac{H(\zeta)}{p - j\zeta}\,d\zeta = \frac{1}{\pi}\int \frac{\operatorname{Re} H(\zeta)}{p - j\zeta}\,d\zeta$$

$$= \frac{j}{\pi}\int \frac{\operatorname{Im} H(\zeta)}{p - j\zeta}\,d\zeta \qquad \text{for} \quad \operatorname{Re} p > 0, \tag{1.1a}$$

and

$$\int \frac{H(\zeta)}{p - j\zeta}\,d\zeta = 0 \qquad \text{for} \quad \operatorname{Re} p < 0, \tag{1.1b}$$

where $H(\zeta)$ is the Fourier transform of h or the boundary value of $H(p)$; and

$$H(\omega) = (1/\pi j)H(\omega) * pv(1/\omega), \tag{1.2}$$

or

$$\operatorname{Re} H(\omega) = (1/\pi)\operatorname{Im} H(\omega) * pv(1/\omega) \tag{1.3a}$$

and

$$\operatorname{Im} H(\omega) = -(1/\pi)\operatorname{Re} H(\omega) * pv(1/\omega), \tag{1.3b}$$

where $pv(1/\omega)$ is a distribution, contained in $\mathscr{D}'_{L_2}$, defined in terms of the principal value integral

$$\int pv\left(\frac{1}{\omega}\right)\varphi(\omega)\,d\omega \equiv pv\int \frac{\varphi(\omega)}{\omega}\,d\omega \qquad \text{for all} \quad \varphi \in C_0^{\infty}.$$

Moreover, the satisfaction of either (1-1b) or (1-2) is sufficient to guarantee that $H(\omega)$ is the transform of a function that is zero for $t < 0$ if $H(\omega) \in \mathscr{D}'_{L_2}$. The first set of results, (1.1a) and (1.1b), are the generalized Cauchy integral representations of analytic functions in terms of their boundary values, whereas the second set, (1.2) and (1.3), are the Hilbert transform relations between the real and imaginary parts of the Fourier transform of "causal" time functions. The above results have classical counterparts when $H(\omega)$ is a magnitude square integrable function, i.e., $H(\omega) \in L_2$ and are discussed in [Til]. However, as simple an example as $H(p) = 1/p$ is enough to convince one that the classic results are not generally applicable. For, in this example, $H(p) = (1/p) \to (1/j\omega)$ for almost all values (a.e.) of ω, and, in particular, $\operatorname{Re} H(p) \to 0$ as $\operatorname{Re} p \to 0$ a.e. However, distributionally, $(1/p) \overset{w}{\to} \pi\delta(\omega) + pv(1/j\omega)$ in the $\mathscr{S}'$ topology and we may see directly that the above formulas will only be satisfied when this expression is used as $H(\zeta)$.

A specific result related to the aforementioned but far deeper in its significance is the following, which we present without proof but which is described in [St1] and [Be1]:

Theorem 1.7. If two functions F_1 and F_2 are analytic in $\operatorname{Re} p > 0$ and satisfy

$$\lim_{\sigma = \operatorname{Re} p \to 0} \int [F_1(\sigma + j\omega) - F_2(\sigma + j\omega)]\varphi(\omega)\, d\omega = 0$$

for all infinitely differentiable $\varphi(\omega)$ that are zero outside of some fixed interval of nonzero length, i.e., $F_1 = F_2$ on the interval, then $F_1 = F_2$ for all $\operatorname{Re} p > 0$.

Thus equality of distributional boundary values on only a finite interval of the boundary implies equality in the entire domain of analyticity. An immediate consequence of the theorem is that a function analytic in $\operatorname{Re} p > 0$ cannot have zero boundary values over any interval of nonzero length without being zero in all $\operatorname{Re} p > 0$. In the classical counterparts of this theorem one must assume that these boundary values are attained pointwise and in a continuous manner, whereas the distributional analog allows for a much more erratic mode of attaining boundary values which may in fact be distributions. We will use the following modification of Theorem 1.7 in a later chapter:

Corollary 1.2. Assume that two given functions F_1 and F_2 are analytic in $\operatorname{Re} p > 0$ and are such that their $\mathscr{S}'$ boundary values $F_1(\omega)$ and $F_2(\omega) \in \mathscr{S}'$. If, in addition, $\operatorname{Re} F_1(\omega) = \operatorname{Re} F_2(\omega)$ for all ω, then for some finite m

$$F_1(p) - F_2(p) = \sum_{n=0}^{m} a_n p^n$$

for all p.

Proof. Assume that Re $\mathscr{F}(\varphi) = 0$, and then note from the definition of the Fourier transform (Appendix I) that

$$\int \mathscr{F}(\varphi)\beta(\omega)\, d\omega = j\int \operatorname{Im} \mathscr{F}(\varphi)\beta(\omega)\, d\omega = \int \varphi(t)\, dt \int \beta(\omega)e^{-j\omega t}\, d\omega,$$

and thus

$$-j\int \operatorname{Im} \mathscr{F}(\varphi)\beta(\omega)\, d\omega = \int [\mathscr{F}(\varphi)\beta^*(\omega)]^*\, d\omega = \int \varphi^*(-t)\, dt \int \beta(\omega)e^{-j\omega t}\, d\omega$$

or $-\varphi^*(-t) = \varphi(t)$. Now if we further assume that $\varphi(t)$ is zero for $t < 0$, then the above implies that $\varphi(t)$ must be zero for all t except perhaps at $t = 0$, i.e., it is a distribution with point support at the origin. However, it is known (see, for example, [Be1]), that all such distributions are of the form

$$\varphi(t) = \sum_{n=0}^{m} a_n \frac{d^n\, \delta(t)}{dt^n},$$

where m is finite. We now apply these arguments to the function $F_1(p) - F_2(p)$ and arrive at the conclusion that

$$F_1 - F_2 = \mathscr{L}\left[\sum a_n \frac{d^n\, \delta(t)}{dt^n}\right] = \sum_{n=0}^{m} a_n p^n.$$

The requirement that $F_1(\omega)$ and $F_2(\omega)$ be the $\mathscr{S}'$ boundary values of the corresponding functions $F_1(p)$ and $F_2(p)$ was implicitly used in the above argument to guarantee the existence of functions $f_1(t)$ and $f_2(t) \in \mathscr{S}'$ that will be zero for $t < 0$ and, for example, $\mathscr{F}(f_1) = F_1(\omega)$, $\mathscr{L}(f_1) = F_1(p)$, and $F_1(p) \xrightarrow{w} F_1(\omega)$. QED

1.3. THE NETWORK VIEWPOINT—SOME ILLUSTRATIONS

In the previous sections we discussed some of the mathematical techniques available for the study of linear operators. The connection between these operators and the physical systems which they represent was the basis for our first definition—namely, the definition of "system." At the heart of this definition, and, in fact, the basis of all of mathematical physics, is the assumption that one can identify certain parameters as those which uniquely characterize a particular physical phenomenon. In many instances it is not clear which set of independent and dependent parameters, or inputs and outputs, should be selected. In this section we will discuss two types of systems, indicate how such a selection may be made, and then explore some of the basic properties of the systems using the aforementioned mathematical techniques of linear system theory. We will also discuss what we have termed the network viewpoint. Here, in lieu of a detailed consideration of all the parameters which might be required to describe a physical system, we focus our attention on only a few of these variables, assuming of course that they contain all the independent parameters of the system. This is a part of the network viewpoint, although some may call it the system viewpoint, or the "black box" approach. What generally distinguishes network theory from system theory is that a network consists of an interconnection of subsystems of a certain type or types. Moreover, if these subsystems have certain properties in common, then a network theorist is interested in determining how these properties are reflected in the performance of the composite or overall network. This aspect of the problem of network theory is one of analysis. The other aspect is represented by the problem of finding an interconnection of subsystems of a certain type that will have some prescribed overall performance; this is the inverse problem or the synthesis problem.

We will discuss two classes of physical systems to which the network viewpoint has been applied with considerable success, and we will devote individual chapters to a more detailed investigation of each of the two classes.

1.3.1. Lumped or Rational Networks

The first class of systems to be considered consists of those networks composed of the elements shown in Fig. 1.1.† Each of these elements represents a physical system whose internal behavior is not considered directly; instead, we identify parameters called voltage and current to describe the overall behavior of the device. For example, the resistor is described by a voltage $v(t)$, a current $i(t)$, and the mapping

$$v(t) = R(t)i(t).$$

The current and voltage are actually vectors whose physical directions are essentially one dimensional, being related to the element directly. By convention, the voltage $v(t)$ corresponds to a vector that is directed from the minus symbol toward the plus symbol on the element; the vector direction of current is shown directly by the arrow on the element. For any of the subsystems or elements in the first column of Fig. 1.1 (called two-terminal or one-port elements) we may identify either $i(t)$ or $v(t)$ as input or independent variable and the other variable then becomes the output or dependent variable. The elements in the second column have four terminals, which may be grouped into two sets forming two ports. For these elements various combinations

† We exclude from our discussion of lumped networks such well-known lumped elements as "controlled sources," since their study is more pertinent if active networks are being investigated.

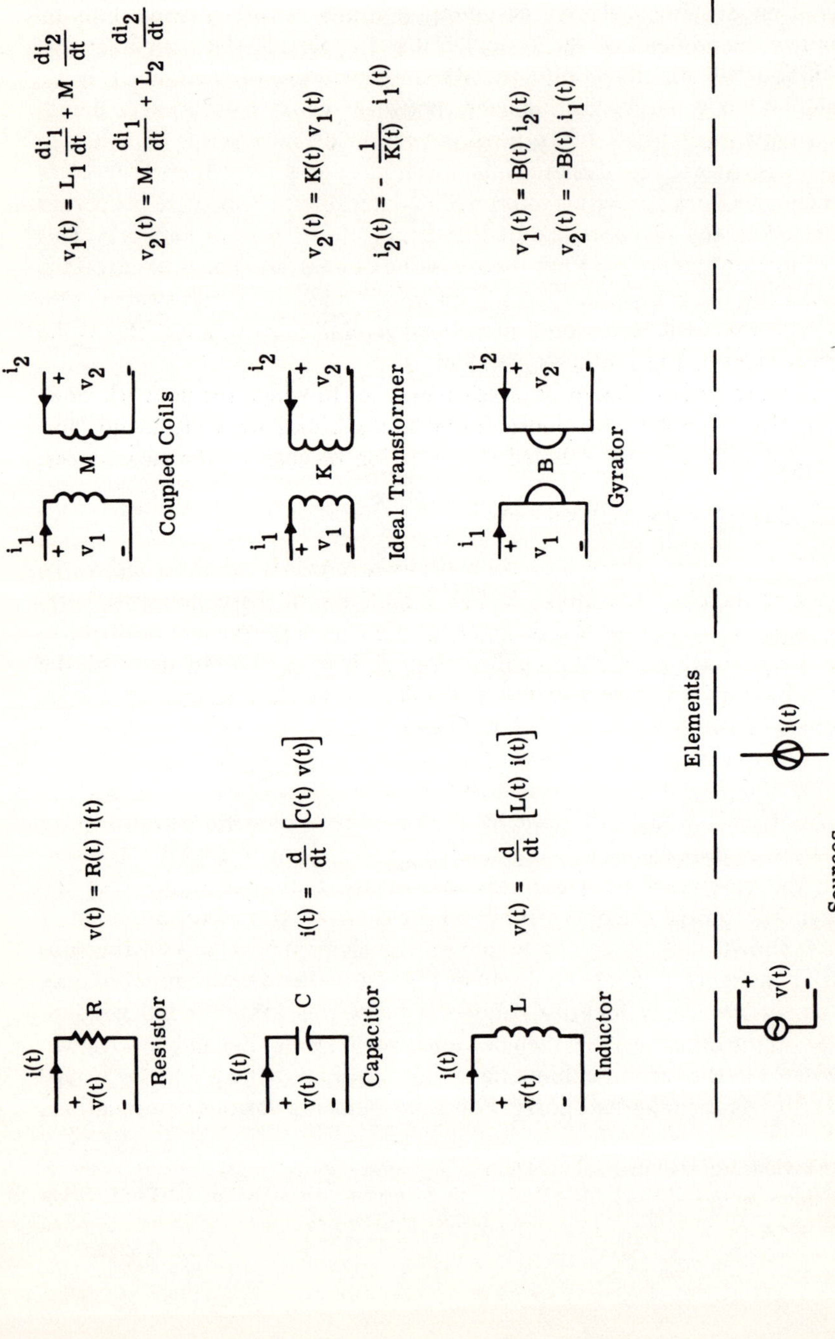

FIG. 1.1 Basic lumped elements and sources.

of excitation (v_1, v_2), (i_1, i_2), (v_1, i_2), etc. are possible, except that the definitions of an ideal transformer or gyrator allow for only certain of these combinations. One may now form an interconnection of all of these elements, a network, and then represent the overall excitation of such a network by the voltage and current sources shown in Fig. 1.1. In particular, if such a network can be excited by n sources, the network is referred to as an n-port. The two laws governing the resultant voltages and currents which are caused to flow in the network by the independent sources are known as Kirchhoff's laws and may be stated as

$$\sum_{\text{node}} i_j \alpha_j = 0 \qquad \text{and} \qquad \sum_{\text{loop}} v_j \beta_j = 0.$$

The first statement is applicable at any node or point in the network where elements are connected together; α_j is $+1$ if the direction of current in a particular element is toward the node and -1 if directed away from the node, and the sum is to be taken over all currents either entering or leaving the particular node. The second statement is applicable around any closed directed path or loop through the network on which one moves from one node to another ending at the same node at which the path originates; β_j is $+1$ if the vector direction of the voltage is in the same direction of the path and -1 if opposed, and the sum extends over all voltages encountered in the loop. By applying these two laws to a specific network we arrive at a set of first-order differential equations in the unknown currents and voltages with the prescribed source terms as forcing functions of the equations. Questions relating to systematic methods for arriving at a sufficient number of independent equations such that unique solutions are possible will not be discussed here, but we refer the reader to, e.g., [Ca2]. Now, if all the elements in a particular network are time-invariant, i.e., $R(t)$, $C(t)$, etc. are constants, then the set of first-order differential equations describing the currents and voltages are not only linear but have constant coefficients. By applying Fourier transforms we may then reduce the solution of these equations to an algebraic problem, since each derivative operation transforms into a multiplication by $j\omega$; $\sigma + j\omega$ if the Laplace transform is used. In particular, we obtain from these equations a characterization of the "n-port" in terms of its system matrix, i.e.,

$$\mathbf{G}(\omega) = [H(\omega)]\mathbf{F}(\omega),$$

where $\mathbf{F}$ are the transforms of the inputs and $\mathbf{G}$ the transforms of the outputs and $[H(\omega)]$ is a matrix all of whose elements are ratios of polynomials in $j\omega$. The fact that $[H(j\omega)]$ is rational in $j\omega$ has prompted some to refer to these networks as rational ones. Now, in those cases where $\mathbf{F} = \mathbf{I}$ and $\mathbf{G} = \mathbf{V}$, or vice versa, then the system matrix is called an immittance matrix. In particular, when $\mathbf{V} = [Z]\mathbf{I}$ or $\mathbf{I} = [Y]\mathbf{V}$, then the immittance matrices $[Z]$ and $[Y]$ are referred to, respectively, as the impedance and admittance matrices of the network. Unfortunately, an arbitrary network may not possess an immittance matrix representation—for example, a single transformer. However, if we assume that the excitation of the transformer is of the form $e_1(t) = v_1(t) + i_1(t)$ and $e_2(t) = v_2(t) + i_2(t)$, which corresponds to the series combination of a voltage source and a resistor of unit value connected to both ports of

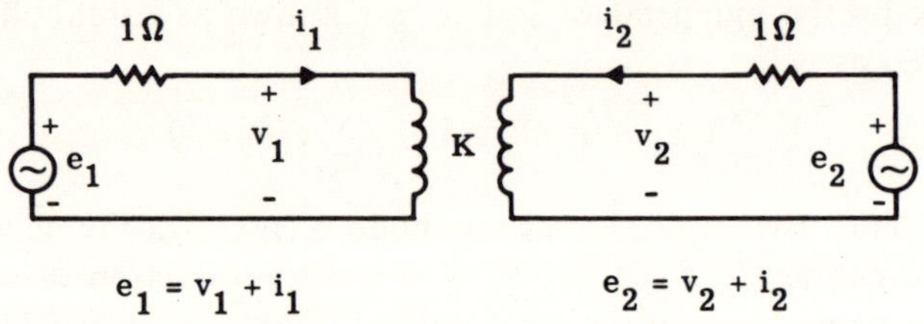

FIG. 1.2 Ideal transformer with arbitrary excitation.

the transform (see Fig. 1.2), then we find that its performance is characterized by

$$\begin{bmatrix} i_1 \\ i_2 \end{bmatrix} = \begin{bmatrix} \dfrac{K^2}{K^2+1} & \dfrac{-K}{K^2+1} \\ \dfrac{-K}{K^2+1} & \dfrac{1}{K^2+1} \end{bmatrix} \begin{bmatrix} e_1 \\ e_2 \end{bmatrix}.$$

We will find in Chapter III that such a characterization is always possible for those networks composed of the elements shown in Fig. 1.1 if all the variables defining the elements are real constants and all the R's, L's, C's, and $(L_1L_2 - M^2)$ terms are positive numbers. In Chapter III we will study the properties of such networks, describe some of their applications, and also note that a complete realizability theory is available for them, expressed solely in terms of their terminal performance. The basic property which leads to a complete characterization of such networks is that for each element of the network

$$\int_{-\infty}^{t} [(v_1 + i_1)^2 - (v_1 - i_1)^2 + (v_2 + i_2)^2 - (v_2 - i_2)^2]\, d\eta$$
$$= 4 \int_{-\infty}^{t} v_1 i_1 + v_2 i_2 \, d\eta > 0$$

for all t, where in the case of the one-port elements $v_2 = i_2 = 0$. We note that the physical implication of the nonnegativity of these integrals is that it implies that these elements are not capable of generating electrical energy. Because of this we will refer to them as passive elements.

1.3.2. Distributed or Irrational Networks

There are many physical phenomenon whose description depends upon physical spatial coordinates in addition to time. The classic example in electrical phenomena is the propagation of electromagnetic fields. In these cases the physical phenomenon requires sets of partial differential equations for its description. In particular, the dependence of the system parameters on spatial coordinates prompts us to classify them as distributed systems, as opposed to the lumped systems of Section 1.3.1, whose description is given entirely as a function of time. Because the study of electromagnetic fields is of great practical importance, we will use it as representative of the class of distributed systems.

To be specific, we consider a region of space inside a perfectly conducting hollow surface or envelope. In addition, we assume that we have hollow pipes or waveguides, also of perfectly conducting material, piercing the envelope. Through these waveguides we will introduce electromagnetic fields into the envelope, they will interact, and we will measure the resulting fields that are reflected back out of the pipes. It will be assumed that the interaction of the electromagnetic fields with the material present in the region enclosed by the envelope and the waveguides leading into it is characterized by Maxwell's equations in the form

$$\nabla \times \mathbf{E}(\mathbf{r}, t) = -\frac{\partial}{\partial t}[\mu H(\mathbf{r}, t)],$$

$$\nabla \times \mathbf{H}(\mathbf{r}, t) = \sigma \mathbf{E}(\mathbf{r}, t) + \frac{\partial}{\partial t}[\varepsilon \mathbf{E}(\mathbf{r}, t)]$$

where $\mathbf{E}(\mathbf{r}, t)$ and $\mathbf{H}(\mathbf{r}, t)$ are the three-dimensional electric and magnetic field intensities as functions of a spatial coordinate $\mathbf{r}$ and time t, and the material is characterized by the parameters μ, σ, and ε (its permeability, conductivity, and permittivity, respectively) which may be functions of both position and time. In a specific problem both the geometry of the envelope and the waveguides and the variation of the material properties (μ, σ, ε) are specified. One must also describe the form of the excitation used to cause the electromagnetic fields ($\mathbf{E}$ and $\mathbf{H}$) to exist in the region. If these sources of excitation are external to the envelope and their effects are introduced to it through the hollow waveguides, then we might expect to appy the network viewpoint to the study of the interaction of such sources if we can identify certain parameters as inputs and others as outputs. In the following discussion of the allowable fields inside a hollow waveguide we will see that certain parameters may be so identified in a natural manner.

We consider as a very general model of these waveguides a region of space surrounded by a cylindrical, perfectly conducting surface with the property that we may choose a coordinate system in which the z axis lies along the cylinder such that the cross section of the waveguide or its intersection with planes perpendicular to the z axis does not change as a function of z. Moreover, we will assume for simplicity that the material filling the waveguide is described by parameters μ, σ, and ε that are independent of time and may be functions only of z. We may now expand the most general electromagnetic field inside such a waveguide into a set of eigenfunctions of the following type, with $\mathscr{F}(\,)$ the Fourier transform (this is also referred to as a model analysis of the fields):

1. TEM modes, in which $\mathscr{F}(\mathbf{E}) = \mathscr{E}(z, \omega)\mathbf{e}(x, y)$ and $\mathscr{F}(\mathbf{H}) = \mathscr{H}(z, \omega)$ $\mathbf{h}(x, y)$.
2. TE modes, in which $\mathscr{F}(\mathbf{E}) = \mathscr{E}(z, \omega)\mathbf{e}(x, y)$ and $\mathscr{F}(\mathbf{H}) = \mathscr{H}(z, \omega)\mathbf{h}(x, y)$ $+\mathbf{z}_0 H_z$.
3. TM modes, in which $\mathscr{F}(\mathbf{E}) = \mathscr{E}(z, \omega)\mathbf{e}(x, y) + \mathbf{z}_0 E_z$ and $\mathscr{F}(\mathbf{H}) =$ $\mathscr{H}(z, \omega)\mathbf{h}(x, y)$.

The interrelationship between the variables defining each mode may be obtained in essentially the same manner as we will now describe for the case of TE modes. Thus with $\mathscr{F}(\mathbf{E}) = \mathscr{E}(z, \omega)\mathbf{e}(z, y)$ and $\mathscr{F}(H) = \mathscr{H}(z, \omega)\mathbf{h}(x, y)$ $+ \mathbf{z}_0 H_z$ we return to Maxwell's equations and after taking Fourier transforms we obtain by direct substitution

$$\nabla \times \mathscr{F}(\mathbf{E}) = \mathbf{z}_0 \times \mathbf{e}\frac{\partial \mathscr{E}}{\partial z} + \mathscr{E}\nabla \times \mathbf{e} = -j\omega\mu\mathbf{h}\mathscr{H} - j\omega\mu\mathbf{z}_0 H_z$$

and

$$\nabla \times \mathscr{F}(\mathbf{H}) = \mathbf{z}_0 \times \mathbf{h}\frac{\partial \mathscr{H}}{\partial z} + \mathscr{H}\nabla \times \mathbf{h} + \nabla \times \mathbf{z}_0 H_z = (\sigma + j\omega\varepsilon)\mathbf{e}\mathscr{E}.$$

But since the terms $\nabla \times \mathbf{e}$, $\nabla \times \mathbf{h}$, and $\mathbf{z}_0 H_z$ have only $\mathbf{z}_0$ components, whereas the other terms have no $\mathbf{z}_0$ components, the above equations break into

$$\mathbf{z}_0 \times \mathbf{e}\frac{\partial \mathscr{E}}{\partial z} = -j\omega\mu\mathbf{h}\mathscr{H}, \qquad \mathscr{E}\nabla \times \mathbf{e} = -j\omega\mu\mathbf{z}_0 H_z,$$

$$\mathbf{z}_0 \times \mathbf{h}\frac{\partial \mathscr{H}}{\partial z} + \nabla \times \mathbf{z}_0 H_z = (\sigma + j\omega\varepsilon)\mathbf{e}\mathscr{E}, \qquad \text{and} \qquad \nabla \times \mathbf{h} = 0.$$

From the first equation we note that $\mathbf{z}_0 \times \mathbf{e} = \mathbf{h}$ and $\partial\mathscr{E}/\partial z = -j\omega\mu\mathscr{H}$, and since this implies that $\mathbf{z}_0 \times \mathbf{h} = -\mathbf{e}$, we also obtain

$$-\mathbf{e}\frac{\partial \mathscr{H}}{\partial z} + \nabla \times \mathbf{z}_0 H_z = (\sigma + j\omega\varepsilon)\mathbf{e}\mathscr{E}.$$

Furthermore, by taking the curl we observe that

$$\nabla \times [\mathscr{E}\nabla \times \mathbf{e}] \equiv \mathbf{z}_0 \times (\nabla \times \mathbf{e})\frac{\partial \mathscr{E}}{\partial z} + \mathscr{E}\nabla \times \nabla \times \mathbf{e}$$

$$\equiv \mathscr{E}\nabla \times \nabla \times \mathbf{e} = -j\omega\nabla \times [\mu \mathbf{z}_0 \mathscr{H}_z] \equiv -j\omega\mu\nabla \times [\mathbf{z}_0 H_z],$$

where we have noted that since $\nabla \times \mathbf{e}$ has only $\mathbf{z}_0$ components, $\mathbf{z}_0 \times (\nabla \times \mathbf{e}) = 0$, and since $\mu = \mu(z)$, i.e., not a function of x or y, $\nabla \times [\mu \mathbf{z}_0 \mathbf{H}_z] = \mu\nabla \times [\mathbf{z}_0 H_z]$. If we use the last expression as a representation for $\nabla \times [\mathbf{z}_0 H_z]$, we may substitute into the previous equation and obtain

$$\nabla \times \nabla \times \mathbf{e} = \mathbf{e}\,\frac{[+\mu\varepsilon\omega^2\mathscr{E} - j\omega\mu\sigma\mathscr{E} - j\omega\mu(\partial\mathscr{H}/\partial z)]}{\mathscr{E}}.$$

Since the term $[\mu\varepsilon\omega^2\mathscr{E} - j\omega\mu\sigma\mathscr{E} - j\omega\mu(\partial\mathscr{H}/\partial z)]/\mathscr{E}$ depends only on z, whereas $\mathbf{e}$ depends only on x and y, the last equation will possess a solution valid for all x, y, and z only if

$$\nabla \times \nabla \times \mathbf{e} = k_e \mathbf{e}$$

and

$$\frac{\mu\varepsilon\omega^2\mathscr{E} - j\omega\mu\sigma\mathscr{E} - j\omega\mu(\partial\mathscr{H}/\partial z)}{\mathscr{E}} = k_e \quad \text{or} \quad \frac{\partial\mathscr{H}}{\partial z} = -\left[\sigma + j\omega\varepsilon + \frac{k_e}{j\omega\mu}\right]\mathscr{E}.$$

In summary, we have found that a TE mode will exist only if

$$\mathbf{z}_0 \times \mathbf{e} = \mathbf{h},$$

$$\nabla \times \mathbf{h} = 0,$$

$$\nabla \times \nabla \times \mathbf{e} = k_e \mathbf{e},$$

$$\partial\mathscr{E}/\partial z = -j\omega\mu\mathscr{H},$$

$$\partial\mathscr{H}/\partial z = -[\sigma + j\omega\varepsilon + (k_e/j\omega\mu)]\mathscr{E},$$

and

$$\mathbf{z}_0 H_z = (\mathscr{E}\nabla \times \mathbf{e})/(-j\omega\mu).$$

If we note that Maxwell's equations also imply that

$$\nabla \cdot [\varepsilon\mathscr{F}(\mathbf{E})] = \nabla \cdot (\varepsilon\mathscr{E}\mathbf{e}) = \varepsilon\mathscr{E}\nabla \cdot \mathbf{e} = 0,$$

where the manipulation follows since ε and $\mathscr{E}$ are only functions of z and $\mathbf{e}$ contains no $\mathbf{z}_0$ component, we note that $\nabla \cdot \mathbf{e} = 0$, so that the equation

$$\nabla \times \nabla \times \mathbf{e} = k_e \mathbf{e}$$

may also be written as $\nabla^2\mathbf{e} + k_e\mathbf{e} = 0$ since $\nabla \times \nabla \times \mathbf{e} = \nabla(\nabla \cdot \mathbf{e}) - \nabla^2\mathbf{e}$. In the reduction of the equations to their present form we are left with the problem of solving the eigenvalue problem

$$\nabla^2 \mathbf{e} + k_e \mathbf{e} = 0$$

subject to the boundary conditions that the tangential electric field, i.e., $\mathbf{n} \times \mathbf{e}$, vanish on the curve which defines the perfectly conducting wall of the waveguide. One final point is that since $\mathbf{h} = \mathbf{z}_0 \times \mathbf{e}$, we observe that $\nabla \times \mathbf{h} = \mathbf{z}_0 \nabla \cdot \mathbf{e}$, so that specifying $\nabla \cdot \mathbf{e} = 0$ implies directly that $\nabla \times \mathbf{h} = 0$.

Similar considerations with the other two modes yield the various equations listed in Table 1.1. In particular, the z dependence of all modes is governed by the differential equations

$$\partial \mathscr{E}(z, \omega)/\partial z = -Z_0(z, \omega)\mathscr{H}(z, \omega),$$

$$\partial \mathscr{H}(z, \omega)/\partial z = -Y_0(z, \omega)\mathscr{E}(z, \omega),$$

where

$$\begin{aligned} Z_0(z, \omega) &= j\omega\mu(z) && \text{TEM and TE} \\ &= j\omega\mu(z) + \frac{k_m}{\sigma(z) + j\omega\varepsilon(z)} && \text{TM} \end{aligned}$$

and

$$\begin{aligned} Y_0(z, \omega) &= \sigma(z) + j\omega\varepsilon(z) && \text{TEM and TM} \\ &= \sigma(z) + j\omega\varepsilon(z) + \frac{k_e}{j\omega\mu(z)} && \text{TE.} \end{aligned}$$

TABLE 1.1

MODAL FIELD RELATIONS

TEM		
$\mathbf{z}_0 \times \mathbf{e} = \mathbf{h}$,	$\nabla \times \mathbf{e} = \nabla \times \mathbf{h} = 0$	
TE		We may require that
$\mathbf{z}_0 \times \mathbf{e} = \mathbf{h}$,	$\mathbf{H}_z = \mathscr{E}(z, \omega)\dfrac{\nabla \times \mathbf{e}}{-j\omega\mu}$	$\int_S \mathbf{h}_i \cdot \mathbf{h}_i^* \, ds = \int \mathbf{e}_i \cdot \mathbf{e}_i^* \, dS$
$\nabla \cdot \mathbf{e} = 0$		
$\nabla^2 \mathbf{e} + k_e \mathbf{e} = 0$		where S is the cross section, but for all combinations of modes
TM		
$\mathbf{z}_0 \times \mathbf{e} = \mathbf{h}$,	$\mathbf{E}_z = \mathscr{H}(z, \omega)\dfrac{\nabla \times \mathbf{h}}{\sigma + j\omega\varepsilon}$	$\int_S (\mathbf{e}_i \times \mathbf{h}_j^*) \cdot \mathbf{z}_0 \, dS = 0$
$\nabla \cdot \mathbf{h} = 0$		when $i \neq j$, with $\mathbf{z}_0$ the unit vector in the z direction
$\nabla^2 \mathbf{h} + k_m \mathbf{h} = 0$		

The partial differential equation given in Table 1.1 involving **e** or **h** must be solved subject to the boundary conditions that the tangential components of either **e** or $\mathbf{z}_0 \times \mathbf{h}$ vanish at the surface of the perfectly conducting waveguide. One may show that these partial differential equations will only possess such solutions for discrete k_e or k_m (thus the name "eigenvalue") and, moreover, that these constants are real and nonnegative. We have also noted in Table 1.1 that $\int_S (\mathbf{e}_i \times \mathbf{h}_j^*) \cdot \mathbf{z}_0 \, dS = 0$, where the subscripts refer to any two modes and S is the cross section of the guide. This fact may be demonstrated by first noting that

$$\begin{aligned} \nabla \cdot (\mathbf{e}_b^* \times \nabla \times \mathbf{e}_a) - \nabla \cdot (\mathbf{e}_a \times \nabla \times \mathbf{e}_b^*) &= \mathbf{e}_a \cdot \nabla \times \nabla \times \mathbf{e}_b^* \\ &\quad - \mathbf{e}_b^* \cdot \nabla \times \nabla \times \mathbf{e}_a \\ &= (k_b^* - k_a)\mathbf{e}_a \cdot \mathbf{e}_b^*, \end{aligned}$$

since for all modes $\nabla \times \nabla \times \mathbf{e} = k\mathbf{e}$, where k may be zero when the mode is TM or TEM. Now, by employing the divergence theorem, $\int_V \nabla \cdot \mathbf{A} \, dv = \oint_S \mathbf{A} \cdot \mathbf{n} \, dS$, with **n** the outward normal to the surface S surrounding the volume V, we have

$$\begin{aligned} \int_V (k_b^* - k_a)\mathbf{e}_a \cdot \mathbf{e}_b^* \, dv &= \int_V [\nabla \cdot (\mathbf{e}_b^* \times \nabla \times \mathbf{e}_a) - \nabla \cdot (\mathbf{e}_a \times \nabla \times \mathbf{e}_b^*)] \, dv \\ &= \oint_S [\mathbf{e}_b^* \times \nabla \times \mathbf{e}_a - \mathbf{e}_a \times \nabla \times \mathbf{e}_b^*] \cdot \mathbf{n} \, dS \\ &\equiv \int_S [\mathbf{n} \times \mathbf{e}_b^* \cdot \nabla \times \mathbf{e}_a - \mathbf{n} \times \mathbf{e}_a \cdot \nabla \times \mathbf{e}_b^*] \, dS. \end{aligned}$$

If we select the surface S to be the wall of the waveguide and the two ends of the waveguide formed on two planes perpendicular to the $\mathbf{z}_0$ axis, then we have $\mathbf{n} \times \mathbf{e} = 0$ from the boundary condition (tangential **e** must vanish) on the wall of the waveguide, and an equal and opposite contribution to the integral at each of the ends (the integrand is not a function of z but the normal **n** must be the outward normal). Therefore

$$\oint_S [\mathbf{n} \times \mathbf{e}_b^* \cdot \nabla \times \mathbf{e}_a - \mathbf{n} \times \mathbf{e}_a \cdot \nabla \times \mathbf{e}_b^*] \, dS = 0$$

and thus

$$(k_b^* - k_a) \int_V \mathbf{e}_a \cdot \mathbf{e}_b^* \, dv = 0.$$

But since $\mathbf{e}_a$ and $\mathbf{e}_b$ are not functions of z, we have

$$(k_b^* - k_a) \int_S \mathbf{e}_a \cdot \mathbf{e}_b^* \, dS = 0,$$

where S is the cross section of the guide. Finally, since we must also satisfy the requirement that $\mathbf{z}_0 \times \mathbf{h} = -\mathbf{e}$, we have

$$(k_b^* - k_a) \int_S (\mathbf{e}_a \times \mathbf{h}_b^*) \cdot \mathbf{z}_0 \, dS = 0,$$

so that

$$\int_S (\mathbf{e}_a \times \mathbf{h}_b^*) \cdot \mathbf{z}_0 \, dS = 0$$

unless $k_b^* = k_a$. The foregoing analysis establishes that the above relationship holds if at least one of the modes is TE, so that $k_b^* - k_a \neq 0$. The alternate case, i.e., where at least one of the two modes is TM, may be treated in essentially the same way using $\mathbf{h}$ in lieu of $\mathbf{e}$ and arriving at the condition $(k_b^* - k_a) \int_V \mathbf{h}_a \cdot \mathbf{h}_b^* \, dv = 0$ which, with $\mathbf{z}_0 \times \mathbf{e} = \mathbf{h}$, yields the same conclusion.

Once the modes have been determined, i.e., the set of eigenfunctions for $\mathbf{e}$ and $\mathbf{h}$ and their corresponding eigenvalues k_e and k_m, the remaining problem is to obtain solutions to the equations

$$\partial \mathscr{E} / \partial z = -Z_0 \mathscr{H} \qquad \text{and} \qquad \partial \mathscr{H} / \partial z = -Y_0 \mathscr{E}.$$

For simplicity, let us consider the situation in which the material parameters $(\mu, \sigma, \varepsilon)$ are all constants. We then note that the solutions to the differential equations for $\mathscr{E}$ and $\mathscr{H}$ are of the form

$$\mathscr{E}_i(z, \omega) = a_i(\omega) \exp\{-(Z_{0i} Y_{0i})^{1/2} z\} + b_i(\omega) \exp\{(Z_{0i} Y_{0i})^{1/2} z\}$$

$$\mathscr{H}_i(z, \omega) = \frac{a_i(\omega)}{(Z_{0i}/Y_{0i})^{1/2}} \exp\{-(Z_{0i} Y_{0i})^{1/2} z\} - \frac{b_i(\omega)}{(Z_{0i}/Y_{0i})^{1/2}} \exp\{(Z_{0i} Y_{0i})^{1/2} z\},$$

where Z_{0i} and Y_{0i} depend upon the specific ith mode in question. In the more general case such solutions cannot be obtained in closed form, since the differential equations for $\mathscr{E}$ and $\mathscr{H}$ have variable coefficients and we must resort to numerical or approximate techniques.

If we now return to the basic problem of determining the fields inside some envelope, then we may first establish a z coordinate system (with a fixed origin) in each of the hollow waveguides leading into the envelope, and then on the basis of the modal decomposition of the fields in the various waveguides identify certain variables as inputs and others as outputs. For example, the set $\mathscr{E}_i(0, \omega)$ in each waveguide could be considered as inputs and the corresponding $\mathscr{H}_i(0, \omega)$ as outputs. We might also designate the set $a_i(\omega)$ as inputs and $b_i(\omega)$ as outputs. In the first case we imply that the electric fields are prescribed at these given coordinate planes in each waveguide, whereas in the second case we imply that the applied fields are some combination of electric and magnetic fields such as $\mathscr{E}_i + (Z_{0i}/Y_{0i})^{1/2} \mathscr{H}_i$ which, as noted in the previous example, would specify a_i. In either case the interaction of the fields inside the envelope is then modeled as a system which relates these inputs and outputs, i.e., the system is a mapping between all of the modes in each waveguide. Unless we can assume that all but a finite number of modes in each waveguide are identically zero, the system or operator will have infinitely dimensioned input and output vectors, and the subsequent analysis would be unmanageable. However, there are some physical situations in which we may assume that only a finite number of modes are present, and for these cases the network viewpoint is indeed a fruitful one.

We will see in Chapter IV that certain properties of distributed systems of the type described above may be obtained without a detailed study of the spatial dependence of the variables. The basis for these results is again a consideration of energy. In particular, if we consider the electromagnetic fields in some region of space of volume V, then, using Maxwell's equations together with the identity $\nabla \cdot (\mathbf{A} \times \mathbf{B}) = \mathbf{B} \cdot (\nabla \times \mathbf{A}) - \mathbf{A} \cdot (\nabla \times \mathbf{B})$, we are led to the following statement of Poynting's theorem:

$$\tfrac{1}{2}\,\mathrm{Re}\int_{-\infty}^{t}\left[\int_V \nabla \cdot (E^* \times H)\,dv\right]dt = -\int_{-\infty}^{t}\left[\int_V \sigma|E|^2\,dv\right]dt - \tfrac{1}{2}\int_V \varepsilon|E|^2\,dv - \tfrac{1}{2}\int_V \mu|H|^2\,dv \le 0.$$

We have assumed that μ, σ, and ε are independent of time and are positive real functions of the spatial coordinates. Now if we consider that specific volume of space enclosed by the metallic envelope (discussed above) and those portions of the hollow waveguides interior to the reference planes in each waveguide, then using the fact that $\int_V \nabla \cdot (\mathbf{E}^* \times \mathbf{H})\,dv = \oint_S (\mathbf{E}^* \times \mathbf{H}) \cdot dS$, where S is the closed surface enclosing the volume, we obtain

$$\int_V \nabla \cdot (E^* \times H)\,dv = -\sum_{\text{all guides}}\ \sum_{\text{all modes}} v_j^*(t)\,i_j(t),$$

where $\mathscr{F}[v_j(t)] = \mathscr{E}_j(0, \omega)$ and $\mathscr{F}[i_j(t)] = \mathscr{H}_j(0, \omega)$. Note that we have imposed the boundary condition $\mathbf{n} \times \mathbf{E} = 0$ on the envelope and we have made use of the fact that $\int (\mathbf{e}_i \times \mathbf{h}_j^*) \cdot d\mathbf{S} = 0$ $(i \neq j)$ over the cross section of each hollow waveguide, where $\mathbf{e}_i$ and $\mathbf{h}_j$ correspond to any two of the normal modes. Thus Poynting's theorem implies that

$$\mathrm{Re}\int_{-\infty}^{t}\left[\sum_{\text{all guides}}\ \sum_{\text{all modes}} v_j^*(\eta)\,i_j(\eta)\right]d\eta \ge 0$$

for all t, and we will again say that the system is passive, i.e., the electromagnetic fields are not gaining energy from the material filling the volume V of the envelope which consitutes the system under consideration.

1.4. SUMMARY

We have presented some of the basic axioms of system theory by giving precise definitions of such terms as system, linearity, time-invariance, causality, and continuity. Moreover, we have shown that the theory of distributions or generalized functions provides a rigorous formalism in which to state these definitions and then study the resulting structure of such systems.

An attempt was made to illustrate what we will refer to as the "network viewpoint." Although some may argue that there is nothing so unusual with this viewpoint that warrants the prefix "network," we will continue to refer to it by that name. In any event, the approach we have indicated to the study of physical phenomenon, by whatever name it is called, is one in which attention is directed to the external behavior of a system, and is facilitated by a proper definition of independent and dependent parameters which describe the system.

Our primary goal in this book will be the development of a theory of passive systems. Our approach will consist of exploring the logical conclusions of the requirement that the input and output variables satisfy a constraint of the form

$$\int_{-\infty}^{t} fg \, d\eta = \tfrac{1}{4} \int_{-\infty}^{t} (f+g)^2 - (f-g)^2 \, d\eta \geq 0.$$

The next chapter is devoted to a study of such systems with little or no structure assumed other than linearity and time-invariance. We will return in later chapters to a detailed consideration of the specific systems discussed in Section 1.3.

II

Passive Systems

INTRODUCTION

Our purpose in this chapter is to discuss that particular class of linear systems which we shall call passive. Since we wish to obtain our results in as general a form as is possible, we will not restrict ourselves to specific systems such as the ones discussed in Section 1.3. The motivation for our current study is the fact that in both the lumped and distributed systems discussed in that section we found that the inputs and outputs satisfied constraints of the form $\int fg \, d\eta = \frac{1}{4} \int (f+g)^2 - (f-g)^2 \, d\eta \geq 0$. In this chapter we will consider that subset of linear systems whose inputs and outputs are so constrained, and we refer to these systems as passive systems because in many instances the physical significance of such integrals is that they are measures of energy absorbed by the system. In particular, the word passive denotes the fact that a system is incapable of generating such energy, and thus it can only absorb energy from the sources used to excite it. However, there are many systems for which constraints of this type exist even though the integrals do not directly correspond to measures of energy. Our approach in this chapter will still be applicable to such systems.

From a purely mathematical viewpoint the two integrals noted above are equivalent. However, in our subsequent studies we will see that they can lead to essentially different formalisms, since in one case we identify f as the input of the system and g as the output, whereas in the other case $f + g$ is considered as the input and $f - g$ as the output. We will compare the formalisms that arise out of both forms of the energy statement, and in addition deduce certain properties of those systems or operators that satisfy them. Moreover, we will obtain representations of such operators and illustrate the practical significance of such results.

2.1. CHARACTERIZATION OF LINEAR, TIME-INVARIANT, PASSIVE SYSTEMS

We will adopt as our definition of a passive system the following:

Definition 7 [Wo10]. A passive system is one whose domain D_T contains all n-dimensional vectors $\mathbf{a}$ such that their elements a_i are real functions of time contained in $C_0{}^\infty$; moreover, for any such input the corresponding output $\mathbf{b}$ is n dimensional, with b_i real functions of time satisfying

$$\sum_{i=1}^{n} \int_{-\infty}^{\infty} [a_i{}^2(t) - b_i{}^2(t)]\, dt = \int_{-\infty}^{\infty} [\mathbf{a}^T\mathbf{a} - \mathbf{b}^T\mathbf{b}]\, dt \geq 0.$$

We note first that the integral condition of this definition is less restrictive than requiring that $\int_{-\infty}^{t} \mathbf{a}^T\mathbf{a} - \mathbf{b}^T\mathbf{b}\, d\eta \geq 0$ for all t. Moreover, we have adopted this form of an energy constraint over the alternate $\int fg\, dt \geq 0$. With regard to the first point we see that the above definition is independent of the first six definitions of Chapter I—namely, linearity, time-invariance, continuity, causality, and reciprocity. In particular, the simple example of a linear time-invariant system with an impulse response given by $\delta(t + \alpha)$, $\alpha > 0$, demonstrates that a system may be passive in the sense of Definition 7 without being causal. We will see later that if the definition of passivity were given by means of the more restrictive integral—namely, $\int_{-\infty}^{t} \geq 0$—then any linear passive system would by necessity be causal. Since one of the major aims of any axiomatic theory is to have independent axioms, we have adopted the above as the definition of passivity. The alternative would have been to leave causality out as an explicit assumption, whereas the experience in all of the areas of mathematical physics indicates that this is probably the most basic of all physical assumptions. We will see in the case of linear time-invariant systems that the foregoing distinction is only one of pedagogical interest, since both definitions will lead to the same conclusions. However, in those instances where the system is not time-invariant or nonlinear this may not be the case, and so for the general case an independent axion for passivity is highly desirable. Finally, we will simply note at this point that the other integral form —namely, $\int fg\, dt$—will be discussed in Section 2.2 and then compared to the definition that has been adopted here.

The first result in this chapter delineates the class of linear, time-invariant, causal systems that are in addition passive. Thus we have:

Theorem 2.1. A linear, continuous, time-invariant, causal operator that is passive in the sense of Definition 7 satisfies

$$\int_{-\infty}^{t} [\mathbf{a}^{*T}\mathbf{a} - \mathbf{b}^{*T}\mathbf{b}]\, d\eta \geq 0$$

for all inputs $\mathbf{a}$ such that a_i are complex functions contained in L_2.

Proof. We must first show that the domain of the operator may be extended to include L_2. Thus, since the operator is linear, continuous, and time-invariant, we may appeal to Corollary 1.1 to conclude that

$$\mathbf{b} = [h] * \mathbf{a}$$

for all $\mathbf{a}$ such that $a_i \in C_0^\infty$. In particular, if we assume that $\mathbf{a} = \varphi\mathbf{y}$, where $\varphi \in C_0^\infty$ is a real function and $\mathbf{y}$ is an n-dimensional vector of complex constants, we obtain

$$\mathbf{b} = [h * \varphi]\mathbf{y} \qquad \text{and} \qquad \mathbf{b}^{*T} = \mathbf{y}^{*T}[h * \varphi]^T,$$

where we have noted that since by definition the system maps real inputs into real outputs, the impulse response matrix $[h]$ is such that all its elements are real. Now, since the system is passive, we see that

$$\int_{-\infty}^{\infty} [\mathbf{a}^{*T}\mathbf{a} - \mathbf{b}^{*T}\mathbf{b}]\, dt = \int_{-\infty}^{\infty} [\mathrm{Re}\,\mathbf{a}]^T[\mathrm{Re}\,\mathbf{a}] - [\mathrm{Re}\,\mathbf{b}]^T[\mathrm{Re}\,\mathbf{b}]\, dt$$
$$+ \int_{-\infty}^{\infty} [\mathrm{Im}\,\mathbf{a}]^T[\mathrm{Im}\,\mathbf{a}] - [\mathrm{Im}\,\mathbf{b}]^T[\mathrm{Im}\,\mathbf{b}]\, dt \geq 0$$

since, for example, $\mathrm{Re}\,\mathbf{b} = [h] * \mathrm{Re}\,\mathbf{a}$. Thus in the case where $\mathbf{a} = \mathbf{y}\varphi$ we have $\int_{-\infty}^{\infty} \mathbf{b}^{*T}\mathbf{b}\, dt \leq \infty$, since with $\varphi \in C_0^\infty$, $0 \leq \int_{-\infty}^{\infty} \mathbf{a}^{*T}\mathbf{a} < \infty$. If we now set each element of $\mathbf{y}$ except y_r equal to zero, we obtain

$$\int_{-\infty}^{\infty} \mathbf{b}^{*T}\mathbf{b}\, dt = |y_r|^2 \int_{-\infty}^{\infty} \sum_{j=1}^{n} (h_{jr} * \varphi)^2\, dt < \infty$$

and we conclude that

$$\int_{-\infty}^{\infty} (h_{jr} * \varphi)^2\, dt < \infty$$

for all real $\varphi \in C_0^\infty$. Thus (see Appendix I) we know that $h_{jr} \in \mathcal{D}'_{L_2}$, i.e., it may be written as a finite sum of generalized derivatives of L_2 functions; moreover, its Fourier transform $H_{jr}(\omega)$ exists and may be written as the product of a polynomial and an L_2 function. In particular, we may apply Plancherel's equality (see Appendix I) to conclude on the basis of passivity that

$$\int_{-\infty}^{\infty} \mathbf{a}^{*T}\mathbf{a} - \mathbf{b}^{*T}\mathbf{b}\, dt = \frac{1}{2\pi} \int_{-\infty}^{\infty} \mathbf{y}^{*T}[1_n - H^{*T}(\omega)H(\omega)]\mathbf{y}\, |\Phi(\omega)|^2\, d\omega \geq 0,$$

where $H(\omega)$ is the $n \times n$ matrix of Fourier transforms of the impulse response matrix $[h]$, $\Phi(\omega)$ is the transform of $\varphi(t)$, and 1_n is an $n \times n$ matrix with all elements zero except for its diagonal elements, which are unity. It is not surprising that one may then prove that

$$\mathbf{y}^{*T}[1_n - H^{*T}(\omega)H(\omega)]\mathbf{y} \geq 0$$

for all ω, as we will show rigorously by employing a basic result of distribution theory. The difficulty with making this conclusion directly is that even though $\varphi(t)$ is an arbitrary member of C_0^{∞} so that $\Phi(\omega)$, its transform, is an arbitrary C^{∞} function, $\Phi(\omega)$ is in general not a member of C_0^{∞}, i.e., it does not vanish outside some finite interval, and we must resort to a more detailed argument. To this end we use the fact that

$$\int_{-\infty}^{\infty} |a_i(t)|^2\, dt = \int_{-\infty}^{\infty} \delta(t)\, dt \int_{-\infty}^{\infty} a_i(\tau)a_i^*(\tau - t)\, d\tau = \int_{-\infty}^{\infty} \delta(t) a_i * \tilde{a}_i\, dt,$$

where we use the notation $\tilde{a}$ to denote $a^*(-t)$, and, further, that

$$\int_{-\infty}^{\infty} |(a * u)|^2\, dt = \int \delta(t)(a^* * u^*) * (a * u)\, dt = \int (u * \tilde{u})(a * \tilde{a})\, dt$$

to show directly that

$$\int_{-\infty}^{\infty} [\mathbf{a}^{*T}\mathbf{a} - \mathbf{b}^{*T}\mathbf{b}]\, dt = \int_{-\infty}^{\infty} \mathbf{y}^{*T}[\delta(t)1_n - h^T * \tilde{h}]\mathbf{y}(\varphi * \tilde{\varphi})\, dt,$$

where $h^T * \tilde{h}$ is an $n \times n$ matrix whose elements are obtained as in the ordinary matrix product except that in lieu of multiplication of terms one substitutes convolution. Note, however, that such a matrix is well defined, since each element of $[h]$ was shown to be a member of $\mathscr{D}'_{L_2}$ and, as noted in Appendix I, convolution is defined between arbitrary members of $\mathscr{D}'_{L_2}$. Now we appeal to a basic theorem of Bochner and Schwartz [Sc3, Vol. II, p. 132] which states that if $\tilde{f} = f$ and $\int f\varphi * \tilde{\varphi}\, dt \geq 0$ for all $\varphi \in \mathscr{D}$, then $\mathscr{F}[f]$ is real and $\int \mathscr{F}[f]\beta(\omega)\, d\omega \geq 0$ for all $\beta(\omega) \in \mathscr{D}$ such that $\beta(\omega) \geq 0$ for all ω, i.e., $\mathscr{F}[f]$ is a real nonnegative distribution. Applying this result, we conclude that

$$\mathscr{F}[\mathbf{y}^{*T}(\delta(t)1_n - h^T * \tilde{h})\mathbf{y}] = \mathbf{y}^{*T}[1_n - H^{*T}H]\mathbf{y}$$

must be a nonnegative distribution. However, as we noted above, each element of H is an ordinary function, so that $\mathbf{y}^{*T}[1_n - H^{*T}H]\mathbf{y} \geq 0$ for almost all ω, since a nonnegative distribution f is ≥ 0 pointwise if it is in fact an ordinary function. Now we may conclude the first portion of the proof by letting all elements of $\mathbf{y}$ except y_r equal zero to obtain

$$1 - \sum_{j=1}^{n} |H_{jr}|^2 \geq 0,$$

or $|H_{jr}(\omega)| \leq 1$ for almost all ω. Then, since $\mathscr{F}[\mathbf{b}] = \mathbf{B}(\omega) = H(\omega)\mathbf{A}(\omega)$, we see that $B_i \in L_2$ when all $A_i = \mathscr{F}[a_i] \in L_2$, or $a_i \in L_2$. Thus we may extend the domain of the operator to L_2, since for such inputs, the outputs $\mathbf{b}$ are well defined as L_2 functions, i.e., $b_i \in L_2$, since $\mathscr{F}[b_i] = B_i \in L_2$. Moreover, we see that for $a_i \in L_2$

$$\int_{-\infty}^{\infty} \mathbf{a}^{*T}\mathbf{a} - \mathbf{b}^{*T}\mathbf{b}\,dt = \int_{-\infty}^{\infty} \mathbf{A}^{*T}(\omega)[1_n - H^{*T}H]\mathbf{A}(\omega)\,d\omega \geq 0,$$

since the integrand of the second integral will be a positive function of ω, as we know, having shown that

$$\mathbf{y}^{*T}[1_n - H^{*T}H]\mathbf{y} \geq 0$$

for an arbitrary complex constant vector $\mathbf{y}$.

We are now in a position to bring the proof to a close by defining

$$\begin{aligned} \mathbf{a}_{t_0} &= \mathbf{a} \qquad \text{for} \quad t \leq t_0, \quad a_i \in L_2, \\ &= 0 \qquad \text{for} \quad t > t_0, \end{aligned}$$

and noting that the causality of the system implies that the response of the system to $\mathbf{a}_{t_0}$—namely, $\mathbf{b}_{t_0}$—must be equal to its response to $\mathbf{a}$ for all $t \leq t_0$. Thus

$$\begin{aligned} \int_{-\infty}^{t_0} \mathbf{a}^{*T}\mathbf{a} - \mathbf{b}^{*T}\mathbf{b}\,dt &= \int_{-\infty}^{t_0} \mathbf{a}_{t_0}^{*T}\mathbf{a}_{t_0} - \mathbf{b}_{t_0}^{*T}\mathbf{b}_{t_0}\,dt \\ &\geq \int_{-\infty}^{\infty} \mathbf{a}_{t_0}^{*T}\mathbf{a}_{t_0} - \mathbf{b}_{t_0}^{*T}\mathbf{b}_{t_0}\,dt \geq 0. \end{aligned}$$

Since $\mathbf{a}_{t_0}$ is zero for $t > t_0$, then $\int_{t_0}^{\infty} \mathbf{b}_{t_0}^{*T}\mathbf{b}_{t_0}\,dt \geq 0$, and we are perfectly justified in the classical treatment of the integral, since with $\mathbf{a} \in L_2$, $\mathbf{a}_{t_0}$, $\mathbf{b}_{t_0}$, and $\mathbf{b}$ will all be members of L_2. QED

It is of interest to note that the conclusion of Theorem 2.1 was used as the basic definition of passivity in the classic paper of Youla *et al.* on passive networks [Yo7]. Thus our definition of passivity must lead to precisely the same conclusions as they obtained when applied to time-invariant linear systems, and we will describe these conclusions subsequently. As we noted earlier, we did not adopt this as a definition since we claimed that it implied causality. This observation was first made by Youla, and we give a proof of it using essentially his arguments. Thus we state:

Theorem 2.2. If we assume that a linear, time-invariant, continuous system with a domain containing all $\mathbf{a}$ such that $a_i \in C_0^{\infty}$ satisfies

$$\int_{-\infty}^{t} \mathbf{a}^{T}\mathbf{a} - \mathbf{b}^{T}\mathbf{b}\,d\eta \geq 0$$

for all t and all real $\mathbf{a}$, then the system must be causal, i.e., it must satisfy the conditions of Definition 5.

Proof. For simplicity we consider immediately the scalar case, to which the general problem may be reduced using the arguments of the proof of Theorem 1.1. Now, consider two inputs contained in C_0^∞ of the form $a_1 = a_0$ and $a_2 = a_0 + \alpha\hat{a}$, where α is an arbitrary constant, $\hat{a}$ is zero for $t \leq t_0$, and therefore $a_1 = a_2$ for $t \leq t_0$. By the assumed linearity of the system we have as the corresponding outputs

$$b_1 = b_0 \qquad \text{and} \qquad b_2 = b_0 + \alpha\hat{b},$$

where b_0 is the response to a_0 and $\hat{b}$ the response to $\hat{a}$. Since the system was assumed to satisfy $\int_{-\infty}^{t} a^2 - b^2\, d\eta$ for all $a \in C_0^\infty$ and all t, then

$$\begin{aligned}\int_{-\infty}^{t} a_2^2 - b_2^2\, d\eta &= \int_{-\infty}^{t} a_0^2 - [b_0 + \alpha\hat{b}]^2\, d\eta \\ &= \int_{-\infty}^{t} [a_0^2 - b_0^2]\, d\eta - \alpha \int_{-\infty}^{t} [2b_0\hat{b} + \alpha\hat{b}^2]\, d\eta \geq 0\end{aligned}$$

for all $t \leq t_0$. However, with a_0 (and thus b_0) fixed we could always violate this inequality by picking $|\alpha|$ large enough unless $\int_{-\infty}^{t} \hat{b}^2\, d\eta = 0$ for $t \leq t_0$, or, since the integrand of the last integral is a positive infinitely differentiable function, we see that this would require that $\hat{b} = 0$ for $t \leq t_0$. But this implies that $b_1 = b_2$ for $t \leq t_0$, and since by assumption the corresponding inputs are arbitrary inputs that are equal for $t \leq t_0$, we see that the system satisfies the conditions of Definition 5, i.e., it is causal. QED

Note that we have appended the assumptions that the system is continuous and time-invariant so that the integral $\int_{-\infty}^{t} a^2 - b^2\, d\eta$ will have meaning. In this situation it certainly will, since, as noted in Chapter I, the output of such systems will actually be infinitely differentiable (in the ordinary sense) functions when the inputs are contained in C_0^∞.

Let us now return to the major thread of our argument and proceed to the establishment of the basic characterization of linear, time-invariant, passive systems, specifically:

Theorem 2.3. The necessary and sufficient conditions such that an $n \times n$ matrix $[S(p)]$ be the Laplace transform of the impulse response matrix of a linear, continuous, time-invariant, causal, and passive system are that:

1. Each element of $[S(p)]$ be analytic in $\operatorname{Re} p > 0$.
2. $[S(p^*)] = [S^*(p)]$.
3. $1_n - [S^*(p)]^T[S(p)]$ be a nonnegative definite matrix for all $\operatorname{Re} p > 0$, i.e., $\mathbf{y}^{*T}[1_n - [S^*(p)]^T[S(p)]]\mathbf{y} \geq 0$ in $\operatorname{Re} p > 0$ for all constant n-vectors $\mathbf{y}$.

Such a matrix will be called bounded-real.

Proof. Let the input of such a system be $\mathbf{a} = \mathbf{y}e^{pt}u_0(t_0 - t)$, where $\mathbf{y}$ is a constant n-vector, $p = \sigma + j\omega$, and $u_0(t)$ is the Heaviside step function, then one may compute the corresponding output $\mathbf{b}$ as $\mathbf{b} = [h] * \mathbf{y}e^{pt}$ for $t \leq t_0$, where $[h]$ is the system impulse response and we have used the fact that the system is causal. Specifically, we find that

$$\mathbf{b} = e^{pt}[H(p)]\mathbf{y},$$

since for all $\varphi \in C_0^\infty$

$$\begin{aligned}\int [h] * e^{pt}\mathbf{y}\varphi \, dt &= \int [h]\mathbf{y} \, dt \int e^{px}\varphi(t + x) \, dx \\ &= \int [h]\mathbf{y}e^{pt} \, dt \int e^{-px}\varphi(x) \, dx \\ &= [H(p)]\mathbf{y} \int e^{-px}\varphi(x) \, dx\end{aligned}$$

and the manipulations are well defined for all p such that $[H(p)]$ exists. However, as we showed in the proof of Theorem 2.1, each element of $[h]$ is contained in $\mathscr{D}'_{L_2}$, so that $H_{jk}(p)$ is defined and is an analytic function of p in $\operatorname{Re} p > 0$. Now, since each element of $\mathbf{a} = \mathbf{y}e^{pt}u_0(t_0 - t) \in L_2$ for $\operatorname{Re} p > 0$, we may appeal to Theorem 2.1 to obtain

$$\int_{-\infty}^{t_0} \mathbf{a}^{*T}\mathbf{a} - \mathbf{b}^{*T}\mathbf{b} \, dt = \mathbf{y}^{*T}[1_n - [S^*(p)]^T[S(p)]]\mathbf{y} \int_{-\infty}^{t_0} e^{2\sigma t} \, dt \geq 0$$

for all $\operatorname{Re} p > 0$, which implies that

$$\mathbf{y}^{*T}[1_n - [S^*(p)]^T[S(p)]]\mathbf{y} \geq 0$$

for $\operatorname{Re} p > 0$. Finally, since passive systems by their definition map real inputs into real outputs, the impulse response matrix consists of real distributions, so that their Laplace transforms satisfy $H_{jk}(p^*) = H_{jk}^*(p)$.

In order to establish the converse, we note first that if $1_n - [S^*(p)]^T[S(p)]$ is nonnegative-definite, then by setting all the elements of an n-vector except y_r equal to zero we obtain

$$|y_r|^2\left[1 - \sum_{k=1}^{n} |S_{kr}(p)|^2\right] \geq 0 \qquad \text{for} \quad \operatorname{Re} p > 0,$$

or each element of $[S(p)]$ must be less than one in magnitude for all $\operatorname{Re} p > 0$. Thus each element is analytic and bounded by one in $\operatorname{Re} p > 0$, so that, appealing to Theorem 1.6, we may conclude that $[S(p)]$ is the Laplace transform of some $n \times n$ matrix all of whose elements are distributions that are zero for $t < 0$. By identifying an operator with such a matrix using the convolution form, we observe that we may create a linear, time-invariant, continuous, and causal operator or system whose impulse response is the given matrix. Finally, we note that since by assumption

$$\mathbf{y}^{*T}[1_n - [S^*(p)]^T[S(p)]]\mathbf{y} \geq 0$$

for all Re $p > 0$, then

$$\lim_{\text{Re}\, p=\sigma\to 0} \mathbf{y}^{*T}[1_n - [S^*(p)]^T[S(p)]]\mathbf{y} \geq 0$$

if its exists. However, by a classic theorem of Fatou [Hil], we know that the limit exists pointwise a.e., since each element of $[S(p)]$, being bounded and analytic in Re $p > 0$, has such a limit. Thus

$$\mathbf{y}^{*T}[1_n - [S^*(\omega)]^T[S(\omega)]]\mathbf{y} \geq 0$$

for almost all ω. Therefore we may conclude that

$$\int_{-\infty}^{\infty} \mathbf{a}^{*T}\mathbf{a} - \mathbf{b}^{*T}\mathbf{b}\, dt = \int_{-\infty}^{\infty} \mathbf{A}^{*T}[1_n - [S^*(\omega)]^T[S(\omega)]]\mathbf{A}(\omega)\, d\omega \geq 0$$

for any $\mathbf{A}(\omega) = \mathscr{F}[\mathbf{a}]$ such that $a_i \in C_0^{\infty}$. QED

The utility of this theorem is that it uniquely characterizes all linear, passive, time-invariant systems by placing them in one-to-one correspondence with the set of bounded-real matrices. In particular, such matrices may be easily categorized: Conditions 1 and 2 are self-evident and we know that an $n \times n$ matrix $M(p)$, satisfying $M^{*T}(p) = M(p)$, is nonnegative-definite for some fixed value of p if and only if the determinant of every submatrix of $M(p)$ obtainable by deleting some combination of rows and their corresponding columns is nonnegative for that value of p.

Before terminating this portion of our study of passive systems we would also like to define that subset of passive systems which we designate as lossless:

Definition 8 [Yo7]. A system mapping real inputs into real outputs is said to be lossless if†

$$\int_{-\infty}^{\infty} \mathbf{a}^T\mathbf{a} - \mathbf{b}^T\mathbf{b}\, dt = 0$$

for all real **a** such that $a_i \in C_0^{\infty}$.

The physical significance of this definition is that it implies that the system is conservative, or that it can neither dissipate nor generate energy. In some sense this class is then the borderline between passive systems and nonpassive or active systems. The characterization of such systems can be obtained as a

† Note that this implies that $\int_{-\infty}^{\infty} \mathbf{a}^{*T}\mathbf{a} - \mathbf{b}^{*T}\mathbf{b}\, dt = 0$ for arbitrary **a** if $a_i \in C_0^{\infty}$.

special case of the results of Theorem 2.3, i.e., the Laplace transform of the impulse response matrix of a lossless system must be a bounded-real matrix. However, the precise characterization can only be obtained by statements made when $p = j\omega$. Thus we have:

Theorem 2.4. The necessary and sufficient condition that an $n \times n$ bounded-real matrix be the Laplace transform of the impulse response matrix of a lossless system is that its pointwise a.e. boundary value matrix $[S(\omega)]$ consist of measurable functions such that $[S^*(\omega)]^T[S(\omega)] = 1_n$ a.e.

Proof. The sufficiency of the theorem is established by employing Plancherel's equality to obtain for $a_i \in C_0^\infty$:

$$\int_{-\infty}^{\infty} \mathbf{a}^T\mathbf{a} - \mathbf{b}^T\mathbf{b}\, dt = \frac{1}{2\pi}\int_{-\infty}^{\infty} \mathbf{A}^{*T}(\omega)[1_n - [S^*(\omega)]^T[S(\omega)]]\mathbf{A}(\omega)\, d\omega = 0,$$

where $\mathbf{A}(\omega) = \mathscr{F}[\mathbf{a}]$.

The necessity may be argued by first noting that for $\mathbf{a} = \mathbf{y}\varphi$

$$\begin{aligned}\int_{-\infty}^{\infty} \mathbf{a}^{*T}\mathbf{a} - \mathbf{b}^{*T}\mathbf{b}\, dt &= \int_{-\infty}^{\infty} (\mathbf{y}^{*T}\varphi^*\mathbf{y}\varphi - \mathbf{y}^{*T}[s]^T * \varphi^*[s] * \varphi\mathbf{y})\, dt \\ &= \int_{-\infty}^{\infty} \mathbf{y}^{*T}[\delta(t)1_n - [s]^T * [\tilde{s}]]\mathbf{y}(\varphi * \tilde{\varphi})\, dt \\ &= 0,\end{aligned}$$

where $\varphi \in C_0^\infty$, $[s]$ is the impulse response matrix of the system, and we have used the arguments developed in the proof of Theorem 2.1. But if $\int_{-\infty}^{\infty} x(t)\varphi * \tilde{\varphi}\, dt = 0$, then $x * \varphi * \tilde{\varphi}$ must vanish identically, as we may show by the following reasoning. Consider

$$\begin{aligned}\int_{-\infty}^{\infty} (x * \varphi * \tilde{\varphi})(\varphi * \tilde{\varphi})\, dt &= \int_{-\infty}^{\infty} x(t)\, dt \int_{-\infty}^{\infty} (\varphi * \tilde{\varphi})(\varphi * \tilde{\varphi})_{y+t}\, dy \\ &= \int_{-\infty}^{\infty} x(t)[\overbrace{(\varphi * \tilde{\varphi}) * (\varphi * \tilde{\varphi})}]\, dt \\ &= \int_{-\infty}^{\infty} x(t)[(\varphi * \tilde{\varphi}) * (\tilde{\varphi} * \varphi)]\, dt\end{aligned}$$

and observe that if $\int_{-\infty}^{\infty} x(t)\varphi * \tilde{\varphi}\, dt = 0$ for all $\varphi \in C_0^\infty$, then, since the last integral may be written as $\int_{-\infty}^{\infty} x(t)\beta * \tilde{\beta}\, dt$, where $\beta = \varphi * \tilde{\varphi} \in C_0^\infty$, we conclude that $\int_{-\infty}^{\infty} (x * \varphi * \tilde{\varphi})(\varphi * \tilde{\varphi})\, dt = 0$ also. Applying the Bochner theorem (discussed in Theorem 2.1), we conclude that $x * \varphi * \tilde{\varphi}$ is a nonnegative-definite distribution, i.e., its Fourier transform is a nonnegative distribution. But since $\varphi \in C_0^\infty$, it must also be a continuous function. Thus $x * \varphi * \tilde{\varphi}$ is a nonnegative-definite function, and, as is known classically, such functions must be bounded by their value at $t = 0$.† Finally, we note that since

$$\int_{-\infty}^{\infty} x\varphi * \tilde{\varphi}\, dt = x * \varphi * \tilde{\varphi}\Big|_{t=0} = 0,$$

we may conclude that $x * \varphi * \tilde{\varphi} = 0$ for all t. By employing this result, we obtain directly that $\mathbf{y}^{*T}[\delta 1_n - [s]^T * [s]]\mathbf{y} = 0$, and, by taking Fourier transforms, this is equivalent to $\mathbf{y}^{*T}[1_n - [S^*(\omega)]^T[S(\omega)]]\mathbf{y} = 0$, a.e., and since $\mathbf{y}$ is arbitrary, we see that this implies that

$$1_n = [S^*(\omega)]^T[S(\omega)] \quad \text{a.e.} \qquad \text{QED}$$

2.2. PASSIVE IMMITTANCE OPERATORS

We note by comparison with the particular physical systems considered in Section 1.3.1 that the definition of passivity advanced in the previous section considers as inputs to the system $\mathbf{a} = \frac{1}{2}(\mathbf{v} + \mathbf{i})$ and as outputs $\mathbf{b} = \frac{1}{2}(\mathbf{v} - \mathbf{i})$, while the systems considered in Section 1.3.1 were defined in terms of currents $\mathbf{i}$ and voltages $\mathbf{v}$. We note from Fig. 2.1 that $\mathbf{v} + \mathbf{i}$ corresponds to the voltages of those sources applied to the system that have unit resistors in series with them. Alternatively, we may say that we are applying these sources to an augmented n-port that is formed from the original one, in which the voltages $\mathbf{v}$ and the currents $\mathbf{i}$ are defined, by adding unit resistors in series with each of its n-ports. The question which might be raised is: why not deal directly with the original n-port if indeed that is the physical systems under investigation? In particular, we might consider using as a definition of passivity

$$\int_{-\infty}^{t} \mathbf{v}^T\mathbf{i}\, d\eta \equiv \tfrac{1}{4}\int_{-\infty}^{t} (\mathbf{v} + \mathbf{i})^T(\mathbf{v} + \mathbf{i}) - (\mathbf{v} - \mathbf{i})^T(\mathbf{v} - \mathbf{i})\, d\eta$$

$$\equiv \int_{-\infty}^{t} \mathbf{a}^T\mathbf{a} - \mathbf{b}^T\mathbf{b}\, dt \geq 0,$$

but now with $\mathbf{v}$ as the independent parameters. In this section we will describe

† From the Bochner–Schwartz theorem $\mathscr{F}[x]$ is real and nonnegative; therefore in the case where $\mathscr{F}[x]$ is an ordinary function we see using the Fourier transform that

$$|x * \varphi * \tilde{\varphi}| = \frac{1}{2\pi}\left|\int \mathscr{F}[x]\,|\Phi(\omega)|^2 e^{j\omega t}\, d\omega\right| \leq \frac{1}{2\pi}\int \mathscr{F}[x]|\Phi(\omega)|^2\, d\omega$$

$$= x * \varphi * \tilde{\varphi}\,|_{t=0},$$

where $\Phi(\omega) = \mathscr{F}[\varphi]$. The more general case requires a more elaborate proof (see [Sc3]).

such a theory, compare it to the one previously developed, and then note some of its shortcomings.

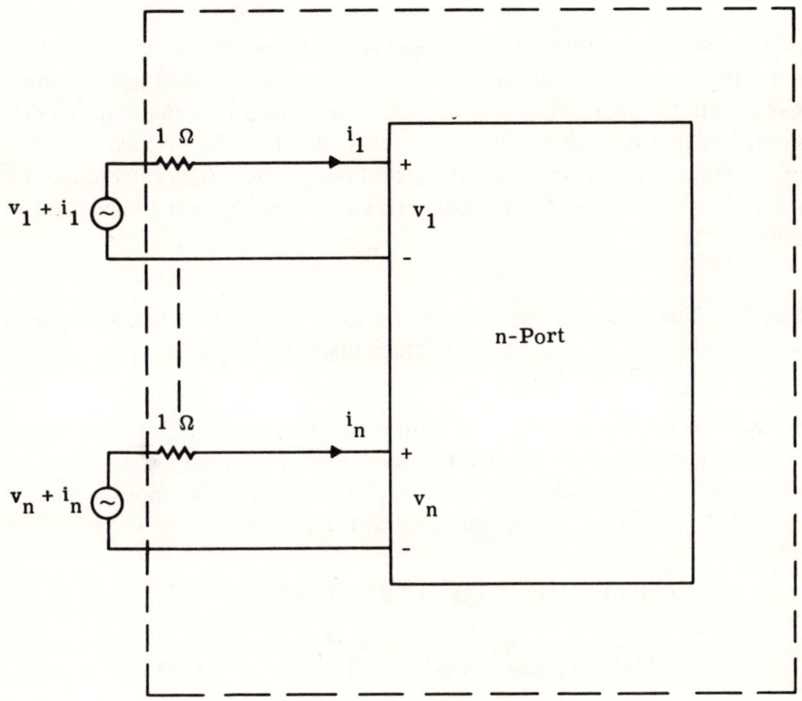

FIG. 2.1 Physical n-port and the augmented n-port.

We start by defining exactly what we mean by a passive immittance operator:

Definition 9. A linear, continuous, translation-invariant operator whose domain contains all real n-vectors $\mathbf{f}$ such that $f_i \in C_0^\infty$ is said to be a *passive immittance operator* if with arbitrary values of t

$$\int_{-\infty}^{t} \mathbf{f}^T\mathbf{g}\, d\eta \geq 0$$

for all $f_i \in C_0^\infty$, where $\mathbf{g}$, the corresponding outputs, are also real n-vectors.

It should be noted that since such operators map real $\mathbf{f}$ into real $\mathbf{g}$, then the definition also implies that

$$\int_{-\infty}^{t} \mathbf{f}^{*T}\mathbf{g} + \mathbf{g}^{*T}\mathbf{f}\, d\eta \geq 0$$

for all complex n-vectors $\mathbf{f} \in C_0^{\infty}$, i.e., n-vectors whose elements are complex-valued functions contained in C_0^{∞}. We will obtain a characterization of passive emittance operators using the proof developed by Zemanian in [Ze2]. However, we first note that using arguments similar to those described in the proof of Theorem 2.2, one may show that *passive immittance operators must be causal.* For the details of our arguments we proceed by first considering two lemmas:

Lemma 2.1. The impulse response matrix of a passive immittance operator is contained in $\mathscr{D}'_{L_\infty}$, i.e., each of its elements is contained in $\mathscr{D}'_{L_\infty}$.

Proof. Let $\mathbf{f} = \mathbf{y}\varphi$, where $\mathbf{y}$ is a constant n-vector and $\varphi \in C_0^{\infty}$. The convolution representation of the operator (we have used its linearity, continuity, and time-invariance) implies that $\mathbf{g} = [h] * \varphi\mathbf{y}$, where $[h]$ is the impulse response matrix. The "passivity" condition implies for $t \to \infty$ that

$$\int_{-\infty}^{\infty} [(\mathbf{y}^{*T}[h]^T\mathbf{y}) * \varphi]\varphi^* + [(\mathbf{y}^{*T}[h]\mathbf{y}) * \varphi^*]\varphi\, dt$$

$$= \int_{-\infty}^{\infty} \delta(t)[(\mathbf{y}^{*T}[h]^T\mathbf{y}) * \varphi * \tilde{\varphi}]\, dt + \int_{-\infty}^{\infty} \delta(t)[(\mathbf{y}^{*T}[h]\mathbf{y}) * \varphi] * \varphi\, dt$$

$$= \int_{-\infty}^{\infty} \delta(t)\{[\mathbf{y}^{*T}([\tilde{h}]^T + [h])\mathbf{y}] * (\varphi * \tilde{\varphi})\}\, dt$$

$$= \int_{-\infty}^{\infty} \{\mathbf{y}^{*T}([\tilde{h}]^T + [h])\mathbf{y}\}\varphi * \tilde{\varphi}\, dt \geq 0,$$

where we have used the identity, valid when either u or $v \in C_0^{\infty}$,

$$\int_{-\infty}^{\infty} uv^*\, dt = \int_{-\infty}^{\infty} \delta(t)u * \tilde{v}\, dt$$

and also the fact that $[h]$ is an $n \times n$ matrix of real distributions. We have already noted in the proof of Theorem 2.4 that if

$$\int_{-\infty}^{\infty} x(t)\varphi * \tilde{\varphi}\, dt \geq 0 \quad \text{for all} \quad \varphi \in C_0^{\infty}, \quad \text{then} \quad \int_{-\infty}^{\infty} (x * \varphi * \tilde{\varphi})\beta * \tilde{\beta}\, dt \geq 0$$

for all $\beta \in C_0^{\infty}$, and, using Bochner's theorem, the conclusion is that $x * \varphi * \tilde{\varphi}$ is a nonnegative-definite distribution. However, it is a continuous function (in fact, it is infinitely differentiable, but we will not need that property) and we may conclude that it must be bounded uniformly for all t —in fact, by its value at $t = 0$, as noted earlier in the proof of Theorem 2.4. Finally, by means of the identity

$$4(\alpha * \beta) = (\alpha + \tilde{\beta}) * (\tilde{\alpha} + \beta) - (\alpha - \tilde{\beta}) * (\tilde{\alpha} - \beta) \\ + j(\alpha + j\tilde{\beta}) * (\tilde{\alpha} - j\beta) - j(\alpha - j\tilde{\beta}) * (\tilde{\alpha} + j\beta),$$

we see that $x * \alpha * \beta$ must also be bounded for any α and $\beta \in C_0^\infty$, so that by a repeated application of the fact that $u * \varphi \in L_\infty$ for all $\varphi \in C_0^\infty$ implies $u \in \mathscr{D}'_{L_\infty}$ (see Appendix I), we see that $x \in D'_{L_\infty}$. Thus we have established that $\mathbf{y}^{*T}([\tilde{h}]^T + [h])\mathbf{y} \in \mathscr{D}'_{L_\infty}$ for all constant n-vectors y. Now if we select a $\mathbf{y}$ such that all its elements except y_r and y_k are zero we obtain

$$\mathbf{y}^{*T}([\tilde{h}]^T + [h])\mathbf{y} = y_r y_r^*(h_{rr} + \tilde{h}_{rr}) + y_r^* y_k(h_{rk} + \tilde{h}_{kr}) \\ + y_k^* y_r(h_{kr} + \tilde{h}_{rk}) + y_k y_k^*(h_{kk} + \tilde{h}_{kk}) \in \mathscr{D}'_{L_\infty}.$$

Setting $y_k = 0$ or $y_r = 0$ yields the conclusion that $h_{rr} + \tilde{h}_{rr} \in \mathscr{D}'_{L_\infty}$ and $h_{kk} + \tilde{h}_{kk} \in \mathscr{D}'_{L_\infty}$. Therefore we may also conclude from the above that $y_r^* y_k(h_{rk} + \tilde{h}_{kr}) + y_k^* y_r(h_{kr} + \tilde{h}_{rk}) \in \mathscr{D}'_{L_\infty}$. Now, if $y_r = y_k = 1$, then $h_{rk} + \tilde{h}_{rk} + h_{kr} + \tilde{h}_{kr} \in \mathscr{D}'_{L_\infty}$, whereas if $y_r = 1$ and $y_k = j$, then $h_{rk} - \tilde{h}_{rk} - h_{kr} + \tilde{h}_{kr} \in \mathscr{D}'_{L_\infty}$. Combining the latter two statements shows that $h_{rk} + \tilde{h}_{kr}$ is also contained in $\mathscr{D}'_{L_\infty}$. Therefore, using the fact that passive immittance operators must be causal, we know that each element of $[h]$ is zero for $t < 0$ and thus the sums, for example $h_{rk} + \tilde{h}_{kr}$, reflects the entire behavior of the distribution except possibly for its behavior at $t = 0$, which may consist of delta functions and their derivatives, which could possibly cancel in the sum. However, such delta functions are also members of $\mathscr{D}'_{L_\infty}$, and we may conclude finally that the four elements h_{rr}, h_{kk}, h_{kr}, and $h_{rk} \in \mathscr{D}'_{L_\infty}$. Since k and r were arbitrary we have established the desired result. QED

Lemma 2.2. A passive immittance operator mapping real $\mathbf{f}$ into real $\mathbf{g}$ yields

$$\int_{-\infty}^{t} \mathbf{f}^{*T}\mathbf{g} + \mathbf{g}^{*T}\mathbf{f}\, d\eta \geq 0$$

for all $\mathbf{f}$ such that $f_i \in \mathscr{S}$ and for all t.

Proof. Lemma 2.1 established the fact that each element of the impulse response matrix $[h]$ of such operators is contained in $\mathscr{D}'_{L_\infty}$. Thus $[h] * \mathbf{f}$ is defined for all $\mathbf{f}$ such that $f_i \in \mathscr{S}$, as we may conclude in the following way. Consider the scalar convolution $u * v$, where $u \in \mathscr{D}'_{L_\infty}$ (thus u may be written as a finite sum of generalized derivatives of bounded functions) and $v \in \mathscr{S}$. Then

$$u * v = \sum (d^n\hat{u}_n/dt^n) * v = \hat{u}_n * \sum (d^n v/dt^n)$$

is well defined, since $\hat{u}_n \in L_\infty$, $\sum (d^n v/dt^n) \in \mathscr{S}$ and is thus an L_1 function when $v \in \mathscr{S}$, and the convolution results in a bounded function, i.e., if $x \in L_p$, and $y \in L_g$, then $x * y \in L_r$, where $1/r = (1/p) + (1/g) - 1$. Therefore we conclude that passive immittance operators are well defined for all inputs $\mathbf{f}$ such that $f_i \in \mathscr{S}$ and, moreover, the corresponding outputs $\mathbf{g}$ will be such that g_i are infinitely differentiable bounded functions of time.

Now consider an arbitrary $f_i \in \mathscr{S}$ and note that there exists an $\alpha(t) \in C_0{}^\infty$ such that $\alpha f_i = f_i$ for all t in some arbitrary interval, and, moreover, $\alpha f_i \in C_0{}^\infty$ [in Appendix I we give an example of an infinitely differentiable function that is equal to zero for $t \geq b + \varepsilon$ and equal to 1 for $t \leq b$, so that $\alpha(t)$ may be constructed by a product of such functions]. By the "passivity" of the operator we then have

$$\int_{-\infty}^{t} (\alpha \mathbf{f}^{*T})\hat{\mathbf{g}} + \hat{\mathbf{g}}^{*T}(\alpha \mathbf{f})\, d\eta \geq 0,$$

where $\hat{\mathbf{g}}$ is the response to the altered input $\alpha\mathbf{f}$. By allowing the interval on which $\alpha(t) = 1$ to become arbitrarily large, we see that $\alpha\mathbf{f}$ converges pointwise to any arbitrary $\mathbf{f} \in \mathscr{S}$. But the corresponding sequence of integrals formed with $\alpha\mathbf{f}$ are all nonnegative, so that if they converge, the result must be nonnegative. Finally, one shows, using the fact that when $f_i \in \mathscr{S}$ then $[h] * \mathbf{f}$ is an n-vector all of whose elements are bounded functions of time, that the sequence of integrals does indeed converge, and the result is established. QED

We are now in a position to state the desired characterization of passive immittance operators as:

Theorem 2.5. Let the $n \times n$ matrix of distributions, $[w(t)]$, be the impulse response of a passive immittance operator; then its Laplace transform $[W(p)]$ is such that:

1. Each element of $[W(p)]$ is analytic in $\operatorname{Re} p > 0$.
2. $[W(p^*)] = [W^*(p)]$.
3. $[W^*(p)]^T + [W(p)]$ is a nonnegative-definite matrix for all $\operatorname{Re} p > 0$.

Such a matrix will be called positive-real.

Proof. If one considers as inputs $\mathbf{f} = \mathbf{y}e^{pt}\alpha(t_0 - t)$, where $\alpha(t) \in C^\infty$ is one for $t \leq 0$ and zero for $t \geq \varepsilon$, then for $\operatorname{Re} p > 0$, $f_i \in S$, and we may appeal to Lemma 2.2 to justify the use of essentially the same arguments employed in the proof of Theorem 2.3 to establish the conclusion of the present theorem. QED

The coverse of this theorem—namely, that every positive-real matrix is the Laplace transform of the impulse response matrix of some passive immittance operator—has also been established, using distributional arguments in an essential manner, by Zemanian [Ze2] but the presentation of these arguments would take us too far afield.

However, we are now ready to compare the two approaches to a definition of passivity—Definitions 7 and 9. First we note that the definition of a linear, passive immittance operator implies causality, and, as we have noted previously, this might be a disadvantage in an axiomatic theory for more general systems, such as nonlinear or time-variable ones, since it is not distinct from the very basic physical assumption of causality. If one attempts to rectify this problem by using as a definition

$$\int_{-\infty}^{\infty} \mathbf{f}^T \mathbf{g}\, dt \geq 0,$$

then a simple example such as the system with impulse response $-(d/dt)\,\delta(t)$ shows that this approach is not fruitful; even though such a system would be linear, continuous, time-invariant, causal, and would satisfy the integral constraint with the upper limit at $t = \infty$, it would not satisfy

$$\int_{-\infty}^{t} fg\, d\eta \geq 0,$$

since $W(p) = \mathscr{L}[-d\delta(t)/dt] = -p$, which is not a positive-real function as required by Theorem 2.5. If we assume that the system under study is such that the inputs $\mathbf{f}$ correspond to voltages $\mathbf{v}$ and the outputs to the resultant currents $\mathbf{i}$, then we see that the difficulty with the definition, from a physical standpoint, is that with $v_i \in C_0^{\infty}$ then $\mathbf{v}^T\mathbf{i}$ will be zero for t larger than some finite value of time, say t_0, and thus in the integral

$$\int_{-\infty}^{\infty} \mathbf{v}^T \mathbf{i}\, dt$$

there will be no contribution for $t > t_0$ even though $\mathbf{i}$ is not zero in this range. Thus, using the above integral, we have no way of determining whether or not the system would have been capable of generating energy during the time $t > t_0$. Now, in the first definition of passivity the inputs considered correspond to $\mathbf{v} + \mathbf{i}$ and, as noted in Fig. 2.1, this corresponds to voltage sources connected to the system through unit resistors. In this case even though $\mathbf{v} + \mathbf{i}$ vanishes for $t > t_0$, the $\mathbf{v}$ and $\mathbf{i}$ will not in general vanish, since we see from the figure that during this period of time the unit resistors are effectively connected across each of the ports of the network. Thus, by measuring the resultant $\mathbf{v}$

and **i** (specifically **v** − **i**), we can determine the energy flow on the interval $t \geq t_0$ and compare it to flow during the interval $t \leq t_0$. In particular for $t \geq t_0$ we note that the contributions to the total measure of energy, i.e.,

$$\int_{-\infty}^{\infty} (v + i)^2 - (v - i)^2 \, dt$$

will be evidenced through the term $-\int_{t_0}^{\infty} (v - i)^2 \, dt$. A related difficulty with the passive immittance formulation is its inability to distinguish lossless systems directly, since for these systems one must phrase a definition in which **v** + **i** are considered as the inputs, and this is essentially a restatement of Definition 8.

The final comparison between these two approaches to a definition of passivity is perhaps the most informative, in that it shows that there will exist systems that may be studied using the first definition (Definition 7), i.e., that are representable by bounded-real matrices, but for which one cannot define an immittance operator. However, we will show that if the system possesses a passive immittance operator representation, then the system will always be describable by a bounded-real matrix and thus satisfy the conditions of Definition 7. We state this as:

Theorem 2.6. Given that in at least one of the representations, $\mathbf{V} = [Z(p)]\mathbf{I}$ or $\mathbf{I} = [Y(p)]\mathbf{V}$, the corresponding matrix either $[Z]$ or $[Y]$, exists and is positive-real, then there will exist a bounded-real matrix $[S(p)]$ such that

$$\mathbf{B}(p) = \tfrac{1}{2}(\mathbf{V}(p) - \mathbf{I}(p)) = [S(p)]\tfrac{1}{2}(\mathbf{V}(p) + \mathbf{I}(p)).$$

Moreover if, for example, $\mathbf{V} = [Z(p)]\mathbf{I}(p)$, then

$$[S(p)] = [Z(p) - 1_n][Z(p) + 1_n]^{-1},$$

where 1_n is the unit $n \times n$ matrix. Conversely, given a bounded-real matrix $[S(p)]$ such that

$$\mathbf{B}(p) = [S(p)]\mathbf{A}(p),$$

then there exists a positive-real matrix $[Z(p)]$ such that $\mathbf{V} = [Z]\mathbf{I}$ if and only if $\det[1_n - S] \neq 0$ in $\operatorname{Re} p > 0$, and there exists a positive-real matrix $[Y(p)]$ such that $\mathbf{I} = [Y]\mathbf{V}$ if and only if $\det[1_n + S] \neq 0$ in $\operatorname{Re} p > 0$. If, for example $\det[1_n - S] \neq 0$, then $Z(p) = [1_n - S]^{-1}[1_n + S]$. But it is possible that neither $[Z]$ nor $[Y]$ exists, i.e., it is possible that

$$\det[1_n \pm S(p)] = 0 \qquad \text{in all} \quad \operatorname{Re} p > 0$$

even though $[S(p)]$ is bounded-real.

Proof. Given that $[S(p)]$ is bounded-real and that $\mathbf{B} = [S]\mathbf{A}$ we see that $\mathbf{B} = \frac{1}{2}(\mathbf{V} - \mathbf{I}) = [S]\mathbf{A} = [S]\frac{1}{2}(\mathbf{V} + \mathbf{I})$, so that, for example,

$$\mathbf{V} = [1_n - S]^{-1}[1 + S]\mathbf{I}$$

if and only if $\det(1_n - S) \neq 0$. Now, if we define $Z = [1_n - S]^{-1}[1_n + S]$, then we have $[S][Z + 1_n] = [Z - 1_n]$, and thus $[Z + 1_n]^{*T}[S]^{*T} = [Z - 1_n]^{*T}$. Thus, by multiplying these matrices, we have

$$\begin{aligned}[Z + 1]^{*T}[S]^{*T}[S][Z + 1_n] &= [Z - 1_n]^{*T}[Z - 1_n]\\ &= [Z + 1_n]^{*T}[Z + 1_n] - 2[Z^{*T} + Z],\end{aligned}$$

or

$$[Z + 1_n]^{*T}\{1_n - [S]^{*T}[S]\}[Z + 1_n] = 2[Z^{*T} + Z],$$

so that

$$(\mathbf{y}^{*T}[Z + 1_n]^{*T})\{1_n - [S]^{*T}[S]\}([Z + 1_n]\mathbf{y}) = 2\mathbf{y}^{*T}[Z^{*T} + Z]\mathbf{y}.$$

But with the assumption that $\det(1_n - S) \neq 0$ in $\operatorname{Re} p > 0$, $[Z]$ exists, and we see that $\mathbf{x} = [Z + 1_n]\mathbf{y}$ is well defined in all $\operatorname{Re} p > 0$ and for arbitrary complex n-vectors $\mathbf{y}$. Now, by assumption, $[S]$ is bounded-real, so that

$$\mathbf{x}^{*T}\{1_n - [S]^{*T}[S]\}\mathbf{x} \geq 0$$

for all $\operatorname{Re} p > 0$, and we see that this implies that $[Z]^{*T} + [Z]$ will be non-negative-definite in $\operatorname{Re} p > 0$. Finally, given that $[S]$ is bounded-real, we know that each of its elements is analytic in $\operatorname{Re} p > 0$, so that if $\det(1_n - S) \neq 0$ in $\operatorname{Re} p > 0$, then each element of the matrix $[Z] = [1_n - S]^{-1}[1_n + S]$ is analytic in $\operatorname{Re} p > 0$. In addition, we may show directly that $[S(p^*)] = [S^*(p)]$ implies $[Z(p^*)] = [Z^*(p)]$, and we may finally conclude that the matrix $[Z(p)]$ is positive-real. A similar argument applies to the case when $\det[1_n + S] \neq 0$ in $\operatorname{Re} p > 0$, showing that the matrix

$$[Y(p)] = [1_n + S]^{-1}[1_n - S]$$

is then positive-real. However, the bounded-real matrix

$$[S(p)] = \begin{bmatrix} 0 & 1 \\ 1 & 0 \end{bmatrix}$$

provides an example in which $\det[1_n \pm S] = 0$ for all $\operatorname{Re} p > 0$ and for which neither $[Z]$ nor $[Y]$ can be defined.

Now we consider the first half of the theorem. Here we find that if $\mathbf{V} = [Z]\mathbf{I}$, then the corresponding $[S]$ is given as

$$[S] = [Z - 1_n][Z + 1_n]^{-1}$$

if $\det[Z + 1_n] \neq 0$ in $\operatorname{Re} p > 0$. However, the assumption that $\det[Z + 1_n] = 0$

at some p_0 in Re $p > 0$ implies that there exists some nonzero n-vector $\mathbf{y}$ such that

$$[Z + 1_n]\mathbf{y} = 0_n,$$

where 0_n is an n-vector with all zero entries, so that we would have $\mathbf{y}^{*T}[Z + 1_n]\mathbf{y} = 0$, or $\mathbf{y}^{*T}[Z]\mathbf{y} = -\mathbf{y}^*\mathbf{y}^T$. Thus one would arrive at the conclusion that there exists some nonzero n-vector $\mathbf{y}$ such that

$$\mathbf{y}^{*T}[Z^{*T} + Z]\mathbf{y} = -2\mathbf{y}^{*T}\mathbf{y} < 0,$$

which contradicts the fact that $[Z(p)]$ has been assumed to be a positive-real matrix. Thus $[S] = [Z - 1_n][Z + 1_n]^{-1}$ exists given that $[Z]$ is positive-real and, moreover, each element in $[S]$ is analytic in Re $p > 0$, since each element of $[Z]$ is analytic there.

Finally, we will use the previous arguments regarding the nonnegative-definiteness of these matrices. First we note that if we consider the expression $\mathbf{x} = [Z + 1_n]\mathbf{y}$, where $\mathbf{x}$ is to be arbitrary, then we may find a corresponding $\mathbf{y} = [Z + 1_n]^{-1}\mathbf{x}$, since $\det[Z + 1_n] \neq 0$. Therefore we may appeal to the previously derived expressions to conclude that if $[Z(p)]$ is positive-real, then

$$\mathbf{x}^{*T}\{1_n - [S]^{*T}[S]\}\mathbf{x} \geq 0,$$

where $\mathbf{x}$ is an arbitrary n-vector. We have thus shown that $[S]$ will be a bounded-real matrix.

Similar arguments apply in the case where $\mathbf{I} = [Y]\mathbf{V}$ and $[Y]$ is given as a positive-real matrix. QED

On the basis of the above theorem and the discussion preceding it we conclude that for the study of linear, time-invariant systems the more general definition of passivity is that of Definition 7, and in subsequent sections we will assume that this definition is applicable whenever we refer to a system as being passive. The matrices discussed in Theorem 2.6 are given specific names in network theory. Thus $[Z]$ is called the impedance matrix, $[Y]$ the admittance matrix (generically both are called immittance matrices), and $[S]$ is called the scattering matrix. Since we will have occasion to use all three in subsequent chapters we will use the next two sections to discuss some useful representations of such matrices. Before ending this section we would like to further illustrate the fact that bounded-real and positive-real matrices are intimately connected by stating:

Lemma 2.3. If $[S(p)]$ is a bounded-real matrix and $[W(p)]$ is a positive-real matrix, then

1. $[W_1]+[W_2]$, $1_n \pm [S(p)]$, $[W]^T$, and $[W(p)]^{-1}$ (assuming $\det[W(p)] \neq 0$ for at least one p in Re $p > 0$) are positive-real matrices.
2. $[S]^T$, and $[S_1][S_2]$ are bounded-real matrices.

Proof. The demonstrations are generally quite straightforward. In particular, we have, given that $[S]$ is bounded-real,

$$\begin{aligned}(1_n \pm [S]^{*T}) + (1_n \pm [S]) &= 2_n \pm ([S]^{*T} + [S]) \\ &= 1_n - [S]^{*T}[S] + (1_n \pm [S]^{*T})(1_n \pm [S])\end{aligned}$$

and

$$\begin{aligned}&\mathbf{y}^{*T}\{1_n - [S]^{*T}[S] + (1_n \pm [S]^{*T})(1_n \pm [S])\}\mathbf{y} \\ &\quad = \mathbf{y}^{*T}\{1_n - [S]^{*T}[S]\}\mathbf{y} + [(1_n \pm [S])\mathbf{y}]^{*T}[(1_n \pm [S])\mathbf{y}] \geq 0.\end{aligned}$$

We also may show that

$$\mathbf{y}^{*T}\{1_n - [S_2]^{*T}[S_1]^{*T}[S_1][S_2]\}\mathbf{y} \geq \mathbf{y}^{*T}\{1_n - [S_2]^{*T}[S_2]\}\mathbf{y} \geq 0,$$

since $[S_1]$ and $[S_2]$ are bounded-real, and thus

$$\mathbf{y}^{*T}[S_2]^{*T}[S_1]^{*T}[S_1][S_2]\mathbf{y} = \mathbf{x}^{*T}[S_1]^{*T}[S_1]\mathbf{x} < \mathbf{x}^{*T}\mathbf{x},$$

where $\mathbf{x} = [S_2]\mathbf{y}$. As one last demonstration we note that

$$\mathbf{y}^{*T}\{[W]^{-1*T} + [W]^{-1}\}\mathbf{y} = \mathbf{y}^{*T}[W]^{-1*T}\{[W] + [W]^{*T}\}[W]^{-1}\mathbf{y} \geq 0,$$

given that $[W]$ is positive-real and that $\det[W] \neq 0$. However, one may show as a consequence of the maximum modulus theorem that if $[W(p)]$ is positive-real, then either $\det[W(p)] = 0$ for all $\operatorname{Re} p > 0$ or $\det[W(p)] \neq 0$ for any p in $\operatorname{Re} p > 0$, and the statement of the theorem follows directly. QED

2.3. REPRESENTATIONS OF BOUNDED-REAL AND POSITIVE-REAL MATRICES

The characterizations of passive systems which we have obtained provide concise requirements on the system function or impulse response matrix of such systems—namely, that the Laplace transform of such matrices be either bounded-real or positive-real. However, one must test both the analyticity of each element in such a matrix and the nonnegativity of related matrices in the entire half plane $\operatorname{Re} p > 0$ to establish the fact that a given matrix is either bounded-real or positive-real. There is then some interest in obtaining alternate conditions which would allow such a test to be performed more easily. Below we discuss such alternate conditions, some of which will be phrased on the boundary of the half plane, i.e., for $p = j\omega$. We will then discuss a transformation between bounded-real matrices, which we will call complex normalization, that will prove of great use in our subsequent studies. The normalization was introduced by Youla [Yo2] and it has proven to be one of the most unifying concepts developed in network theory.

2.3.1. Positive-Real Matrices

The starting point for our discussion will be a representation theorem for arbitrary positive-real matrices. This theorem is based on the following result concerning functions whose real parts are positive in the half plane $\operatorname{Re} p > 0$, and was first stated by Cauer [Ca5] and is a special case of a theorem due to Herglotz (see, for example, [Ne1]):

Theorem 2.7. A function $f(p)$ is analytic and has a positive-real part in the half plane $\operatorname{Re} p > 0$ if and only if

$$f(p) = \int_{-\infty}^{\infty} \frac{1 - pj\tau}{p - j\tau}\, d\beta(\tau) + ap + jg,$$

where $\beta(\tau)$ is a real nondecreasing function of bounded variation on $(-\infty, \infty)$, a is a nonnegative real constant, and g is a real constant. In the case when $f(p^*) = f^*(p)$, i.e., $f(p)$ is a positive-real function, then $\beta(\tau) = -\beta(-\tau)$ and $g = 0$.

We will make use of this result to obtain the desired representation of positive real matrices in essentially the same manner as Youla presented it in [Yo1].

Theorem 2.8. The necessary and sufficient condition that an $n \times n$ matrix $[W(p)]$ be positive-real is that for $\operatorname{Re} p > 0$

$$[W(p)] = [Q] + [A]p + \int_{-\infty}^{\infty} \frac{1 - pj\tau}{p - j\tau}\, d[M(\tau)],$$

where $[Q]$ is a real, constant, skew-symmetric ($[Q]^T = -[Q]$), $n \times n$ matrix; $[A]$ is a real, constant, symmetric, nonnegative-definite matrix; and $[M]$ is an $n \times n$ matrix each of whose elements are of bounded variation on the interval $(-\infty, \infty)$, satisfying $[M]^{*T} = [M]$ and $-[M(-\tau)]^* = [M(\tau)]$ and with $\mathbf{y}^{*T}[M(\tau)]\mathbf{y}$ a real, bounded, nondecreasing function of τ for any complex n-vector $\mathbf{y}$.

Proof. Since $[W(p)]$ is positive-real, $\mathbf{y}^{*T}\{[W(p)] + [W^*(p)]^T\}y \equiv 2 \operatorname{Re}\{\mathbf{y}^{*T}[W(p)]\mathbf{y}\} \geq 0$ in $\operatorname{Re} p > 0$ and for all complex constant vectors $\mathbf{y}$. In particular, if every element of $\mathbf{y}$ except y_r and y_k is zero, then

$$\operatorname{Re}\{y_r^* y_r W_{rr}(p) + y_r^* y_k W_{rk}(p) + y_r y_k^* W_{kr}(p) + y_k^* y_k W_{kk}(p)\} \geq 0.$$

Thus the function $a_1(p) = W_{rr}(p) + W_{rk}(p) + W_{kr}(p) + W_{kk}(p)$ has a positive real part, as we note when $y_r = y_k = 1$. However, since each element satisfies $\chi(p^*) = \chi^*(p)$ and in addition is holomorphic in $\operatorname{Re} p > 0$, we conclude that

$a_1(p)$ is itself a positive-real function. Similarly, if we let $y_r = -y_k = 1$, we observe that $a_2(p) = W_{rr}(p) - [W_{rr}(p) + W_{kr}(p)] + W_{kk}(p)$ is also a positive-real function. In addition, if we let $y_r = j, y_k = 1$, and then $y_r = j, y_k = -1$, we observe that $b_1(p) = W_{rr}(p) - j[W_{rk}(p) - W_{kr}(p)] + W_{kk}(p)$ and $b_2(p) = W_{rr}(p) + j[W_{rk}(p) - W_{kr}(p)] + W_{kk}(p)$ have positive real parts and are holomorphic in $\operatorname{Re} p > 0$ (they are not positive-real, since they fail the symmetry property). Therefore $W_{rk}(p) + W_{kr}(p) = \frac{1}{2}[a_1(p) - a_2(p)]$ is the difference between two positive-real functions, and $j[W_{rk}(p) - W_{kr}(p)] = \frac{1}{2}[b_2(p) - b_1(p)]$ is the difference between two analytic functions with positive real parts. Employing Theorem 2.7, we obtain the representation

$$W_{rk}(p) + W_{kr}(p) = ap + \int_{-\infty}^{\infty} \frac{1 - pj\tau}{p - j\tau}\, d\beta(\tau),$$

where a is real (not necessarily positive) and $\beta(\tau) = -\beta(-\tau)$ is a bounded function (not necessarily nondecreasing). In addition,

$$W_{rk}(p) - W_{kr}(p) = q - jap - j\int_{-\infty}^{\infty} \frac{1 - pj\tau}{p - j\tau}\, d\theta(\tau),$$

where q and a are real and $\theta(\tau)$ is bounded. However, since $W_{rk}(p^*) - W_{kr}(p^*) = W_{rk}^*(p) - W_{kr}^*(p)$, we observe that $a = 0$ and $\theta(-\tau) = \theta(\tau)$, or

$$W_{rk}(p) - W_{kr}(p) = q - j\int_{-\infty}^{\infty} \frac{1 - pj\tau}{p - j\tau}\, d\theta(\tau).$$

Thus, considering each element of the matrix $[W(p)]$ separately, we may conclude that if $[W(p)]$ is positive-real,

$$[W(p)] + [W(p)]^T = [A]p + \int_{-\infty}^{\infty} \frac{1 - pj\tau}{p - j\tau}\, d[\beta(\tau)],$$

$$[W(p)] - [W(p)]^T = [Q] + \int_{-\infty}^{\infty} \frac{1 - pj\tau}{p - j\tau}\, d[-j\theta(\tau)],$$

where $[A]$ is a real symmetric $n \times n$ matrix of constants, $[Q]$ is a real skew-symmetric $n \times n$ matrix of constants, $[\beta(\tau)]$ is a symmetric $n \times n$ matrix of odd-bounded functions, and $[\theta(\tau)]$ is a skew-symmetric $n \times n$ matrix of even-bounded functions. By addition, $[\beta(\tau)] - j[\theta(\tau)] = [M(\tau)]$ is an $n \times n$ matrix of bounded functions satisfying $[M]^{*T} = [M]$ and $-[M(-\tau)]^* = [M(\tau)]$. The theorem follows finally by adding the two matrix expressions and noting that

$$\mathbf{y}^{*T}([W]^{*T} + [W])\mathbf{y} = 2(\mathbf{y}^{*T}[A]\mathbf{y})p + \int_{-\infty}^{\infty} \frac{1 - pj\tau}{p - j\tau}\, d\{\mathbf{y}^{*T}([M]^{*T} + [M])\mathbf{y}\} \geq 0$$

in $\operatorname{Re} p > 0$ implies via Theorem 2.7 that $\mathbf{y}^{*T}[A]\mathbf{y}$ must be nonnegative, i.e.,

$[A]$ is nonnegative-definite, and that $\mathbf{y}^{*T}([M]^{*T} + [M])\mathbf{y} = 2\mathbf{y}^{*T}[M(\tau)]\mathbf{y}$ must be a real, bounded, nondecreasing function of τ for all complex vectors $\mathbf{y}$. QED

One might inquire as to the significance of the matrix $[M(\tau)]$ and particularly its connection with $[W(p)]$. The following theorem shows the connection and in addition provides another concise representation of positive-real matrices, although it requires a distributional interpretation:

Theorem 2.9. [Wo2]. The necessary and sufficient condition that an $n \times n$ matrix of distributions $[W(\omega)]$ be the boundary behavior in the $\mathscr{S}'$ topology of a positive-real matrix is that

$$[W(\omega)] = [Q] + j\omega\left[[A] + \int_{-\infty}^{\infty} d[M(\tau)]\right] + \pi(1+\omega^2)\left[\frac{dM(\omega)}{d\omega}\right]$$

$$+ (1+\omega^2)\left(\left[\frac{dM(\omega)}{d\omega}\right] * pv\,\frac{1}{j\omega}\right),$$

where $[Q]$, $[A]$, and $[M]$ are defined in Theorem 2.8. In particular,

$$[W(\omega)]^{*T} + [W(\omega)] = 2\pi(1+\omega^2)[dM(\omega)/d\omega].$$

Proof. The necessity will be established if we demonstrate that the matrix $[W(\omega)]$ given in the theorem is the $\mathscr{S}'$ boundary value of the representation for $[W(p)]$ given in Theorem 2.8. The first two terms in $[W(p)]$ converge in their present form with $p \to j\omega$. By noting that

$$\frac{1 - pj\tau}{p - j\tau} = p + \frac{1 - p^2}{p - j\tau},$$

the third term may be rewritten in the form

$$p\int_{-\infty}^{\infty} d[M] + (1 - p^2)\int_{-\infty}^{\infty} \frac{d[M]}{p - j\tau},$$

where we have observed that the bounded variation of $[M(\tau)]$ implies that $\int_{-\infty}^{\infty} d[M]$ exists. The third term is then obtained directly. Finally, consider

$$\lim_{\sigma\to 0}\int_{-\infty}^{\infty} (1 - p^2)\int_{-\infty}^{\infty} \frac{d[M]}{p - j\tau}\,\varphi(\omega)\,d\omega$$

$$= \lim_{\sigma\to 0}\int_{-\infty}^{\infty} dM(\tau)\int_{-\infty}^{\infty} \frac{(1 - p^2)\varphi(\omega)}{p - j\tau}\,d\omega$$

for all $\varphi \in \mathscr{S}$, where the classic Fubini theorem justifies the interchange.

However, Plancherel's equality implies that

$$\int_{-\infty}^{\infty} \frac{(1-p^2)\varphi(\omega)}{p-j\tau}\, d\omega = 2\pi \int_{-\infty}^{\infty} \hat{\varphi}_\sigma(t) e^{j\tau t} e^{-\sigma t} H(t)\, dt,$$

where $\hat{\varphi}_\sigma(t) = \mathscr{F}^{-1}[(1-p^2)\varphi(\omega)] \in S$ for every $\sigma \geq 0$, $\mathscr{F}^{-1}[1/(p-j\tau)] = e^{j\tau t} e^{-\sigma t} H(t)$, and $H(t)$ is the Heaviside step function. Therefore

$$\left|\int_{-\infty}^{\infty} \frac{(1-p^2)\varphi(\omega)}{p-j\tau}\, d\omega\right| = 2\pi \left|\int_{-\infty}^{\infty} \hat{\varphi}_\sigma(t) e^{j\tau t} e^{-\sigma t} H(t)\, dt\right|$$
$$\leq 2\pi \int_0^{\infty} |\hat{\varphi}_\sigma| e^{-\sigma t}\, dt \leq C$$

for all $\sigma \geq 0$. Then the dominated convergence theorem allows one to conclude that

$$\lim_{\sigma\to 0} \int_{-\infty}^{\infty} d[M(\tau)] \int_{-\infty}^{\infty} \frac{(1-p^2)\varphi(\omega)}{p-j\tau}\, d\omega = \int_{-\infty}^{\infty} dM \lim_{\sigma\to 0} \int_{-\infty}^{\infty} \frac{(1-p^2)\varphi(\omega)}{p-j\tau}\, d\omega.$$

But with $\varphi \in \mathscr{S}$ one may obtain classically

$$\lim_{\sigma\to 0} \int_{-\infty}^{\infty} \frac{(1-p^2)\varphi(\omega)}{p-j\tau}\, d\omega = (1+\tau^2)\pi\varphi(\tau) + pv \int_{-\infty}^{\infty} \frac{(1+\omega^2)\varphi(\omega)}{j(\omega-\tau)}\, d\omega$$
$$= (1+\tau^2)\pi\varphi(\tau)$$
$$+ pv \int_{-\infty}^{\infty} \frac{[1+(\omega+\tau)^2]\varphi(\omega+\tau)}{j\omega}\, d\omega,$$

or, finally,

$$\lim_{\sigma\to 0} \int_{-\infty}^{\infty} (1-p^2) \int_{-\infty}^{\infty} \frac{d[M]}{p-j\tau}\, \varphi\, d\omega$$
$$= \int_{-\infty}^{\infty} (1+\tau^2)\pi\varphi(\tau)\, d[M] + \int_{-\infty}^{\infty} d[M]\, pv \int_{-\infty}^{\infty} \frac{[1+(\omega+\tau)^2]\varphi(\omega+\tau)}{j\omega}\, d\tau$$
$$= \int_{-\infty}^{\infty} \pi(1+\omega^2)\left[\frac{dM}{d\omega}\right]\varphi(\omega)\, d\omega$$
$$+ \int_{-\infty}^{\infty} (1+\omega^2)\left(\left[\frac{dM}{d\omega}\right] * pv\frac{1}{j\omega}\right)\varphi(\omega)\, d\omega,$$

which may be written

$$(1-p^2)\int_{-\infty}^{\infty} \frac{d[M]}{p-j\tau} \xrightarrow{w} \pi(1+\omega^2)\left[\frac{dM(\omega)}{d\omega}\right] + (1+\omega^2)\left(\left[\frac{dM(\omega)}{d\omega}\right] * pv\frac{1}{j\omega}\right).$$

The sufficiency of the theorem follows from the remark that $[W]^{*T} + [W] = 2\pi(1 + \omega^2)[dM(\omega)/d\omega]$ because this allows one to identify $[M]$ given $[W(\omega)]$; then by inspection one determines $[Q]$ and $[A]$. The conditions of the theorem guarantee that the $[Q]$, $[A]$, and $[M]$ so identified will satisfy Theorem 2.8. Thus the $[W(p)]$ formed from these parameters by means of Youla's representation will be positive-real. The above demonstration proves that this $[W(p)]$ will converge in the $\mathscr{S}'$ topology to the original distribution. QED

The theorem takes on a particularly simple form when the positive-real matrix reduces to a scalar function, i.e., the boundary value of a positive-real function must be of the form

$$W(\omega) = j\left[\omega A + \omega \int_{-\infty}^{\infty} dM - (1 + \omega^2)\left(\frac{dM}{d\omega} * pv\frac{1}{\omega}\right)\right] + \pi(1 + \omega^2)\frac{dM}{d\omega}$$

with $\operatorname{Re} W(\omega) = \pi(1 + \omega^2)(dM/d\omega)$, A a real positive constant, and M a bounded, odd, nondecreasing, real function. Thus $dM/d\omega$ is a nonnegative measure and we see most simply how the positiveness of the real part of the holomorphic function, $W(p)$, is reflected in the behavior of the boundary distribution. In addition, the statement of the theorem can be interpreted as a specialized Hilbert transform relationship for positive-real functions, i.e., the imaginary part (at least in the scalar case) of the boundary distribution is determined, with the exception of the constant A, from the real part. In any event, the theorem characterizes the positive-real class completely and does so by considering only the boundary values $[W(\omega)]$. Other concise characterizations have been obtained, notably one by König and Zemanian [Ko3]. In their formulation direct use is made of the impulse response matrix $[w(t)]$ of the system. Specifically, they proved that $[w(t)]$ is the generalized Green's function of a passive immittance operator if an only if

$$[w(t)] = [A]\frac{d\delta(t)}{dt} + [w_0(t)],$$

where $[A]$ is a real, symmetric, nonnegative-definite matrix and $[w_0]$ is a matrix of real distributions each of whose elements are of zero order, such that $\mathbf{y}^{*T}([w_0] + [\tilde{w}_0{}^T])\mathbf{y}$ is a nonnegative-definite distribution for all constants $n \times 1$ vectors $\mathbf{y}$ and each element of $[w_0(t)]$ is zero for $t < 0$.

In many cases the following theorem provides a much easier means for establishing that a given matrix is positive-real. However, it is only applicable to those matrices whose elements are meromorphic functions, i.e., functions that are analytic in the entire finite part of the p plane except at isolated points where they may have only poles as singularities, the only essential singularities allowed being at the point of infinity.

Theorem 2.10. An $n \times n$ matrix $[W(p)]$ all of whose elements are real meromorphic functions is positive-real if and only if:

1. $W_{jk}(p)$ has no poles in $\operatorname{Re} p > 0$.
2. The poles of $W_{jk}(p)$ on $p = j\omega$ are simple.
3. $[\hat{W}(j\omega)]^{*T} + [\hat{W}(j\omega)]$ is nonnegative-definite for almost all ω and

$$\int_{-\infty}^{\infty} \frac{\mathbf{y}^{*T}([\hat{W}(j\tau)]^{*T} + [\hat{W}(j\tau)])\mathbf{y}}{1+\tau^2}\, d\tau < \infty$$

for all constant n-vectors $\mathbf{y}$, where $[\hat{W}(\omega)]$ is the infinitely differentiable part of the $\mathcal{S}'$ boundary value of $[W(p)]$.

4. If $p = j\omega_0$ is a point where $W_{jk}(p)$ has a pole and $\alpha_{jk}(\omega)$ is the residue of this pole, then the matrix formed with these residues, $[\alpha(\omega_0)]$, must satisfy $[\alpha(\omega_0)]^{*T} = [\alpha(\omega_0)]$, it must be nonnegative-definite, and

$$\sum_i \frac{\mathbf{y}^{*T}[\alpha(\omega_i)]\mathbf{y}}{1+\omega_i^2} < \infty,$$

where the sum extends over all the points $p = j\omega_i$ at which $W_{jk}(p)$ has poles.

5. Asymptotically, $[W(p)] \to [A]p$ in $\operatorname{Re} p > 0$, where $[A]$ is a real, constant, symmetric, nonnegative-definite matrix.

Proof. First we note that if any element of $[W(p)]$ has a pole of order n at the point $p = j\omega_0$, then using the Laurent expansion about this point leads us to the conclusion that in $\operatorname{Re} p > 0$

$$\mathbf{y}^{*T}([W]^{*T} + [W])\mathbf{y} \approx \operatorname{Re} \frac{a_n}{(p - j\omega_0)^n}$$

$$= \frac{\operatorname{Re} a_n \cos n\theta - \operatorname{Im} a_n \sin n\theta}{r^n}$$

in the vicinity of such a pole, where $p - j\omega_0 = re^{j\theta}$ and $-(\pi/2) < \theta < (\pi/2)$. Thus, unless $n = 1$, we see that it is always possible to find a θ for which the expression is negative. But since $[W(p)]$ is, by assumption, positive-real, $[W]^{*T} + [W]$ must be nonnegative-definite, and we conclude that the order of these poles may not exceed 1, i.e., the poles must be simple on $p = j\omega$.

We further note that since each element of $[W(p)]$ is a meromorphic function, then in the vicinity of the poles

$$W_{jk} = \frac{a}{p - j\omega_0} + \hat{W}_{jk},$$

where $\hat{W}_{jk}$ is analytic. Moreover, as $\operatorname{Re} p \to 0$ in $\operatorname{Re} p > 0$, $a/(p - j\omega_0)$

converges distributionally to

$$a\left[\pi\delta(\omega - \omega_0) + pv\,\frac{1}{j(\omega - \omega_0)}\right].$$

Therefore we may always express $W_{jk}(\omega)$ as

$$W_{jk}(\omega) = \hat{W}_{jk}(\omega) + \sum_i a_i\left[\pi\delta(\omega - \omega_i) + pv\,\frac{1}{j(\omega - \omega_i)}\right],$$

where $\hat{W}_{jk}(\omega)$ is an infinitely differentiable function (classically) and the sum extends over all of the poles of W_{jk} on $p = j\omega$.

We are now ready to consider the $\mathscr{S}'$ boundary value of the positive-real matrix as given in Theorem 2.9, in particular, the expression

$$[W(\omega)]^{*T} + [W(\omega)] = 2\pi(1 + \omega^2)[dM(\omega)/d\omega].$$

First, since each element of the matrix $[M]$ is of bounded variations, we know from classical analysis (see, for example, [Na1]) that these elements may be written as a sum of a continuous function of bounded variation and a function, called the saltus function, that is constant except at specific points where it jumps discontinuously. In addition, it is known that if the derivative of the continuous part of such functions, say, $\hat{M}_{jk}(\tau)$, exists everywhere, then

$$\hat{M}_{jk}(\tau) = \int_a^\tau \frac{d\hat{M}_{jk}(\eta)}{d\eta}\,d\eta + \hat{M}_{jk}(a).$$

However, the generalized derivative of the saltus function exists and it is a sum of delta functions each with a multiplier given by the value of the jump, i.e., if $M_{jk}(\tau) = \hat{M}(\tau) + M_s(\tau)$, then

$$dM_s(\tau)/d\tau = \sum \gamma_i\,\delta(\tau - \tau_i),$$

where τ_i are the positions of the jumps in the saltus function M_s and $\gamma_i = \lim_{\varepsilon\to 0}[M(\tau_i + \varepsilon) - M(\tau_i - \varepsilon)]$, with $\varepsilon \geq 0$. Now, since we argued that in the case of a meromorphic $W_{jk}(p)$ the function $W_{jk}(j\omega)$ could be written as an infinitely differentiable function and a sum of deltas, we see that we may write

$$\begin{aligned} R(\omega) &= [W(\omega)]^{*T} + [W(\omega)] \\ &= 2\pi(1 + \omega^2)\left[\frac{d\hat{M}(\omega)}{d\omega}\right] + 2\pi(1 + \omega^2)\left[\frac{dM_s(\omega)}{d\omega}\right], \end{aligned}$$

and then if we are given $R(\omega)$, we may identify in a unique manner both $[\hat{M}(\omega)]$ and $[M_s(\omega)]$. In particular, we see that

$$\hat{M}(\omega) = \int_a^\omega \frac{\hat{R}(\tau)}{2\pi(1 + \tau^2)}\,d\tau + \hat{M}(a),$$

where $\hat{R}(\omega)$ is the infinitely differentiable part of $[W(\omega)]^{*T} + [W(\omega)]$. We note that since from Youla's theorem $\mathbf{y}^{*T}[M]\mathbf{y}$ must be a bounded function, then $\mathbf{y}^{*T}[\hat{M}]\mathbf{y}$ must also be bounded, or we must have

$$\int_{-\infty}^{\infty} \frac{\mathbf{y}^{*T}([\hat{W}(\tau)]^{*T} + [\hat{W}(\tau)])\mathbf{y}}{1+\tau^2}\,d\tau < \infty.$$

In order to relate the saltus function to the residues of the poles of $[W(p)]$ on $p = j\omega$, we return to Youla's theorem and write it in the form

$$[W(p)] = [Q] + [A]p + \int_{-\infty}^{\infty} \frac{1 - pj\tau}{p - j\tau}\,d[\hat{M}] + \int_{-\infty}^{\infty} \frac{1 - pj\tau}{p - j\tau}\,d[M_s].$$

Now, if $M_{slk}(\tau)$ has a jump of value γ_{lk} at $\tau = \tau_{lk}$, then we will have a term in $d[M_s]$ of the form $\gamma_{lk}\,\delta(\tau - \tau_{lk})$, and thus $W_{lk}(p)$ will have a component of the form

$$\gamma_{lk}\left[\frac{1 - pj\tau_{lk}}{p - j\tau_{lk}}\right],$$

which, as we see, is a pole of W_{jk} at $p = j\tau_{lk}$, and its residue is then $\gamma_{lk}(1 + \tau_{lk}^2)$. Thus the saltus function can be reconstructed from the residues of the poles of W_{jk} on $p = j\omega$. We conclude that the entire matrix $[M] = [\hat{M}] + [M_s]$ can be uniquely determined from the smooth part of $[W(\omega)]^{*T} + [W(\omega)]$ and the residues of the poles of $[W(p)]$ on $p = j\omega$. In particular, the matrix of jumps in the saltus function are determined by

$$[\gamma(\omega_i)] = \lim_{\varepsilon \to 0} \{[M(\omega_i + \varepsilon) - M(\omega_i - \varepsilon)]\},$$

with $\varepsilon \geq 0$ and ω_i the position of these jumps, and since $[M(\tau)]^{*T} = [M(\tau)]$, and $\mathbf{y}^{*T}M(\tau)\mathbf{y}$ is a real, bounded, nondecreasing function of τ, we see that

$$[\gamma(\omega_i)]^{*T} = [\gamma(\omega_i)], \qquad \mathbf{y}^{*T}[\gamma(\omega_i)]\mathbf{y} \geq 0, \qquad \text{and} \qquad \sum_i \mathbf{y}^{*T}[\gamma(\omega_i)]\mathbf{y} < \infty,$$

where we have noted that

$$\sum_i \lim_{\varepsilon \to 0} \mathbf{y}^{*T}\{[M(\omega_i + \varepsilon)] - [M(\omega_i - \varepsilon)]\}\mathbf{y} = \sum_i \mathbf{y}^{*T}[\gamma(\omega_i)]\mathbf{y} < \infty,$$

since $\mathbf{y}^{*T}M(\omega)\mathbf{y}$ must be a bounded, nondecreasing function of ω. The conditions stated in the theorem then follow, since the residues of the poles on $p = j\omega$ were found to correspond to these jumps via

$$[\alpha(\omega_i)] = (1 + \omega_i^2)[\gamma(\omega_i)]$$

and we have shown the necessity of all the statements in the theorem.

Now, as to their sufficiency, we note first that the statements allow us to identify a unique $[M(\tau)]$ that will satisfy $[M]^{*T} = [M]$, $-[M(-\tau)]^* = [M]$, and that $\mathbf{y}^{*T}[M(\tau)]\mathbf{y}$ will be an odd, bounded, nondecreasing function of τ. We may then form a positive-real matrix $[X(p)]$ using this $[M]$, the given asymptotic behavior of $[W(p)]$, i.e., $[A]p$, and an arbitrary constant, skew-symmetric matrix $[Q]$. In particular, $[X(p)]$ and the given matrix $[W(p)]$ satisfy $[X(\omega)]^{*T} + [X(\omega)] = [W(\omega)]^{*T} + [W(\omega)]$ by construction. Moreover, both matrices will be analytic in $\operatorname{Re} p > 0$ and have $\mathscr{S}'$ boundary values as $\operatorname{Re} p \to 0$ ($[X]$ will by Theorem 2.9, and we may show that the conditions of the theorem imply that $[W(p)]$ will also). If we form the function $f(p) = \mathbf{y}^{*T}[X(p) - W(p)]\mathbf{y}$ using an arbitrary constant n-vector y, we observe that $f(p)$ will be analytic in $\operatorname{Re} p > 0$, have $\mathscr{S}'$ boundary values for $\operatorname{Re} p \to 0$, and $\frac{1}{2}\{f(p) + f^*(p)\} = 0$ for $p = j\omega$. Appealing to Corollary 1.2 we conclude that for $\operatorname{Re} p > 0$

$$f(p) = \sum_{n=1}^{m} a_n p^n$$

for some finite m. But by construction, $[X(p)]$ and $[W(p)]$ have the same asymptotic behavior in $\operatorname{Re} p > 0$, i.e., $[A]p$, and we see that $f(p)$ must be equal to zero in $\operatorname{Re} p > 0$. Thus

$$f(p) + f(p^*) = \mathbf{y}^{*T}\{[X]^{*T} + [X]\}\mathbf{y} - \mathbf{y}^{*T}\{[W]^{*T} + [W]\}\mathbf{y} = 0$$

and since by construction $[X]$ is positive-real, $\mathbf{y}^{*T}\{[X]^{*T} + [X]\}\mathbf{y} \geq 0$, or

$$\mathbf{y}^{*T}\{[W]^{*T} + [W]\}\mathbf{y} \geq 0$$

in $\operatorname{Re} p > 0$. We have thus shown that the conditions of the theorem guarantee that the matrix $[W(p)]$ satisfies all the properties of positive-real matrices. QED

We note in the case where the elements of $[W(p)]$ are rational functions of p that the conditions of the theorem may be further simplified; since the number of poles is finite, the sum involving the residue matrices will certainly be bounded, and the polynomial-like behavior of the elements of $[W(p)]$ implies that each element in $[\hat{W}]^{*T} + [\hat{W}]$ is bounded for all ω, so that the integral involving this matrix will also be bounded.

2.3.2. Bounded-Real Matrices

As was the case with positive-real matrices, the boundary behavior of bounded-real matrices allows us to obtain concise representations of them. As a preliminary, we state a theorem that relates to scalar functions (for a proof the reader is referred to [Bel]):

Theorem 2.11. A necessary and sufficient condition that a bounded measurable function $f(\omega)$ be the boundary value in the $\mathscr{S}'$ topology, as well as pointwise a.e. of a bounded analytic function in $\operatorname{Re} p > 0$, is that

$$\frac{df(\omega)}{d\omega} = \frac{1}{\pi j} f * \frac{d}{d\omega}\left(pv \frac{1}{\omega}\right),$$

where the generalized derivative is to be understood.

This theorem is, in a sense, a converse of the classic Fatou theorem which states that bounded analytic functions converge pointwise a.e. to bounded measurable functions as $\operatorname{Re} p \to 0$ in $\operatorname{Re} p \geq 0$ [Hil]. We may also use a classic representation theorem for bounded analytic functions [Nel] to obtain a maximum modulus statement. Thus if $f(p)$ is analytic and bounded in $\operatorname{Re} p > 0$, it may be expressed for $\operatorname{Re} p > 0$ as

$$f(p) = \frac{\sigma}{\pi} \int_{-\infty}^{\infty} \frac{f(\zeta)}{(\zeta - \omega)^2 + \sigma^2} \, d\zeta,$$

where $f(\zeta)$ is its boundary value, i.e., $f(\zeta) = \lim_{p \to j\zeta} f(p)$. Therefore we find

$$|f(p)| \leq M \frac{\sigma}{\pi} \int_{-\infty}^{\infty} \frac{1}{(\zeta - \omega)^2 + \sigma^2} \, d\zeta = M,$$

where $|f(\omega)| \leq M$, and we see that the bound M applies uniformly throughout the region of analyticity $\operatorname{Re} p > 0$.

We now proceed to the characterization of bounded-real matrices in terms of their boundary behavior; specifically:

Theorem 2.12 [Wo2]. The necessary and sufficient condition that an $n \times n$ matrix of measurable functions $[S(\omega)]$ be the boundary behavior pointwise a.e. as well as in the $\mathscr{S}'$ topology of a bounded-real matrix are:

1. $1_n - [S(\omega)]^{*T}[S(\omega)]$ be nonnegative-definite a.e.
2. $[dS(\omega)/d\omega] = (1/\pi j)[S(\omega)] * (d/d\omega)(pv\ 1/\omega)$.
3. $[S(-\omega)] = [S^*(\omega)]$.

Proof. The fact that $1_n - [S(p)]^{*T}[S(p)]$ is nonnegative-definite in $\operatorname{Re} p > 0$ implies that each element of $[S(p)]$ is bounded in $\operatorname{Re} p > 0$, so that Condition 2 follows from Theorem 2.11, since, by assumption, each element of $[S(p)]$ is analytic in $\operatorname{Re} p > 0$. The bounded nature of each element also implies, on the basis of Fatou's theorem, that they will converge pointwise a.e. to bounded measurable functions as $\operatorname{Re} p \to 0$. Therefore

$$\lim_{\operatorname{Re} p \to 0} \mathbf{y}^{*T}\{1_n - [S(p)]^{*T}[S(p)]\}\mathbf{y} = \mathbf{y}^{*T}\{1_n - [S(\omega)]^{*T}[S(\omega)]\}\mathbf{y} \geq 0$$

a.e., since $1_n - [S(p)]^{*T}[S(p)]$ is nonnegative-definite in all $\operatorname{Re} p > 0$, so that when the limit exists it must also be nonnegative. The necessity of these statements are thus established by noting finally that $[S(p^*)] = [S^*(p)]$ implies $[S(-\omega)] = [S^*(\omega)]$ with $p = j\omega$.

As to the sufficiency of these statements: 1 and 2 imply, on the basis of Theorem 2.11, that $[S(p)] = \mathscr{L}[\mathscr{F}^{-1}[S(\omega)]]$ is a matrix all of whose elements are bounded holomorphic functions in $\operatorname{Re} p > 0$ with $[S(\omega)]$ as their boundary value. Now consider

$$\mathbf{y}^{*T}\{1_n - [S(p)]^{*T}[S(p)]\}\mathbf{y} = \sum_{k=1}^{n} |y_k|^2 - \sum_{j=1}^{n} \left| \sum_{k=1}^{n} S_{jk}(p) y_k \right|^2 .$$

But since $S_{jk}(p)y_k$ is a bounded function, we may use the representation theorem for such functions to obtain for $\sigma > 0$

$$\sum_{j=1}^{n} \left| \sum_{k=1}^{n} S_{jk}(p) y_k \right|^2 = \sum_{j=1}^{n} \left| \frac{\sigma}{\pi} \int_{-\infty}^{\infty} \frac{\sum_{k=1}^{n} S_{jk}(j\zeta) y_k}{(\zeta - \omega)^2 + \sigma^2} \, d\zeta \right|^2 ,$$

and by means of Schwartz' inequality, $|\int ur\, dt|^2 \le \int |u|^2\, dt \int |r|^2\, dt$, we then find

$$\begin{aligned}
\sum_{j=1}^{n} \left| \frac{\sigma}{\pi} \int_{-\infty}^{\infty} \frac{\sum_{k=1}^{n} S_{jk}(j\zeta) y_k}{(\zeta - \omega)^2 + \sigma^2} \, d\zeta \right|^2 &\le \frac{\sigma}{\pi} \int_{-\infty}^{\infty} \frac{\sum_{j=1}^{n} \left| \sum_{k=1}^{n} S_{jk}(j\zeta) y_k \right|^2}{(\zeta - \omega)^2 + \sigma^2} \, d\zeta \\
&\le \sup\left\{ \sum_{j=1}^{n} \left| \sum_{k=1}^{n} S_{jk}(j\omega) y_k \right|^2 \right\} \\
&\equiv \sum_{j=1}^{n} \left| \sum_{k=1}^{n} S_{jk}(j\omega_0) y_k \right|^2 \\
&\equiv y^{*T}[S(\omega_0)]^{*T}[S(\omega_0)]\mathbf{y},
\end{aligned}$$

where ω_0 is that value of ω at which the maximum is achieved (such an ω will exist, since each $S_{jk}(\omega)$ is a bounded measurable function). Thus we have

$$\mathbf{y}^{*T}\{1_n - [S(p)]^{*T}[S(p)]\}\mathbf{y} \ge \mathbf{y}^{*T}\{1_n - [S(\omega_0)]^{*T}[S(\omega_0)]\}\mathbf{y} \ge 0,$$

since by assumption $1_n - [S(\omega)]^{*T}[S(\omega)]$ is nonnegative-definite. QED

Another characterization may now be obtained on the basis of the previous theorem. In a sense it is an analogue of Youla's representation theorem for positive-real matrices. We state it as:

Theorem 2.13. The necessary and sufficient condition for an $n \times n$ matrix $[S(p)]$ to be bounded-real is that it be expressible in $\operatorname{Re} p > 0$ as

$$[S(p)] = \left[\frac{\sigma}{\pi} \int_{-\infty}^{\infty} \frac{S_{jk}(j\zeta)}{(\zeta - \omega)^2 + \sigma^2} \, d\zeta \right],$$

where $S_{jk}(\omega)$ are the elements of a matrix $[S(\omega)]$ satisfying the conditions of Theorem 2.12.

Proof. Follows from the representations theorem for bounded analytic functions noted above and Theorem 2.12. QED

More detailed results are available in the case of scalar bounded-real functions, and since we will have occasion to use them in later chapters we will state them as:

Theorem 2.14. Let the scalar function $S(p)$ be analytic and bounded by 1 in $\operatorname{Re} p > 0$; moreover, let $S(p^*) = S(p)$; then for $\operatorname{Re} p > 0$

$$S(p) = \pm b(p)e^{-ap}\exp\left\{-\int_{-\infty}^{\infty} \frac{1 - pj\tau}{p - j\tau}\, dm(\tau)\right\},$$

where a is a real, nonnegative constant, $m(\tau)$ is a real, odd, bounded, non-decreasing function, and $b(p)$ is a regular Blaschke product formed with the zeros of $S(p)$ in $\operatorname{Re} p > 0$ as

$$b(p) = \prod_k \frac{p - a_k}{p + a_k^*} \prod_l \frac{a_l - p}{a_l^* + p}\frac{a_l^*}{a_l},$$

with a_k those zeros such that $|a_k| < 1$, and a_l the remaining ones. Moreover,

$$S_m(p) = \frac{S(p)}{\pm b(p)e^{-ap}}$$

is also a bounded (by 1) analytic function in $\operatorname{Re} p > 0$, $|S_m(j\omega)| = |S(j\omega)|$ a.e., and in addition $S_m(p)$ is nonzero in $\operatorname{Re} p > 0$.

In the special case where $S(\omega)$ is a continuous function of ω

$$S_m(p) = \frac{S(p)}{\pm b(p)e^{-ap}} = \exp\left\{\frac{p}{\pi}\int_{-\infty}^{\infty} \frac{\log|S(j\eta)|}{p^2 + \eta^2}\, d\eta\right\}.$$

Proof. We admit without proof (see, for example, [Hil]) the fact that $b(p)$ converges to an analytic function in $\operatorname{Re} p > 0$ and that $S(p)/[\pm b(p)e^{-ap}]$ is also a bounded analytic function in $\operatorname{Re} p > 0$, where a is some real, non-negative constant. Granting these facts, however, we see that we may now define a single-valued log of the function $S_m = S(p)/[\pm b(p)e^{-ap}]$, since S_m is devoid of zeros in $\operatorname{Re} p > 0$. However, with S_m bounded by 1 in $\operatorname{Re} p > 0$ we see that $-\operatorname{Re}[\log S_m] \geq 0$ for all $\operatorname{Re} p > 0$, and thus we have $S_m = \exp[-Z(p)]$, where $Z(p)$ is a positive-real function. The basic representation statement then follows from Theorem 2.8. But $|S_m(\omega)| = |S(\omega)|$ a.e.—i.e., $|b(j\omega)e^{-ja\omega}| = 1$ a.e., so that $\log|S_m(\omega)| = \log|S(\omega)| = -\operatorname{Re} Z(j\omega)$ a.e. Using Theorem 2.8 again we see that $\operatorname{Re} Z(j\omega)$ is in general a distribution given by $\operatorname{Re} Z(j\omega) = \pi(1 + \omega^2)\, dm(\omega)/d\omega$, so that $\pi(1 + \omega^2)\, dm(\omega)/d\omega = -\log|S(\omega)|$ a.e. By employing arguments similar to those discussed in the proof of

Theorem 2.10 (see, for example, [Bo1] or [Le1]), it can be shown that when $S(\omega)$ is a continuous function of ω this identification allows one to write the integral in the representation theorem as

$$-\int_{-\infty}^{\infty} \frac{1 - pj\tau}{p - j\tau}\, dm(\tau) = \int_{-\infty}^{\infty} \frac{1 - pj\tau}{p - j\tau} \frac{\log|S(j\tau)|}{\pi(1 + \tau^2)}\, d\tau,$$

and, since $|S(j\tau)| = |S(-j\tau)|$,

$$\int_{-\infty}^{\infty} \frac{1 - pj\tau}{p - j\tau} \frac{\log|S(j\tau)|}{\pi(1 + \tau^2)}\, d\tau = \frac{1}{2}\left\{\int_{-\infty}^{\infty} \left[\frac{1 - pj\tau}{p - j\tau} + \frac{1 + pj\tau}{p + j\tau}\right] \frac{\log|S(j\tau)|}{\pi(1 + \tau^2)}\, d\tau\right\},$$

which with algebraic simplification is the result stated. QED

For completeness we would also mention the Paley–Wiener theorem for bounded functions; the necessary and sufficient condition such that a bounded measurable function $A(\omega)$ be the magnitude of the boundary value of some function $A(p)$ which is bounded and analytic in $\operatorname{Re} p > 0$ is that

$$\int_{-\infty}^{\infty} \frac{\log A(\omega)}{1 + \omega^2}\, d\omega < \infty.$$

We see a demonstration of the sufficiency portion of this result in Theorem 2.14, i.e., given that

$$\int_{-\infty}^{\infty} \frac{\log|S(j\omega)|}{1 + \omega^2}\, d\omega < \infty$$

we may establish that the integral

$$\int_{-\infty}^{\infty} \frac{1 - pj\tau}{p - j\tau} \frac{\log|S(j\tau)|}{\pi(1 + \tau^2)}\, d\tau = S_m(p)$$

exists for all $\operatorname{Re} p > 0$, is bounded in that half plane, and, moreover,

$$\lim_{\operatorname{Re} p \to 0} |S_m(p)| = |S(j\omega)| \text{ a.e.}$$

The previous theorems dealt with the general class of bounded-real matrices. We noted in Theorem 2.4 that the class of lossless systems (Definition 8) is characterized by a subset of all bounded-real matrices—specifically, those that are unitary for $p = j\omega$, i.e., $[S(\omega)]^{*T}[S(\omega)] = 1_n$. We will have many occasions in subsequent chapters to consider lossless systems, and the following alternate characterizations will prove very useful:

Theorem 2.15 [Yo1]. Given an $n \times n$ bounded-real matrix of meromorphic functions $[S(p)]$ such that $[S(\omega)]^{*T}[S(\omega)] = 1_n$, then for all p, $[S(-p)]^T[S(p)] = 1_n$.

Proof. First we note from $[S(\omega)]^{*T}[S(\omega)] = 1_n$ that $|\det[S(\omega)]| = 1$ for almost all ω. Therefore $\det[S(p)]$, which is an analytic function in $\operatorname{Re} p > 0$ when $[S(p)]$ is bounded-real, cannot vanish identically in $\operatorname{Re} p > 0$. Therefore we know that $[S(p)]^{-1}$ exists, and its elements are then meromorphic functions given that the elements of $[S(p)]$ are meromorphic. Now, let $[A(-p)]^T = [S(p)]^{-1}$ and note that the meromorphic nature of the elements of $[A(p)]$ and $[S(p)]$ together with the fact that $|\det[S(\omega)]| = 1$ implies that we can always find a region of the p plane such that some finite interval of the $p = j\omega$ axis is contained in its interior, and, moreover, throughout this region $[A(p)]$ and $[S(p)]$ are analytic. However,

$$\lim_{\substack{\operatorname{Re} p \to 0 \\ (\operatorname{Re} p \leq 0)}} A(p) \equiv A(j\omega) = \lim_{\substack{\operatorname{Re} p \to 0 \\ (\operatorname{Re} p \geq 0)}} [S(-p)]^{-1T} = \{[S(j\omega)]^{*T}\}^{-1}.$$

But by assumption $[S(j\omega)]^{*T}[S(j\omega)] = 1_n$, so that

$$\lim_{\substack{\operatorname{Re} p \to 0 \\ (\operatorname{Re} p \leq 0)}} [A(p)] = [S(j\omega)],$$

and we conclude that $[A(p)]$ is the analytic continuation of the matrix $[S(p)]$ into $\operatorname{Re} p < 0$; we have shown that two functions analytic in some common domain have the same value on all points of some arc contained in the interior of the domain, and thus that they must be analytic continuations of each other (see, for example, [Hil]). Therefore for $\operatorname{Re} p < 0$, $[S(p)] = [[S(-p)]^T]^{-1} = [A(p)]$. QED

The following result is of general interest, since it shows that the explicit assumption that the elements of the matrix be meromorphic functions need not be made but may be replaced with equivalent statements:

Corollary 2.1. Let $[S(p)]$ be a bounded-real matrix satisfying $1_n = [S(\omega)]^{*T} [S(\omega)]$ a.e. The sufficient conditions guaranteeing that each element in $[S(p)]$ is actually meromorphic, and thus $[S(-p)]^T[S(p)] = 1_n$ for all p, are either (1) each element in $[S(p)]$ be a continuous function when $p = j\omega$, or (2) there exist positive constants K and ε such that

$$1/\det[S(p)] \leq K \text{ for all } \varepsilon > \operatorname{Re} p > 0.$$

Proof. To the author's knowledge the proof of Condition 1 was first given by Youla in [Yo1] and proceeds as follows. As in the proof of Theorem 2.15 we define $[A(-p)]^T = [S(p)]^{-1}$, and since $[S(p)]$ is analytic in $\operatorname{Re} p > 0$, we see that $[A(-p)]^T$ is defined except for isolated singularities—namely, poles in $\operatorname{Re} p > 0$. Again, as before,

$$\lim_{\substack{\operatorname{Re} p \to 0 \\ (\operatorname{Re} p \leq 0)}} [A(p)] = \lim_{\substack{\operatorname{Re} p \to 0 \\ (\operatorname{Re} p \geq 0)}} [S(p)]$$

pointwise a.e., since by Fatou's theorem each element of $[S(p)]$, being analytic and bounded in $\operatorname{Re} p > 0$, has such a boundary value. But by assumption, each element of $[S(j\omega)]$ is a continuous function of ω, so that the above limits hold for every value of ω. Now we appeal to a sharper version of the classic analytic continuation theorem [Hil] which states that if two functions analytic in domains with a common boundary have equal boundary values along some arc of the common boundary and, moreover, these boundary values are continuous functions, then the two functions must be analytic continuations of each other. The result then follows as in the proof of Theorem 2.15.

With regard to the second sufficient condition we note that the matrix $[A(p)]$ is defined as before, but now with the assumption concerning $\det[S(p)]$

$$|A_{jk}(p)| \leq C$$

we have for $-\varepsilon < \operatorname{Re} p < 0$. Thus we have, using a classic dominated convergence argument, that for all $\varphi \in C_0^{\infty}$

$$\lim_{\substack{\operatorname{Re} p \to 0 \\ (\operatorname{Re} p \leq 0)}} \int [A(\sigma + j\omega)]\varphi(\omega)\, d\omega = \lim_{\substack{\operatorname{Re} p \to 0 \\ (\operatorname{Re} p \geq 0)}} \int [S(\sigma + j\omega)]\varphi(\omega)\, d\omega,$$

since, as before $[A(j\omega)] = [S(j\omega)]$ a.e. Appealing to a simple modification of Theorem 1.7 than allows us to conclude that $[S(p)]$ and $[A(p)]$ are analytic continuations of each other, and the result follows again as in the proof of Theorem 2.15. QED

The simple function $e^{-1/p}$, which is bounded-real but not meromorphic, shows that the above result is nontrivial, since this function is neither continuous for $p = j\omega$ nor can one find the required bound on $1/e^{-1/p}$, and it is certainly not meromorphic

2.3.3. Complex Normalization on Bounded-Real Matrices

In many applications the bounded-real matrices represent a system in which the inputs **a** and the outputs **b** are in fact combinations of other vectors **V** and **I**—e.g., $\mathbf{a} = \frac{1}{2}(\mathbf{V} + \mathbf{I})$ and $\mathbf{b} = \frac{1}{2}(\mathbf{V} - \mathbf{I})$; the matrix $[S(p)]$ is then a mapping between $\mathbf{V} + \mathbf{I}$ and $\mathbf{V} - \mathbf{I}$. In particular, when one considers boundary conditions of the form $V_j = -I_j, j = 2, \ldots, n$, then $a_j = 0, j \neq 1$, and we obtain

$$\frac{V_1 - I_1}{V_1 + I_1} = S_{11}(p).$$

For example, when **V** and **I** correspond to the voltages and currents of the n-port of Fig. 2.1 we see that such boundary conditions correspond to the presence of unit resistors across ports 2 to n and

$$S_{11}(p) = \frac{V_1 - I_1}{V_1 + I_1} = \frac{Z_{in} - 1}{Z_{in} + 1},$$

where $Z_{in} = V_1/I_1$ is the resulting input impedance of the system with these resistors in place. It will become evident in many of the applications discussed in Chapters III and IV that this connection with at least one element of the [S] matrix is of direct utility. We would like to extend this result so that it holds for other than unit resistor terminations, but the extension is to be done in a way that will still allow us to deal with bounded-real matrices as representations of the system. Such a procedure was proposed by Youla [Yo2], and we will describe it as a transformation between bounded-real matrices. The technique was extended by Rohrer [Ro1] to n-port normalizations, but we will restrict our discussion to the simpler case.

We will say that an $n \times n$ matrix $[S_Z(p)]$ is normalized to n functions of p, $Z_i(p)$, $i = 1, \ldots, n$, if $\mathbf{b}_Z = [S_Z]\mathbf{a}_Z$, and

$$2[H(p)]\mathbf{a}_Z = \mathbf{V} + [Z]\mathbf{I}$$

and

$$2[H(-p)]\mathbf{b}_Z = \mathbf{V} - [Z(-p)]\mathbf{I}$$

relate the variables $\mathbf{V}$ and $\mathbf{I}$ to $\mathbf{a}_Z$ and $\mathbf{b}_Z$. Here $[Z]$ is an $n \times n$ matrix all of whose elements are zero except the diagonal ones, with $Z_{kk}(p) = Z_k(p)$ one of the given functions of p. The matrix $[H]$ is also diagonal, with

$$h_{kk}(p)h_{kk}(-p) = \tfrac{1}{2}\{Z_k(p) + Z_k(-p)\}$$

and $h_{kk}(p)$ defined as that factorization of the expression in which $h_{kk}(p)$ is analytic in $\operatorname{Re} p > 0$ and nonzero in $\operatorname{Re} p < 0$. We note that in general such a factorization may not be possible, and we therefore define for future use the following class of functions:

Definition 10. $Z(p)$ is a regular normalizing function if:

1. $Z(p)$ is a meromorphic positive-real function.
2. The expression $h(p)h(-p) = \frac{1}{2}\{Z(p) + Z(-p)\}$ determines a unique $h(p)$ such that $h(p)$ is (a) a meromorphic function analytic in $\operatorname{Re} p > 0$, (b) nonzero in $\operatorname{Re} p < 0$ but having the same zeros as $Z(p) + Z(-p)$ in $\operatorname{Re} p \geq 0$, and (c) both $h(p)$ and $h(p)/h(-p)$ are bounded in $\operatorname{Re} p > 0$ such that (d) $\lim_{|p| \to \infty} [\log h(p)/p] = 0$ in $\operatorname{Re} p > 0$ and (e) they satisfy $h(\sigma) \geq 0$ when $\sigma = \operatorname{Re} p < 0$.

The prototype for this definition is the class of rational positive-real functions, $Z(p)$, such that $Z(p) + Z(-p)$, or, equivalently, $\operatorname{Re} Z(j\omega)$, is not identically zero, since the factorization of such functions may always be performed in a way that meets all the requirements stated in the definition. In general, however, an arbitrary meromorphic positive-real function may not be "regular" either because the function $h(p)$ may not exist or the five conditions noted in part 2 of the definition are not sufficient to determine $h(p)$ uniquely. With regard to the uniqueness, we note that if $h(p)$ satisfied the five conditions, we could find another function $\hat{h}(p) = \pm h(p)b(p)e^{-ap}$ [where $b(p)$ is an arbitrary Blaschke function] which still satisfied the expression

$$\begin{aligned} h(p)h(-p) &= (\pm h(p)b(p)e^{-ap})(\pm h(-p)b(-p)^{+ap}) = h(p)h(-p) \\ &= \tfrac{1}{2}\{Z(p) + Z(-p)\}. \end{aligned}$$

However, the new function $\hat{h}(p)$ would not satisfy all the conditions, since from Condition (e) we note that the arbitrary ($\pm$) ambiguity would be resolved [this condition is no more restrictive, since by Condition (b), $h(p)$ is nonzero in $\operatorname{Re} p < 0$], Condition (b) would not allow the arbitrary Blaschke product, and finally, Condition (d) would require that $a = 0$. Thus the conditions do imply that a unique factorization is determined at least with regard to arbitrary multipliers of the type $(\pm b(p)e^{-ap})$.

Before we determine some of the properties of matrices normalized in the way indicated above, which we will refer to as complex normalization, we note that such a technique has at least succeeded in relating one element of the resultant scattering matrix to a corresponding input impedance of the network, since we see that

$$S_{11}(p) = \frac{Z_{\text{in}}(p) - Z_1(-p)}{Z_{\text{in}}(p) + Z_1(p)} \frac{h_1(p)}{h_1(-p)}$$

when the boundary conditions, $V_j = -Z_j(p)I_j$, are imposed on ports 2 through n and Z_1 is the normalizing impedance at port 1.

In order to phrase the complex normalization as a transformation between scattering matrices, let us assume that we are given two scattering matrices $[S_1]$ and $[S_2]$, where $\mathbf{b}_1 = [S_1]\mathbf{a}_1$, with

$$2[H_1]\mathbf{a}_1 = \mathbf{V} + [Z_1]\mathbf{I},$$

$$2[H_1(-p)]\mathbf{b}_1 = \mathbf{V} - [Z_1(-p)]\mathbf{I},$$

and

$$\mathbf{b}_2 = [S_2]\mathbf{a}_2,$$

with

$$2[H_2]\mathbf{a}_2 = \mathbf{V} + [Z_2]\mathbf{I}$$

and

$$2[H_2(-p)]\mathbf{b}_2 = \mathbf{V} - [Z_2(-p)]\mathbf{I};$$

however, **V** and **I** are the same in both cases. We may solve for **V** and **I** as, for example,

$$\mathbf{I} = [H_2(-p)]^{-1}\mathbf{a}_2 - [H_2]^{-1}\mathbf{b}_2$$

and

$$\mathbf{V} = [Z_2(-p)][H_2(-p)]^{-1}\mathbf{a}_2 + [Z_2][H_2]^{-1}\mathbf{b}_2$$

and substitute these expressions into the defining equations for $\mathbf{a}_1$ and $\mathbf{b}_1$ to obtain

$$2[H_1]\mathbf{a}_1 = \{([Z_2(-p)] + [Z_1])[H_2(-p)]^{-1} + ([Z_2] - [Z_1])[H_2]^{-1}[S_2]\}\mathbf{a}_2$$

and

$$\begin{aligned}2[H_1(-p)]\mathbf{b}_1 = \{&([Z_2(-p)] - [Z_1(-p)])[H_2(-p)]^{-1}\\ &+([Z_2] + [Z_1(-p)])[H_2]^{-1}[S_2]\}\mathbf{a}_2 .\end{aligned}$$

But if $\mathbf{b}_1 = [S_1]\mathbf{a}_1$, we see, using the fact that $\mathbf{a}_2$ is an arbitrary n-vector, that

$$\begin{aligned}[S_1] = [H_1(-p)]^{-1}&\{([Z_2(-p)] - [Z_1(-p)])[H_2(-p)]^{-1}\\ &+([Z_2] + [Z_1(-p)])[H_2]^{-1}[S_2]\}\\ &\times\{([Z_2(-p)] + [Z_1])[H_2(-p)]^{-1} + ([Z_2] - [Z_1])[H_2]^{-1}[S_2]\}^{-1},\end{aligned}$$

assuming of course that the latter matrix does possess an inverse. Thus we see that a given scattering matrix with a specific normalization may be transformed to another scattering matrix with some other normalization. A pertinent question is whether these transformed matrices are still bounded-real if the system being studied is passive. The answer to this question is yes, and we state this as:

Theorem 2.16. Let $[S_{1_n}]$ be an $n \times n$ bounded-real matrix normalized to $[Z] = [1_n]$, and assume that all its elements are meromorphic functions of p. Let $[S_Z]$ be the transformed matrix obtained by selecting an arbitrary normalizing matrix $[Z]$ all of whose elements are regular normalizing functions. Then $[S_Z]$ will be a bounded-real matrix all of whose elements are meromorphic functions of p. Moreover, if $[S_{1_n}]^T = [S_{1_n}]$, then $[S_Z]^T = [S_Z]$, or if $[S_{1_n}(-p)]^T[S_{1_n}(p)] = 1_n$, then $[S_Z(-p)]^T[S_Z(p)] = 1_n$.

Proof. We first establish that $[S_Z]$ will exist, and in the process of accomplishing this we will need the following: there exists a positive-real matrix $[Y_a]$ such that $\mathbf{I} = [Y_a](\mathbf{V} + [Z]\mathbf{I})$. If we let $\mathbf{a}_1$ and $\mathbf{b}_1$ be the variables associated with $[S_{1_n}]$, then from $\mathbf{V} = \mathbf{a}_1 + \mathbf{b}_1$ and $\mathbf{I} = \mathbf{a}_1 - \mathbf{b}_1$ we obtain

$$\mathbf{V} + [Z]\mathbf{I} = \mathbf{a}_1 + \mathbf{b}_1 + [Z](\mathbf{a}_1 - \mathbf{b}_1) = \{1_n + [S_{1_n}] + [Z](1_n - [S_{1_n}])\}\mathbf{a}_1 .$$

But we see that $\mathbf{I} = (1_n - [S_1])\mathbf{a}_1$, so that if $\{1_n + [S_{1_n}] + [Z](1_n - [S_{1_n}])\}^{-1}$ exists, we will have

$$\mathbf{I} = (1_n - [S_{1_n}])\{1 +_n [S_{1_n}] + [Z](1_n - [S_{1_n}])\}^{-1}(\mathbf{V} + [Z]\mathbf{I}).$$

However,

$$1_n + [S_{1_n}] + [Z](1_n - [S_{1_n}]) = ([Z] + 1_n) \times \{1_n - ([Z] + 1_n)^{-1}([Z] - 1_n)[S_{1n}]\},$$

and since $[Z]$ is by assumption positive-real, the determinant of $([Z] + 1_n)$ will not vanish in $\operatorname{Re} p > 0$ (see the proof of Theorem 2.6), so that the expression for $\mathbf{I}$ will follow if $\det\{1_n - ([Z] + 1_n)^{-1}([Z] - 1_n)[S_{1_n}]\} \neq 0$. If we assume that it did vanish at some point $p = p_0$, then there must exist some constant n-vector $\mathbf{y}$ such that

$$\mathbf{y} = ([Z] + 1_n)^{-1}([Z] - 1_n)[S_{1_n}]\mathbf{y}$$

for that value of p, and this implies that

$$\mathbf{y}^{*T}\mathbf{y} = \mathbf{y}^{*T}[S_{1_n}]^{*T}\{([Z] + 1_n)^{-1}([Z] - 1_n)\}^{*T} \times \{([Z] + 1_n)^{-1}([Z] - 1_n\}[S_{1_n}]\mathbf{y}.$$

But since each normalizing function is positive-real, we see that

$$\mathbf{x}^{*T}\mathbf{x} - \mathbf{x}^{*T}\{([Z] + 1_n)^{-1}([Z] - 1_n)\}^{*T}\{([Z] + 1_n)^{-1}([Z] - 1_n)\}\mathbf{x} > 0$$

in $\operatorname{Re} p > 0$ because

$$1_n - \{([Z] + 1_n)^{-1}([Z] - 1_n)\}^{*T}\{([Z] + 1_n)^{-1}([Z] - 1_n)\}$$

is a matrix with nonzero elements on its diagonal, i.e., these diagonal elements are given by

$$1 - \left|\frac{Z_i(p) - 1}{Z_i(p) + 1}\right|^2 > 0$$

(in $\operatorname{Re} p > 0$), where Z_i is the ith normalizing function. Therefore

$$\mathbf{y}^{*T}\mathbf{y} = \mathbf{y}^{*T}[S_{1_n}]^{*T}\{([Z] + 1_n)^{-1}([Z] - 1_n)\}^{*T}\{([Z] + 1_n)^{-1}([Z] - 1_n)\} \times [S_{1_n}]\mathbf{y} < \mathbf{y}^{*T}[S_{1_n}]^{*T}[S_{1_n}]\mathbf{y} \leq \mathbf{y}^{*T}\mathbf{y},$$

where the last inequality follows from the given fact that $[S_{1_n}]$ is bounded-real. We have arrived at a contradiction, so that we may assume that the determinant will not vanish in $\operatorname{Re} p > 0$, and we have shown that the matrix $[Y_a]$ exists. Moreover, we note that we have also established that each element in $[Y_a]$ is an analytic function of p for $\operatorname{Re} p > 0$, since, by assumption $[S_{1_n}]$ is bounded-real and $[Z]$ is a regular normalizing matrix. Now we may conclude this digression by showing that $[Y_a]^{*T} + [Y_a]$ is nonnegative-definite in $\operatorname{Re} p > 0$ or that finally $[Y_a]$ is positive-real. Thus

$$\begin{aligned}
&(\mathbf{V} + [Z]\mathbf{I})^{*T}\mathbf{I} + \mathbf{I}^{*T}(\mathbf{V} + [Z]\mathbf{I}) \\
&\quad = (\mathbf{V} + [Z]\mathbf{I})^{*T}[Y_a](\mathbf{V} + [Z]\mathbf{I}) + (\mathbf{V} + [Z]\mathbf{I})^{*T}[Y_a]^{*T}(\mathbf{V} + [Z]\mathbf{I}) \\
&\quad = (\mathbf{V} + [Z]\mathbf{I})^{*T}\{[Y_a]^{*T} + [Y_a])\}(\mathbf{V} + [Z]\mathbf{I}) \\
&\quad = \mathbf{V}^{*T}\mathbf{I} + \mathbf{I}^{*T}\mathbf{V} + \mathbf{I}^{*T}\{[Z]^{*T} + [Z]\}\mathbf{I} \\
&\quad = 2[\mathbf{a}_1{}^{*T}\mathbf{a}_1 - \mathbf{b}_1{}^{*T}\mathbf{b}_1] + \mathbf{I}^{*T}\{[Z]^{*T} + [Z]\}\mathbf{I} \geq 0
\end{aligned}$$

since $[S_{1_n}]$ being bounded-real yields $\mathbf{a}_1{}^{*T}\mathbf{a}_1 - \mathbf{b}_1{}^{*T}\mathbf{b}_1 \geq 0$ and $[Z]$ is a regular normalizing matrix and thus positive-real.

Now we return to the definition of $\mathbf{b}_z$ and $\mathbf{a}_z$ (which denote the variables normalized to $[Z]$) to obtain

$$\begin{aligned}
\mathbf{b}_Z &= \tfrac{1}{2}[H(-p)]^{-1}(\mathbf{V} - [Z(-p)]\mathbf{I}) \\
&= \tfrac{1}{2}[H(-p)]^{-1}\{\mathbf{V} + [Z]\mathbf{I} - ([Z(-p)] + [Z])\mathbf{I}\} \\
&= \tfrac{1}{2}[H(-p)]^{-1}\{\mathbf{V} + [Z]\mathbf{I} - 2[H][H(-p)][Y_a](\mathbf{V} + [Z]\mathbf{I})\} \\
&= \{[H][H(-p)]^{-1} - 2[H][Y_a][H]\}\mathbf{a}_Z
\end{aligned}$$

so that

$$[S_Z] = [H][H(-p)]^{-1} - 2[H][Y_a][H]$$

exists. Moreover since $[Y_a]$ was shown to be positive-real and $[H][H(-p)]^{-1}$ as we note from the definition of regular normalizing impedances will be analytic and bounded (together with $[H]$) in $\operatorname{Re} p > 0$, then $[S_Z]$ will have elements that are meromorphic functions analytic in $\operatorname{Re} p > 0$ and asymptotically bounded by a polynomial (see the representation of Theorem 2.8) in $\operatorname{Re} p > 0$.

Now, by direct calculation we find, since $[S_{1_n}]$ is bounded-real, that for $\operatorname{Re} p = 0$

$$\tfrac{1}{2}[\mathbf{V}^{*T}\mathbf{I} + \mathbf{I}^{*T}\mathbf{V}] = \mathbf{a}_Z^{*T}\{1_n - [S_Z]^{*T}[S_Z]\}\mathbf{a}_Z = \mathbf{a}_1^{*T}\{1_n - [S_{1_n}]^{*T}[S_{1_n}]\}\mathbf{a}_1 \geq 0$$

for all constant n-vectors $\mathbf{a}_1$. But we also find that

$$\mathbf{a}_1 = 2\{1_n - ([Z] + 1_n)^{-1}([Z] - 1_n)[S_{1_n}]\}^{-1}([Z] + 1_n)^{-1}[H]\mathbf{a}_Z,$$

which implies that given an arbitrary $\mathbf{a}_Z$, we can find for almost all ω a corresponding $\mathbf{a}_1$, i.e., with $[S_{1_n}]$ and $[Z]$ meromorphic we see that the inverse matrix will exist for almost all ω since we have established that it exists for all $\operatorname{Re} p > 0$. Thus we may conclude that

$$\mathbf{a}_Z^{*T}\{1_n - [S_Z]^{*T}[S_Z]\}\mathbf{a}_Z \geq 0$$

for arbitrary constant n-vectors $\mathbf{a}_Z$ for almost all values of ω. This implies that each element of $[S_Z]$ is bounded by 1 for all $p = j\omega$, and thus that they are analytic functions on $p = j\omega$, since they will be, as we demonstrated above, meromorphic functions of p. We may then bring this portion of the proof to a close by employing the same arguments used in Theorem 2.12 to establish that

$$\mathbf{y}^{*T}\{1_n - [S_Z]^{*T}[S_Z]\}\mathbf{y} \geq 0$$

for all $\operatorname{Re} p > 0$ with $\mathbf{y}$ an arbitrary constant n-vector. These arguments are appropriate, since the fact that each element of $[S_Z]$ is meromorphic, analytic, and bounded by 1 for $p = j\omega$ and bounded asymptotically by a polynomial in $\operatorname{Re} p > 0$ implies that these elements are bounded functions for all $\operatorname{Re} p > 0$.

Finally, we note from the expression $[S_Z] = [H][H(-p)]^{-1} - 2[H] \times [Y_a][H]$ that $[S_Z]^T = [H][H(-p)]^{-1} - 2[H][Y_a]^T[H]$, since $[H]$ is diagonal. Moreover, we have shown that

$$[Y_a] = (1_n - [S_{1_n}])\{1_n + [S_{1_n}] + [Z](1_n - [S_{1_n}])\}^{-1},$$

so that when $[S_{1_n}]^T = [S_{1_n}]$ we have

$$[Y_a]^T = \{1_n + [S_{1_n}] + (1_n - [S_{1_n}])[Z]\}^{-1}(1_n - [S_{1_n}]).$$

But we have the identity

$$\begin{aligned}(1_n - [S_{1_n}])\{1 + [S_{1_n}] + [Z](1_n - [S_{1_n}])\}\\ \equiv \{1_n + [S_{1_n}] + (1_n - [S_{1_n}])[Z]\}(1_k - [S_{1_n}]),\end{aligned}$$

so we see that $[Y_a]^T = [Y_a]$, and thus $[S_Z]^T = [S_Z]$ when $[S_{1_n}]^T = [S_{1_n}]$. QED

We might note that the physical interpretation of the fact that $[S_Z]^T = [S_Z]$ when $[S_{1_n}]^T = [S_{1_n}]$ is that the reciprocity of the system is invariant under changes of the normalization, i.e., reciprocity is still in one-to-one correspondence with the symmetry of the scattering matrix.

We will use this theorem in some of the applications described in Chapter IV, but for the discussion in Chapter III we will need what amounts to a converse of the above result. In effect, we will show that a given matrix, even though it be bounded-real, may not be the scattering matrix of some passive system normalized to a given $[Z]$. We will establish the desired result under the assumption that the matrices are rational, although similar results can be obtained in the more general case of meromorphic matrices. Thus we will establish:

Theorem 2.17 [Wo3]. The necessary and sufficient conditions that an $n \times n$ matrix $[S_Z]$ of rational functions be the transformation of some rational bounded-real matrix $[S_{1_n}]$, where the regular normalizing matrix $[Z]$

consists of rational functions, are:

1. $1_n - [S_Z]^{*T}[S_Z]$ be nonnegative-definite for $p = j\omega$.
2. $[Y] = \frac{1}{2}[H]^{-1}\{[H][H(-p)]^{-1} - [S_Z]\}[H]^{-1}$ be analytic in $\operatorname{Re} p > 0$,

and either:

3(a). $\det\{1_n - ([Z] - 1_n)[Y]\} \neq 0$ in $\operatorname{Re} p > 0$, or 3(b) the matrix $\{1_n - ([Z] - 1_n)[Y]\}([Z] + 1_n)$ have simple poles when $p = j\omega$, and the matrix formed with the residues at these poles be nonnegative-definite.

Moreover, if $[S_Z]^T = [S_Z]$, then $[S_{1_n}]^T = [S_{1_n}]$.

Proof. In this case we may use the transformation between the matrices in the form

$$\begin{aligned} &1_n - \tfrac{1}{2}([Z] + 1_n)[H]^{-1}([H][H(-p)]^{-1} - [S_Z])[H]^{-1} \\ &\quad = [S_{1_n}]\{1_n - \tfrac{1}{2}([Z] - 1_n)[H]^{-1}([H][H(-p)]^{-1} - [S_Z])[H]^{-1}\}, \end{aligned}$$

or, by means of the identity

$$\begin{aligned} &([Z(-p)] - 1_n)[H(-p)]^{-1}[H]^{-1} + ([Z] + 1_n)[H][H(-p)]^{-1}[H]^{-1}[H]^{-1} \\ &\quad \equiv ([Z(-p)] + 1_n)[H(-p)]^{-1}[H]^{-1} \\ &\quad + ([Z] - 1_n)[H][H(-p)]^{-1}[H]^{-1}[H]^{-1} \\ &\quad \equiv 2, \end{aligned}$$

we have

$$1_n - ([Z] + 1_n)[Y] = [S_{1_n}]\{1_n - ([Z] - 1_n)[Y]\},$$

where

$$[Y] = \tfrac{1}{2}[H]^{-1}([H][H(-p)]^{-1} - [S_Z])[H]^{-1}.$$

But we may also express the transformation between a_Z and a_{1_n} as

$$2[H]a_Z = \mathbf{V} + [Z]\mathbf{I} = \{(1_n + [Z]) + (1_n - [Z])[S_{1_n}]\}\mathbf{a}_{1_n}$$

using $\mathbf{I} = \mathbf{a}_{1_n} - \mathbf{b}_{1_n}$ and $\mathbf{V} = \mathbf{a}_{1_n} + \mathbf{b}_{1_n}$. Combining this statement with the one obtained by expressing $\mathbf{I}$ and $\mathbf{V}$ in terms of $\mathbf{a}_Z$ and $\mathbf{b}_Z$, i.e.,

$$2\mathbf{a}_{1_n} = \{(Z(-p) + 1_n)[H(-p)]^{-1} + ([Z] - 1_n)[H]^{-1}[S_Z]\}\mathbf{a}_Z,$$

we obtain

$$\begin{aligned} &\tfrac{1}{2}\{(1_n + [Z]) + (1_n - [Z])[S_{1_n}]\} \\ &\quad \times \tfrac{1}{2}\{([Z(-p)] + 1_n)[H(-p)]^{-1}[H]^{-1} + ([Z] - 1_n)[H]^{-1}[S_Z][H]^{-1}\} \\ &\quad = 1_n, \end{aligned}$$

which may also be written, using the above identity, as

$$\begin{aligned} &\tfrac{1}{2}\{(1_n + [Z]) + (1_n - [Z])[S_{1_n}]\} \\ &\quad \times \{1_n - \tfrac{1}{2}([Z] - 1_n)[H]^{-1}([H][H(-p)]^{-1} - [S_Z])[H]^{-1}\} = 1_n. \end{aligned}$$

Now we may establish the necessity of the statements of the theorem. Thus from the last equation we have

$$\det \tfrac{1}{2}\{(1_n + [Z]) + (1_n - [Z])[S_{1_n}]\} \times \det\{1_n - \tfrac{1}{2}([Z] - 1_n)[H]^{-1}([H][H(-p)]^{-1} - [S_Z])[H]^{-1}\} = 1,$$

so that if $[S_{1_n}]$ is to be analytic in $\operatorname{Re} p > 0$,

$$\det\{1_n - ([Z] - 1_n)[Y]\} \neq 0$$

in $\operatorname{Re} p > 0$, since $[Z]$, being a regular normalizing matrix, is analytic in $\operatorname{Re} p > 0$. We may also use the first relationship obtained between $[S_{1_n}]$ and $[S_Z]$ to obtain the expression

$$[Y] = \{[S_{1_n}]([Z] - 1_n) - ([Z] + 1_n)\}^{-1}([S_{1_n}] - 1_n).$$

Using the arguments employed in the proof of Theorem 2.16, we see that if $[S_{1_n}]$ is to be bounded-real, then

$$[S_{1_n}]([Z] - 1_n) - ([Z] + 1_n) = \{[S_{1_n}]([Z] - 1_n)([Z] + 1_n)^{-1} - 1_n)\}([Z] + 1_n)$$

must possess an analytic inverse; thus it is necessary that each element in $[Y]$ be analytic in $\operatorname{Re} p > 0$. Also, as we noted in the proof of Theorem 2.16, one has

$$\tfrac{1}{2}[\mathbf{V}^{*T}\mathbf{I} + \mathbf{I}^{*T}\mathbf{V}] = \mathbf{a}_Z^{*T}\{1_n - [S_Z]^{*T}[S_Z]\}\mathbf{a}_Z = \mathbf{a}_{1_n}^{*T}\{1_n - [S_{1_n}]^{*T}[S_{1_n}]\}\mathbf{a}_{1_n}$$

for $p = j\omega$, so that if $[S_{1_n}]$ is to be bounded-real, $1_n - [S_Z]^{*T}[S_Z]$ must be nonnegative-definite for $p = j\omega$. To establish the necessity of the last statement, the alternate Condition to 3(a) we rewrite the second form of the transformation between $[S_{1_n}]$ and $[S_Z]$ as

$$\begin{aligned}&\{1_n - \tfrac{1}{2}([Z] - 1_n)[H]^{-1}([H][H(-p)]^{-1} - [S_Z])[H]^{-1}\}\\ &\quad\times \tfrac{1}{2}\{([Z] + 1_n) + (1_n - [Z])[S_{1_n}]\}\\ &\quad= \{1_n - ([Z] - 1_n)[Y]\}\tfrac{1}{2}([Z] + 1_n)\{1_n - ([Z] + 1_n)^{-1}([Z] - 1_n)[S_{1_n}]\}\\ &\quad= 1_n,\end{aligned}$$

or

$$\{1_n - ([Z] - 1_n)[Y]\}\tfrac{1}{2}([Z] + 1_n) = \{1_n - ([Z] + 1_n)^{-1}([Z] - 1_n)[S_{1_n}]\}^{-1}.$$

But $([Z] + 1_n)^{-1}([Z] - 1_n)$ is bounded real, so that

$$1_n - ([Z] + 1_n)^{-1}([Z] - 1_n)[S_{1_n}]$$

must be positive-real if $[S_{1_n}]$ is to be bounded-real (see Lemma 2.3). The fact that the determinant of the matrix does not vanish, as noted previously, again allows us to appeal to Lemma 2.3 to conclude that its inverse must be positive-real, and the necessity of Condition 3(b) then follows from Theorem 2.10.

The sufficiency of these statements may be established directly. For example, using the first form of the transformation between $[S_{1_n}]$ and $[S_Z]$, we have

$$[S_{1_n}] = \{1_n - ([Z] + 1_n)[Y]\}\{1_n - ([Z] - 1_n)[Y]\}^{-1},$$

which will be bounded-real: Conditions 2 and 3(a) guarantee that each element of $[S_{1_n}]$ will be analytic in $\operatorname{Re} p > 0$; Condition 1 implies that $1_n - [S_{1_n}]^{*T}[S_{1_n}]$ will be nonnegative-definite for $p = j\omega$; since the resulting matrix $[S_{1_n}]$ is rational, then its analyticity in $\operatorname{Re} p > 0$, together with the nonnegative-definitions of $1_n - [S_{1_n}]^{*T}[S_{1_n}]$ on $p = j\omega$, implies that each element will be bounded in $\operatorname{Re} p > 0$, and we may again use the techniques employed in the proof of Theorem 2.12 to show directly that

$$1_n - [S_{1_n}]^{*T}[S_{1_n}]$$

will be nonnegative-definite in all $\operatorname{Re} p > 0$. The sufficiency of the alternate statements, i.e., Conditions 1, 2, and 3(b) may be established by showing that they imply that

$$\{1_n - ([Z] - 1_n)[Y]\}\tfrac{1}{2}([Z] + 1_n) = [M(p)]$$

will be a positive-real matrix that possesses a positive-real inverse. But this means that

$$\det\{1_n - ([Z] - 1_n)[Y]\} \neq 0$$

in $\operatorname{Re} p > 0$, as we noted in Lemma 2.3, and we could then appeal to the first set of conditions to conclude the demonstration. But we see that these conditions do imply that the above matrix is positive-real since the fact that $1_n - [S_Z]^{*T}[S_Z]$ is nonnegative-definite for $p = j\omega$ implies, as we may show directly, that $[M]^{*T} + [M]$ is nonnegative-definite for almost all $p = j\omega$, and we may then appeal to Theorem 2.10 using Condition 3(b). If we can establish that $\det M(p)$ exists for at least one value of p in $\operatorname{Re} p \geq 0$, then, as we also noted in Lemma 2.3, this implies that it will not vanish in all $\operatorname{Re} p > 0$. To accomplish this we write

$$\begin{aligned}&\{1_n - ([Z] - 1_n)[Y]\}([Z] + 1_n)\\ &\quad = \tfrac{1}{2}([Z] + 1_n)[H]^{-1}\{([Z] + 1_n)^{-1}([Z(-p)] + 1_n)[H][H(-p)]^{-1}\\ &\qquad + ([Z] + 1_n)^{-1}([Z] - 1_n)[S_Z]\}[H]^{-1}([Z] + 1_n)\end{aligned}$$

using the identity noted above. But since $\det\{([Z] + 1_n)[H]^{-1}\}$ does not vanish for all $p = j\omega$, it will be sufficient to establish that

$$\begin{aligned}\det\{&([Z] + 1_n)^{-1}([Z(-p)] + 1_n)[H][H(-p)]^{-1}\\ &+ ([Z] + 1_n)^{-1}([Z] - 1_n)[S_Z]\}\end{aligned}$$

does not vanish at at least one point $p_0 = j\omega_0$. Assuming that it did vanish for all $p = j\omega_0$ would imply, using arguments similar to those employed in Theorem 2.16, that for all $p = j\omega$

$$|\det\{([Z] + 1_n)^{-1}([Z] - 1_n)\}| = 1,$$

but this is a contradiction, since the elements of this diagonal matrix are of the form

$$\frac{Z_i(p) - 1}{Z_i(p) + 1},$$

where $Z_i(p)$, being a regular normalizing function, is positive-real, so that

$$\left|\frac{Z_i(j\omega) - 1}{Z_i(j\omega) + 1}\right| < 1$$

unless $Z_i(j\omega) = 0$ or ∞, which will not be the case for all ω. In this contradiction argument we use the fact that each element of

$$([Z] + 1_n)^{-1}([Z(-p)] + 1_n)[H][H(-p)]^{-1}$$

is in magnitude equal to 1 for $p = j\omega$. QED

It is of some interest to single out two special cases for which we may simplify the statements of the theorem:

Corollary 2.2. Let $[Z]$, the normalizing matrix, be rational and analytic on $p = j\omega$; then the conclusions of Theorem 2.17 follow if Conditions 1 and 2 are satisfied when in addition $[H(p)]^{-1}$ is analytic on $p = j\omega$. If $[H(p)]^{-1}$ is not analytic on $p = j\omega$, then we require Conditions 1 and 2 and additionally the stipulation that $[Y]$ have simple poles in $p = j\omega$ with nonnegative-definite residue matrices.

Proof. In the first case we note that when $[Z]$ and $[H(-p)]^{-1}$ are analytic for $p = j\omega$

$$\{1_n - ([Z] - 1_n)[Y]\}([Z] + 1_n)$$

will be analytic for $p = j\omega$, so that Condition 3(*b*) of Theorem 2.17 is trivially satisfied and the conclusion of the theorem follows.

In the second case we note that the residue matrix of $\{1_n - ([Z] - 1_n)[Y]\} \times ([Z] + 1_n)$ reduces to the residue matrix of $(1_n - [Z])[Y]([Z] + 1_n)$ at a pole of $[Y]$, or, as we see, a zero of $[H(p)]$. We then may show directly that such a residue matrix is nonnegative-definite if and only if the residue matrix of $[Y]$ is nonnegative-definite. Thus Condition 3(b) of Theorem 2.17 is again satisfied and the conclusion of the theorem follows. QED

We will see in many of the applications to be considered later that the technique of complex normalization of scattering matrices, or, equivalently, the transformation of such matrices, is a powerful tool in the study of passive systems. In particular, in the next section we will deduce some limitations inherent in the performance of passive systems, and these results will be obtained using this normalization procedure.

2.4. SOME APPLICATIONS

In the previous sections of this chapter we have studied, in a rather abstract setting, those systems that are incapable of generating energy. The resulting class of passive operators which we used to represent such physical systems was shown to be in a one-to-one correspondence with the set of all bounded-real matrices, and we then proceeded to explore various properties of such matrices. We would like to take this opportunity to make some direct use of these studies by deducing some of the limitations inherent in the performance of passive systems. We will be able to obtain certain results that are applicable to the general class of passive systems, i.e., we will not have to impose additional assumptions such as rationality. Since subsequent chapters will be concerned with these specialized systems, we have placed this material in the present chapter to emphasize its wider applicability.

As a preliminary step in our discussion, we will demonstrate that positive-real functions, and the bounded-real functions using the transformation $[f(p) - 1]/[f(p) + 1]$, cannot take on arbitrary values at points in $\operatorname{Re} p > 0$. We state this limitation as follows:

Lemma 2.4. If $f(p)$ is positive-real and not of the form pL or $1/pC$, then with $f(p_0) = R_0 + jX_0$ and $p_0 = \sigma_0 + j\omega_0$ $(\sigma_0 > 0)$

$$\frac{R_0}{\sigma_0} > \left|\frac{X_0}{\omega_0}\right| \qquad \text{and} \qquad \frac{R_0}{\sigma_0} > \left|\frac{df}{dp}\right|_{p=p_0},$$

whereas if $f(p)$ is analytic at some point $p_0 = j\omega_0$ and at this point $R_0 = 0$, then

$$[dX_0/d\omega]_{\omega=\omega_0} > |X(\omega_0)/\omega_0|.$$

Proof. We form the function

$$W(p) = \frac{f(p) - f(p_0)}{f(p) + f^*(p_0)}$$

and note that $|W(p)| \leq 1$, since the fact that $f(p)$ is positive-real implies that

$\operatorname{Re} f(p) \geq 0$ in $\operatorname{Re} p > 0$. Now form the function

$$K(p) = \frac{W(p)}{(p - p_0)/(p + p_0^*)}$$

and observe that since $W(p_0) = 0$, then $K(p)$ is analytic in $\operatorname{Re} p > 0$. But $|(j\omega - p_0)/(j\omega + p_0^*)| = 1$, and we may appeal to the classic maximum modulus theorem to conclude that $|K(p)| < 1$ in $\operatorname{Re} p > 0$ unless $|K(p)| = 1$, or $|W(p)| = |(p - p_0)/(p + p_0^*)|$ in $\operatorname{Re} p > 0$. However, this alternative can easily be shown to imply that $f(p) = pL$ or $1/pc$ when $f(p)$ is positive-real, and since we exclude these, we may then conclude that $|K(p)| < 1$ in $\operatorname{Re} p > 0$. In particular, since

$$|K(p_0^*)| = \left| \frac{x_0}{\omega_0} \left(\frac{\sigma_0 - j\omega_0}{R_0 - jx_0} \right) \right| < 1,$$

we see that $R_0/\sigma_0 > |x_0/\omega_0|$. We also find that

$$|K(p_0)| = \left| \frac{(dW/dp)(p_0)}{1/(p_0 + p_0^*)} \right| = \left| \frac{\sigma_0}{R_0} \frac{df}{dp}(p_0) \right| < 1.$$

Now, if $f(p)$ is analytic at some point $p = j\omega_0$, where $\operatorname{Re} f = R_0 = 0$, then a Taylor series expansion yields

$$f(p) = R + jX = jX(\omega_0) + (p - j\omega_0)(dX/d\omega)(\omega_0) + a_2(p - j\omega_0)^2 + \cdots,$$

or $R = \sigma X'(\omega_0)$ and $X = X(\omega_0) + (\omega - \omega_0)X'(\omega_0)$ to first-order terms for $(\omega - \omega_0)$ sufficiently small. Thus to first order

$$(R/\sigma) \pm (X/\omega) = X'(\omega_0) \pm [X(\omega_0)/\omega],$$

and since $(R/\sigma) \pm (X/\omega) > 0$, as we noted above, for all $\sigma > 0$, then we see that

$$\lim_{\substack{|p - j\omega_0| \to 0 \\ (\operatorname{Re} p > 0)}} \left[\frac{R}{\sigma} \pm \frac{X}{\omega} \right] = X'(\omega_0) \pm \frac{X(\omega_0)}{\omega_0} > 0$$

and the second condition of the lemma follows. QED

2.4.1. Gain in Passive Systems

In this section we will consider the limitations imposed on the transfer function $(V_2/E_s)(p)$ of the system shown in Fig. 2.2 when we assume that the linear, time-invariant, causal, two-port network N is passive and the two impedances Z_L and Z_g are arbitrary positive-real functions of p.†

† We would like to emphasize the fact that Z_L and Z_g are *not* assumed to be regular normalizing functions.

Moreover, we assume that the network N is such that its scattering matrix $[S_{1_n}]$, defined by

$$\mathbf{b} = \tfrac{1}{2}\begin{bmatrix} V_1 - I_1 \\ V_2 - I_2 \end{bmatrix} = [S_{1_n}]\mathbf{a} = [S_{1_n}]\tfrac{1}{2}\begin{bmatrix} V_1 + I_1 \\ V_2 + I_2 \end{bmatrix}$$

is meromorphic. We may now transform $[S_{1_n}]$, which will be bounded-real,

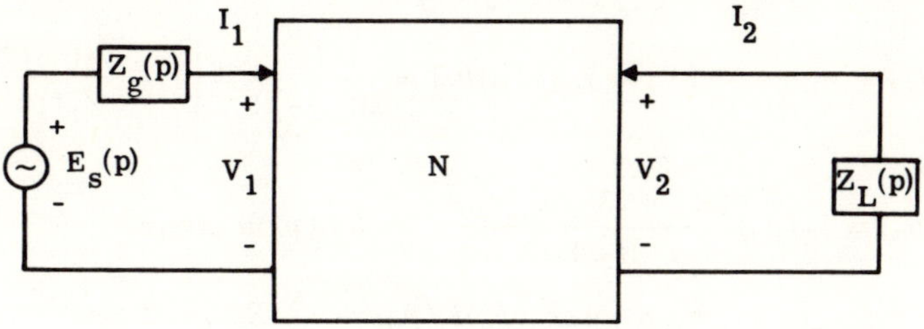

FIG. 2.2 Matching network.

since the two-port N is passive, using the complex normalization technique of Section 2.3, to obtain the matrix $[S_Z]$ where

$$[Z] = \begin{bmatrix} Z_1 & 0 \\ 0 & Z_2 \end{bmatrix},$$

with

$$Z_{1,2} = R_{1,2} - \left|\frac{X_{1,2}}{\omega_0}\right|\sigma_0 + \frac{a_{1,2}}{p} + b_{1,2}\,p,$$

$$\begin{aligned} b_{1,2} &= X_{1,2}/\omega_0 \quad \text{if} \quad X_{1,2}/\omega_0 \quad \text{positive}, \\ &= 0 \quad \text{if} \quad X_{1,2}/\omega_0 \quad \text{negative}, \\ a_{1,2} &= -X_{1,2}/\omega_0({\sigma_0}^2 + {\omega_0}^2) \quad \text{if} \quad X_{1,2}/\omega_0 \quad \text{negative}, \\ &= 0 \quad \text{if} \quad X_{1,2}/\omega_0 \quad \text{positive}, \end{aligned}$$

and $R_{1,2} + jX_{1,2} = Z_{g,L}(p_0)$ and $\operatorname{Re} p_0 > 0$. The notation $Z_{1,2}$, etc., stands for either Z_1 or Z_2. However, the fact that Z_g and Z_L are positive-real (although not necessarily regular normalizing functions) implies on the basis of Lemma 2.4 that Z_1 and Z_2 are positive-real functions, and, moreover, they will be regular normalizing functions, as we may verify using Definition 10. Therefore $[S_Z]$ will be a bounded-real matrix, as we may conclude from Theorem 2.16. Moreover, we see that

$$b_{2Z} = S_{21}a_{1Z} + S_{22}\,a_{2Z},$$

where S_{21} and S_{22} are the appropriate elements of $[S_Z]$, and, by definition of $Z_2(p)$,

$$a_{2Z} = \frac{1}{2}\frac{1}{h_2(p)}(V_2 + Z_2 I_2)$$

will be zero for $p = p_0$, since $V_2 = -Z_L(p_0)I_2 = -Z_2(p_0)I_2$. Here $h_{1,2} = (R_{1,2} - |X_{1,2}/\omega_0|\sigma_0)^{1/2}$, since $\frac{1}{2}\{Z_{1,2}(p) + Z_{1,2}(-p)\} = R_{1,2} - |X_{1,2}/\omega_0|\sigma_0$. Thus

$$b_{2Z} = \frac{1}{2h_2(-p)}[V_2 - Z_2(-p)I_2] = \frac{1}{2h_2(-p)} V_2\left[1 + \frac{Z_2(-p)}{Z_L(p)}\right]$$

and

$$b_{2Z}(p_0) = V_2(p_0)\frac{(R_2 - |X_2/\omega_0|\sigma_0)^{1/2}}{R_2 + jX_2} = S_{21}(p_0)a_{1Z}(p_0)$$

$$= S_{21}(p_0)\frac{V_1(p_0) + Z_1(p_0)I_1(p_0)}{2(R_1 - |X_1/\omega_0|\sigma_0)^{1/2}} = \frac{S_{21}(p_0)E_s(p_0)}{2(R_1 - |X_1/\omega_0|\sigma_0)^{1/2}}$$

and we obtain

$$\left|\frac{V_2}{E_s}(p_0)\right| = \left|\frac{S_{21}(p_0)}{2}\right| \frac{|R_2 + jX_2|}{\{(R_1 - |X_1/\omega_0|\sigma_0)(R_2 - |X_2/\omega_0|\sigma_0)\}^{1/2}}$$

$$\leq \frac{(R_2{}^2 + X_2{}^2)^{1/2}}{2\{(R_1 - |X_1/\omega_0|\sigma_0)(R_2 - |X_2/\omega_0|\sigma_0)\}^{1/2}},$$

since the fact that $[S_Z]$ is bounded-real certainly implies that $|S_{21}(p)| \geq 1$ for all $\mathrm{Re}\, p > 0$. The above result is the limitation imposed on the voltage gain, $|V_2/E_s|$, of a passive system, which, as we see, applies in all $\mathrm{Re}\, p > 0$, since p_0 was some arbitrary point in that region.

2.4.2. Lossless Two-Ports

Assume that we have a linear, time-invariant, causal, and lossless two-port network which is therefore characterized by a 2×2 bounded-real scattering matrix $[S(p)]$; we assume that its normalization is also prescribed, but for our present purposes this is immaterial. Our purpose is to explore the limitations imposed on the system, or, equivalently, its scattering matrix, by the assumption of losslessness.

Since from Theorem 2.4 we know that such matrices must also satisfy $1_n = [S(\omega)]^{*T}[S(\omega)]$ a.e., we find by expanding this matrix statement that

$$|S_{11}|^2 + |S_{21}|^2 = 1 = |S_{22}|^2 + |S_{12}|^2$$

and

$$S_{12}^* S_{11} + S_{22}^* S_{21} = 0$$

for almost all ω. We may obtain alternate conditions by first noting that

$|\det[S(\omega)]| = 1$ a.e., so that

$$[S]^{*T} = [S]^{-1} = \frac{1}{\det[S]} \begin{bmatrix} S_{22} & -S_{12} \\ -S_{21} & S_{11} \end{bmatrix},$$

and, in particular,

$$S_{11}^* = S_{22}/\det S \qquad \text{and} \qquad S_{21}^* = -S_{12}/\det S,$$

so that $|S_{11}| = |S_{22}|$ and $|S_{21}| = |S_{12}|$ a.e. In the special case when either $|S_{11}|$ or $|S_{22}|$ (we see that they are the same) is equal to 1 then $|S_{21}|^2 = |S_{12}|^2 = 1 - |S_{22}|^2 = 0$, and we have the interesting physical conclusion that a lossless two-port such that either $|S_{11}(\omega)|$ or $|S_{22}(\omega)| = 1$ a.e. must be reciprocal, i.e., $S_{12} = S_{21} = 0$. We note that in general, however, a lossless two-port need not be reciprocal, since only $|S_{12}| = |S_{21}|$, and thus their phase angles may differ.

Let us further assume that the scattering matrix $[S(p)]$ is meromorphic; then, as we saw in Theorem 2.15,

$$[S(-p)]^T[S(p)] = 1_n,$$

and thus

$$S_{11}(p)S_{11}(-p) + S_{21}(p)S_{21}(-p) = 1,$$

$$S_{22}(p)S_{22}(-p) + S_{12}(p)S_{12}(-p) = 1,$$

and

$$S_{12}(-p)S_{11}(p) + S_{22}(-p)S_{21}(p) = 0.$$

If we consider the situation in which $S_{11}(p)$ is known, then we see that

$$|S_{21}(j\omega)| = |S_{12}(j\omega)| = \{1 - |S_{11}(j\omega)|^2\}^{1/2}$$

are determined, and from the representation of bounded-real functions given in Theorem 2.14 [since S_{21} and S_{12} are assumed to be meromorphic, then $S_{21}(\omega)$ and $S_{12}(\omega)$ are continuous functions], we have for $\operatorname{Re} p > 0$

$$S_{21}(p) = \pm(\exp -a_{21}p)b_{21}(p)S_m(p)$$

and

$$S_{12}(p) = \pm(\exp -a_{12}p)b_{12}(p)S_m(p),$$

where, also in $\operatorname{Re} p > 0$,

$$S_m(p) = \exp\left\{\frac{p}{\pi}\int_{-\infty}^{\infty} \frac{\log(1 - |S_{11}(j\eta)|^2)^{1/2}}{p^2 + \eta^2}\, d\eta\right\}.$$

But once a_{21}, a_{12}, b_{21}, and b_{12} are prescribed (together with the appropriate signs), then $S_{21}(p)$ and $S_{12}(p)$ may be found for all p, since the condition

$$S_{21}(-p) = \{1 - S_{11}(p)S_{11}(-p)\}/S_{21}(p)$$

uniquely determines $S_{21}(p)$ for $\operatorname{Re} p < 0$. Then

$$S_m(p) = S_{21}(p)/[\pm(\exp -a_{21}p)b_{21}(p)],$$

and so $S_{12}(p)$ is also determined. Finally, we obtain $S_{22}(p)$ as

$$S_{22}(p) = -S_{11}(-p)S_{12}(p)/S_{21}(-p).$$

We state these conclusions for future reference as:

Theorem 2.18. A 2×2 bounded-real matrix whose elements are meromorphic functions and which in addition satisfies $[S(\omega)]^{*T}[S(\omega)] = 1_n$ is uniquely determined by $S_{11}(p)$, two nonnegative constants a_{12} and a_{21}, and two regular Blaschke functions (see Theorem 2.14) $\hat{b}_{12}(p)$ and $\hat{b}_{21}(p)$. In particular,

$$S_{21}(p) = [\exp -a_{21}p]b_{21}(p)S_m(p),$$

$$S_{12}(p) = [\exp -a_{12}p]b_{12}(p)S_m(p),$$

$$S_{22}(p) = -S_{11}(-p)S_{12}p/S_{21}(-p),$$

and $S_m(p)$ is that factorization of $S_m(p)S_m(-p) = 1 - S_{11}(p)S_{11}(-p)$ that is analytic and nonzero in $\operatorname{Re} p > 0$.

We note in the statement of the theorem that the $(\pm)$ ambiguity has been absorbed into the Blascke functions. The theorem is particularly interesting since it illustrates the fact that passivity—specifically, losslessness—is reflected in a very strong structural constraint on the corresponding scattering matrix of the system. If we further assume that the system is reciprocal, so that $S_{12} = S_{21}$, we see that once $S_{11}(p)$ is determined we have only one degree of freedom left in the system—the specification of the function $e^{-ap}b(p)$.

2.4.3. Gain-Bandwidth Limitations

Let us assume that $T(p)$ is the ratio between two input-output parameters of a system. For example, it might represent the ratio of the transform of the output voltage of a system to the corresponding input voltage. By a gain-bandwidth limitation we mean a constraint of the form

$$\int_{-\infty}^{\infty} f(|T(j\omega)|)\, d\omega \le K,$$

where the operator $f(\cdot)$ is known. In this section we will obtain some gain-bandwidth limitations that are inherent in the performance of the system shown in Fig. 2.2. This system is of considerable engineering interest in the case where Z_L and Z_g are prescribed and we may vary the two-port network N so as to achieve some desirable voltage transfer ratio $T(p) = (V_2/E_s)(p)$. The limitations which we will obtain reflect the constraint of assuming that the network N and the two impedances Z_L and Z_g are passive. In fact, the result is related, as we will see, to the constraint obtained in Section 2.4.1 for the same system. First we will consider a procedure for determining the limitations imposed on the performance of an arbitrary system by both passivity and the fact that certain portions of the system are prescribed *a priori*. In a sense, we are attempting to find the invariants of the system's performance imposed by the fixed or prescribed portions of the system.

The starting point for our discussion is the representation theorem for bounded-real functions whose boundary values are continuous functions of ω (Theorem 2.14). In particular, we have

$$\log|T_m(p)| = \mathrm{Re}\left\{\frac{p}{\pi}\int_{-\infty}^{\infty} \frac{\log|T(j\eta)|}{p^2+\eta^2}\, d\eta\right\},$$

where $T(j\omega)$ is the corresponding boundary value of the function $T(p)$ and $T_m(p) = T(p)/[\pm e^{-ap}b(p)]$. The right-hand side of the equation is indeed a gain-bandwidth statement, and the utility of this result follows from the fact that for many systems the equivalent quantity $|T_m(p)|$ can be expressed in a manner which displays the fixed portions of the system, and thus the constraints imposed by these fixed portions may be reflected into constraints on $|T_m(p)|$, or, equivalently, the gain-bandwidth integral. For example, if the fixed portions imply that $|T_m(p_0)| \leq M$, then

$$\mathrm{Re}\left\{\frac{p_0}{\pi}\int_{-\infty}^{\infty} \frac{\log|T(j\eta)|}{{p_0}^2+\eta^2}\, d\eta\right\} \leq \log M.$$

In many cases one wishes to obtain a transfer ratio such that

$$|T(j\omega)| = GA(\omega),$$

where $A(\omega)$ is some prescribed function and the constant G, called the gain constant, is to be as large as possible. Assuming that we have found that $|T_m(p_0)| \leq M$ due to the fixed portions of the system, then we find that

$$\log G \leq \log M - \mathrm{Re}\left\{\frac{p_0}{\pi}\int_{-\infty}^{\infty} \frac{\log A(\omega)}{{p_0}^2+\eta^2}\, d\eta\right\},$$

and we would have determined an upper bound on the gain constant G.

Consider now the specific system shown in Fig. 2.2, in which the network N is used to obtain some voltage transfer ratio $(V_2/E_s)(p)$ between a given "load" Z_L and a source with a fixed "output" impedance Z_g. Assume that the scattering matrix of the two-port and the impedance Z_L and Z_g are all meromorphic and positive-real. Then, by considering that matrix which is normalized to the functions Z_1 and Z_2 introduced in Section 2.4.1, we obtain

$$T(p) = \frac{h_1(p)}{h_2(p)} \frac{S_{21}(p)(Z_2 + Z_2(-p)S_L)}{Q(p)},$$

where

$$Q(p) = Z_1 S_{11}(1 - S_L S_{22}) + Z_1 S_{12} S_{21} S_L + Z_1(-p)(1 - S_L S_{22})$$
$$+ Z_g[(1 - S_{11})(1 - S_L S_{22}) - S_{12} S_{21} S_L],$$

$$S_L = \frac{Z_L - Z_2}{Z_L + Z_2(-p)},$$

and $h_{1,2} = (R_{1,2} - |X_{1,2}/\omega_0|\,\sigma_0)^{1/2}$, with $R_{1,2} + jX_{1,2} = Z_{g,L}(p_0)$. Now we observe that

$$Z_2 + Z_2(-p)S_L = \frac{Z_L 2h_2{}^2}{Z_L + Z_2(-p)},$$

so that $Z_2 + Z_2(-p)S_L$ has no zeros in the finite part of the half plane $\operatorname{Re} p > 0$, i.e., Z_L, being positive-real, is analytic in $\operatorname{Re} p > 0$, whereas $Z_2(-p)$ is by definition (Section 2.4.1) analytic except at either $p = 0$ or $p = \infty$. Moreover, we see that $Q(p)$ is analytic in $\operatorname{Re} p > 0$ except possibly at those points where S_L has poles, and that at these poles, since the numerator of the expression for $T(p)$ also contains S_L, $T(p)$ will not vanish unless $Z_2(-p)$ vanishes. Since this possibility is not compatible with S_L having a pole at the point in question (Z_L, being positive-real, may not vanish in $\operatorname{Re} p > 0$), we may conclude finally that the only zeros of $T(p)$ in $\operatorname{Re} p > 0$ are among the zeros of $S_{21}(p)$.

Now if we use the result of Section 2.4.1 that $T(p)$ is bounded at every point in $\operatorname{Re} p > 0$, we may conclude that $T(p)$ is analytic in $\operatorname{Re} p > 0$. If we further assume that there exists a constant K such that $|T(p)| \leq K$ in $\operatorname{Re} p > 0$, then we may write $T(p) = K\hat{T}(p)$, where $\hat{T}$ is bounded by 1 in $\operatorname{Re} p > 0$, i.e., $\hat{T}(p)$ is bounded-real. We may now use the representation of Theorem 2.14 to state that

$$T(p) = \pm K e^{-ap} \hat{T}_m(p) b(p) = \pm e^{-ap} b(p)[K\hat{T}_m(p)] = \pm e^{-ap} b(p) T_m(p),$$

where $a \geq 0$ and $b(p)$ is a regular Blaschke function which on the basis of our previous discussion is zero in $\operatorname{Re} p > 0$ only at those points where $S_{21}(p)$ vanishes. Now, regarding the constant a, we assume that asymptotically $Z_2 + Z_2(-p)S_L \sim \exp(k_1 p)$, $S_{21} \sim \exp(k_2 p)$, and $Q(p) \sim \exp(k_3 p)$, so that

$T(p) \sim \exp[(k_1 + k_2 - k_3)p]$, and thus $a = k_3 - k_1 - k_2$. We may then conclude that

$$T_m(p) = \frac{T(p)}{\pm e^{-ap} b(p)}$$

$$= \frac{(h_1/h_2)[S_{21}(p)/\hat{b}(p)](Z_2 + Z_2(-p)S_L)\exp[-(k_1 + k_2 - k_3)p]}{Q(p)},$$

where $\hat{b} = \pm b(p)$. However, if $p = p_0$, then by construction, $Z_L = Z_2$ and $Z_g = Z_1$, so that $S_L = 0$ and

$$T_m(p_0) = \frac{[S_{21}(p_0)/\hat{b}(p)]Z_L(p_0)\,[\exp-(k_1 + k_2 - k_3)p_0]}{2h_1 h_2}.$$

But since $\hat{b}(p)$ vanishes only at those points where $S_{21}(p)$ vanishes, we conclude that $|e^{-k_2 p} S_{21}(p)/\hat{b}(p)| \leq 1$ because $|\hat{b}(j\omega)| = 1$ and $S_{21}(p)$ itself is bounded by 1 in $\operatorname{Re} p > 0$, and finally we obtain the constraint (since k_3 is negative)

$$|T_m(p_0)| \leq \frac{|Z_L(p_0)|}{2h_1 h_2} \exp[-k_1 \sigma_0],$$

which is related to the constraint obtained in Section 2.4.1 on the function $|T(p_0)|$. Using this restriction, we obtain as a gain-bandwidth limitation

$$\operatorname{Re}\left\{\frac{p_0}{\pi}\int_{-\infty}^{\infty} \frac{\log|T(j\eta)|}{p_0{}^2 + \eta^2}\, d\eta\right\} \leq \log\left\{\frac{|Z_L(p_0)|}{2[(R_1 - |X_1/\omega_0|\,\sigma_0)(R_2 - |X_2/\omega_0|\,\sigma_0)]^{1/2}}\right\} - k_1 \sigma_0,$$

where $R_{1,2} + jX_{1,2} = Z_{g,L}(p_0)$ and $\operatorname{Re} p_0 > 0$. As indicated previously, the technique has displayed the fact that the performance of the system is inherently constrained by the particular fixed impedances Z_L and Z_g present in the system and the fact that the matching network N is passive.

2.5. LINEAR, TIME-VARYING, PASSIVE SYSTEMS†

In the preceding portions of this chapter we were dealing with those linear passive systems that are time-invariant. We found that the Fourier and Laplace transforms were very effective tools in this study; in particular, they allowed us to state explicit conditions on the impulse response of a linear time-invariant system which completely characterized passive systems. In this

† The author would like to acknowledge that the following discussion of linear passive operators is based on some preliminary investigations that were jointly undertaken with Professor E. J. Beltrami, and to him the author expresses his thanks.

section we will discuss those systems that are not time-invariant, although they will still be assumed to be linear and passive. We will be hampered to a large extent by the fact that these transform techniques are no longer useful, since in general the eigenfunctions of an arbitrary, linear, causal operator need not be of the form e^{pt} and, as noted in Chapter I, this was the critical fact which facilitated the use of the transform techniques.

Before we start our discussion we might note that a primary engineering interest in time-variable systems, at least from a design standpoint, concerns their ability to perform in an active or nonpassive manner. We may then justify our discussion of passive systems on the ground that if a system fails to meet the necessary conditions for passivity, then it by definition will be active.

We will be further restricted in the following discussion by the fact that many of the mathematical tools necessary for the study of time-variable operators, at least the distributional ones, have either only recently been developed or are still under development. For this reason we will describe a theory that hopefully can be fully justified at some later date. We will then proceed on a formal basis and only indicate some of the places where fuller justification is needed. The starting point is the representation for the output $b(t)$ of a linear continuous system (assumed for simplicity to be scalar) in terms of its input $a(t) \in C_0^{\infty}$. Thus from Theorem 1.1 we have

$$b(t) = \int h(t, \tau)a(\tau)\, d\tau,$$

where $h(t, \tau)$ is the response to the input $a(t) = \delta(t - \tau)$. If the system is assumed to be passive, then

$$\int_{-\infty}^{\infty} a^2 - b^2\, dt = \int_{-\infty}^{\infty}\int_{-\infty}^{\infty} a(x)a(y) \times \{\delta(x - y) - \int_{-\infty}^{\infty} h(t, x)h(t, y)\, dt\}\, dx\, dy \geq 0$$

for all $a \in C_0^{\infty}$; we assume that a proper definition has been used for $\int_{-\infty}^{\infty} h(t, x)h(t, y)\, dt$. Now a classic theorem (see, for example, [Co1]) is available which states that a real, symmetric kernel $(K(x, y) = K(y, x))$ is positive in the sense that $\iint K(x, y)\varphi(x)\varphi(y)\, dx\, dy \leq 0$ if and only if its eigenvalues β are real and nonnegative, i.e., all the β for which solutions ψ exist to the equation

$$\beta\psi(y) = \int K(x, y)\psi(x)\, dx$$

must be real and nonnegative constants. Thus we might expect that the comparable condition for the present problem would be that the eigenvalues of the kernel $\delta(x - y) - \int_{-\infty}^{\infty} h(t, x)h(t, y)\, dt$, which we see is real and symmetric, be real and nonnegative. Now, assume that φ is an eigenfunction of the given linear operator; then

$$\int_{-\infty}^{\infty} h(t, \tau)\varphi(\tau)\, d\tau = \lambda\varphi(t),$$

where λ is the corresponding eigenvalue. Moreover,

$$\int_{-\infty}^{\infty}\left\{\delta(x-y) - \int_{-\infty}^{\infty} h(t, x)h(t, y)\, dt\right\}\varphi(x)\, dx = \varphi(y) - \lambda \int_{-\infty}^{\infty} h(t, y)\varphi(t)\, dt.$$

However, $\int_{-\infty}^{\infty} h(t, y)\varphi(t)\, dt$ may be recognized as the adjoint operator's representation, and, as is the case classically, we would expect that

$$\int_{-\infty}^{\infty} h(t, y)\varphi(t)\, dt = \lambda^*\varphi(t),$$

since $\varphi(t)$ is an eigenfunction of the original operator. Granting the previous arguments, one then finds that the eigenvalues β of the operator $\delta(x-y) - \int_{-\infty}^{\infty} h(t, x)h(t, y)\, dt$ are expressible as

$$\beta = 1 - \lambda\lambda^* = 1 - |\lambda|^2,$$

so that the requirement that β be nonnegative demands that the eigenvalues of the operator itself must be bounded in magnitude by 1. This conclusion is certainly borne out when the operator is time-invariant, since in that case the eigenvalues (see Chapter I) become $\lambda = H(p) = \mathscr{L}[h(t)]$ for $\operatorname{Re} p > 0$ and, as we have seen previously, the passivity of such operators demand that $|H(p)| \leq 1$ in $\operatorname{Re} p > 0$. Thus one might expect that the generalization of the previous results concerning linear, time-invariant, passive systems to the class of linear passive systems would be in the form of a requirement that the eigenvalues of such systems be less than or equal to 1 in magnitude.

The above discussion was certainly not rigorous, since, for example, it is not clear that all linear continuous operators possess a complete set of eigenfunctions belonging to some known class of distributions. Moreover, the classic theorem concerning positive kernels has not been verified in a distributional setting. However, this brief discussion does seem to indicate that the natural extensions of the time-invariant results lie along the lines which we have described. In particular, some progress has been made, using a slightly different approach, by Newcomb and Anderson [Ne3]. In particular, they argued that lossless operators, as we might expect from the above, would be characterized by impulse response functions satisfying

$$\delta(x-y) = \int_{-\infty}^{\infty} h(t, x)h(t, y)\, dt.$$

2.6. SUMMARY

We have discussed the characterization of those linear causal systems that we have defined as passive. We have seen that in the case of time-invariant systems this property may be reflected in a one-to-one manner into a requirement that the Laplace transform of the impulse response of such operators, sometimes referred to as the operator's Green's function, be either bounded-real or positive-real.

Because of the wide physical application of the notion of passivity, we have discussed two apparently equivalent definitions that differ only in an assumption concerning the domain of the operator. It was shown that the first definition, which has also been referred to as passivity on a scattering basis because of its connection with problems arising in the theory of electromagnetic and quantum mechanical scattering, may be considered the more basic one, since it was shown in Theorem 2.6 to encompass the second definition at least for linear time-invariant systems. The second definition, also referred to by some as passivity on an immittance basis, was shown to be more restrictive in the n-dimensional case due to the fact that the domain of such operators may not be sufficiently large to allow a unique characterization to be obtained. In both cases we presented representation theorems for the resulting classes of bounded-real and positive-real matrices and then illustrated their utility with some specific applications of engineering interest. In these applications we emphasized the limitations imposed on the behavior of a physical system if it performs in a passive manner. In fact, this is one of the more fundamental reasons for the systematic study of passive operators, since if one can deduce the properties of such operators, then one may expect to find these properties to be reflected into physical limitations inherent in these systems.

In the next two chapters we will study two specific classes of passive systems. The first consists of those systems that are composed of lumped elements, or subsystems, and are therefore characterized by system functions that are rational functions of the Laplace transform variable p. We will see that the properties of such systems have been rather completely delineated, reflecting the fact that such systems have been studied systematically over the past 50 years. The last chapter is by no means as comprehensive, since it deals with distributed systems, which have only received direct attention during recent years. One of the difficulties in the area of distributed systems is the fact that there does not seem to be a finite number of basic subsystems which can be defined, so that a large class of these operators have representations in the form of networks composed of these subsystems. This is in direct contrast with rational systems, whose characterizations have been rather completely obtained. In both chapters we will see that the results of the present chapter are used in an essential manner.

III

Lumped Networks

INTRODUCTION

The purpose of the first systematic study of passive operators was to determine the properties of those networks that were composed of the lumped electrical elements described in Section 1.3.1—namely, resistors, capacitors, inductors, transformers, gyrators, and coupled coils. Our purpose in this chapter is to describe the results of these studies and to indicate some of their applications to problems of engineering interest. At the same time we will be exploring the structure of "rational" passive operators in a purely mathematical way.

As we noted, there are certain results which we will present without proof because we feel that the background material necessary to establish them rigorously is too extensive to be covered. In this chapter we will exercise this option extensively. Fortunately, there are many excellent books now available to which we may refer the interested reader, in particular as general references we mention Newcomb's book [Ne2] on the synthesis problem and the basic texts of Carlin and Giordano [Ca2] and Kuh and Rohrer [Ku1] for questions of analysis.

3.1. REALIZATION OF LUMPED NETWORKS

3.1.1. *n*-Port Networks

We assume that a finite number of the lumped elements defined in Section 1.3.1 are interconnected in some fashion, forming what we call a network, and moreover that there are n points or ports in this network at which we can connect the independent voltage and current sources that represent the inputs to our network. At each of the n ports in our network we define a voltage and a current so that the network is described by means of the n-vectors $\mathbf{v}(t)$ and $\mathbf{i}(t)$ whose elements are these parameters. If we connect a unit resistor to each of the n ports, as illustrated in Fig. 2.1, then we may consider the augmented network thus formed, and, in particular, we may excite it with voltage sources designated by $\mathbf{E}$. Thus, since $\mathbf{E} = \mathbf{v}(t) + \mathbf{i}(t)$, we see the relation between these sources and the parameters of the original n-port under consideration. Now, the physical response to the inputs $\mathbf{E}$ will be the currents $\mathbf{i}(t)$, but since $\mathbf{E} - (\mathbf{v} - \mathbf{i}) = 2\mathbf{i}$, we see that we may consider as the output the quantity $\mathbf{v} - \mathbf{i}$. Our purpose for selecting this quantity as the output will become clear as we proceed; in any event, the network is now represented by some operator T that maps $\mathbf{v} + \mathbf{i}$ into $\mathbf{v} - \mathbf{i}$. Let us now assume that the lumped elements in our network are all time-invariant, and, moreover, that all R's, L's, and C's are nonnegative-real numbers, while the remaining variables defining the gyrator, transformer, and coupled coils are real numbers, and, in the case of the coupled coils the variables satisfy $L_1 > 0$, $L_2 > 0$ and $L_1 L_2 \geq M^2$. Then a systematic application of Kirchhoff's laws (Section 1.3.1), which we assume govern the voltages and currents in our network, will result in a set of ordinary differential equations in the unknown parameters with the inputs $\mathbf{v} + \mathbf{i}$ as forcing functions. An application of the Laplace transform to these equations results in a system function representing the operator T in the form

$$(\mathbf{V}(p) - \mathbf{I}(p)) = [S(p)](\mathbf{V}(p) + \mathbf{I}(p)),$$

where the system function $[S(p)]$ is an $n \times n$ matrix of rational functions of p. As noted earlier, it is this which has prompted the name "rational network." For future reference we designate the specific system function $[S(p)]$ as the scattering matrix of the network, and our first task is to prove that the networks we have described will always possess a scattering matrix representation.

To this end consider the original network before it was augmented with the unit resistors, and note that since the elements are interconnected by wires, and these wires form the only constraint on the physical position of the elements, then we may rearrange our network so that it has the appearance shown in Fig. 3.1. Specifically, the lumped elements are all arranged in order outside some new network M that consists only of wires connecting the elements back to the original n ports, i.e., from the original n-port network we have formed another network M composed only of wires, and we note that M specifies how the lumped elements were interconnected to form the original n-port. In particular, if we had r resistors, c capacitors, l inductors, t transformers, g gyrators, and m coupled coils in the original network, then the new network M will have $n + r + c + l + 2t + 2y + 2m$ ports, and we will define voltages and currents at these ports so that the network is described by inputs

$$\mathbf{a}_M = (\mathbf{v}_M + \mathbf{i}_M) = [\mathbf{a}/\hat{\mathbf{a}}]$$

and outputs

$$\mathbf{b}_M = (\mathbf{v}_M - \mathbf{i}_M) = [\mathbf{b}/\hat{\mathbf{b}}]$$

where $\mathbf{a}$ is the n-vector of inputs to the original n-port, $\hat{\mathbf{a}}$ the $(r + c + l + 2t + 2y + 2m)$-vector of inputs to the additional ports of the network M, and $\mathbf{b}$ and $\hat{\mathbf{b}}$ are the corresponding outputs at the two sets of ports. We could now analyze the network M to show that it possesses a scattering matrix such that

$$\mathbf{b}_M = [S_M]\mathbf{a}_M,$$

where $[S_M]$ is a matrix all of whose elements are real constants, and, more-

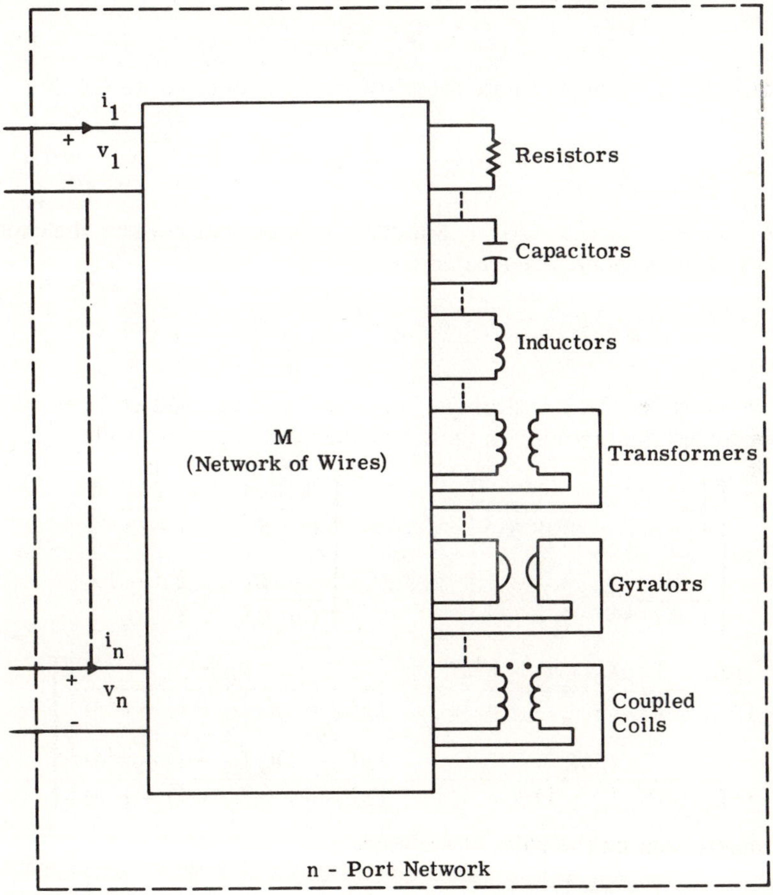

FIG. 3.1 General form of a lumped network.

over, $[S_M]^T = [S_M]$ and $[S_M]^T[S_M] = 1_n$, i.e., $[S_M]$ is a unitary symmetric matrix of real constants. This demonstration, although straightforward, is tedious in detail and we will omit it. Accepting these facts, however, we may then partition the matrix $[S_M]$ as

$$[S_M] = \begin{bmatrix} [S_{11}] & [S_{12}] \\ [S_{21}] & [S_{22}] \end{bmatrix},$$

where $[S_{11}]$ is an $n \times n$ matrix, $[S_{12}]$ is $n \times q$, $[S_{21}]$ is $q \times n$, and $[S_{22}]$ is $q \times q$, with $q = r + c + l + 2t + 2g + 2m$, and obtain the statements

$$\mathbf{b} = [S_{11}]\mathbf{a} + [S_{12}]\hat{\mathbf{a}}$$

and

$$\hat{\mathbf{b}} = [S_{21}]\mathbf{a} + [S_{22}]\hat{\mathbf{a}}.$$

Now, at those ports of M where the resistors are connected we see that

$$\hat{a}_i = \frac{R_i - 1}{R_i + 1}\,\hat{b}_i,$$

with R_i the value of the resistor. Similarly, at those ports where the capacitors C_i and inductors L_i are located either

$$\hat{a}_i = \frac{(1/pC_i) - 1}{(1/pC_i) + 1}\,\hat{b}_i \qquad \text{or} \qquad \hat{a}_i = \frac{pL_i - 1}{pL_i + 1}\,\hat{b}_i.$$

When we consider the transformers, gyrators, and coupled coils, we must view the corresponding ports in pairs and then obtain, respectively,

$$[S_i] = \begin{bmatrix} \dfrac{1 - K_i^2}{1 + K_i^2} & \dfrac{2K_i}{1 + K_i^2} \\[2ex] \dfrac{2K_i}{1 + K_i^2} & \dfrac{K_i^2 - 1}{1 + K_i^2} \end{bmatrix} \quad \text{or} \quad \begin{bmatrix} \dfrac{B_i^2 - 1}{1 + B_i^2} & \dfrac{2B_i}{1 + B_i^2} \\[2ex] \dfrac{-2B_i}{1 + B_i^2} & \dfrac{B_i^2 - 1}{1 + B_i^2} \end{bmatrix} \quad \text{or}$$

$$\begin{bmatrix} \dfrac{(pL_1 - 1)(pL_2 + 1) - p^2M^2}{(pL_1 + 1)(pL_2 + 1) - p^2M^2} & \dfrac{2pM}{(pL_1 + 1)(pL_2 + 1) - p^2M^2} \\[2ex] \dfrac{2pM}{(pL_1 + 1)(pL_2 + 1) - p^2M^2} & \dfrac{(pL_2 - 1)(pL_1 + 1) - p^2M^2}{(pL_1 + 1)(pL_2 + 1) - p^2M^2} \end{bmatrix}$$

as the matrix relating the pairs of variables

$$\begin{bmatrix} a_{1i} \\ a_{2i} \end{bmatrix} = [S_i]\begin{bmatrix} b_{1i} \\ b_{2i} \end{bmatrix}.$$

We may then write

$$\hat{\mathbf{a}} = [Q]\hat{\mathfrak{b}},$$

where $[Q]$ is a matrix all of whose elements are zero except either the diagonal ones [corresponding to the ports where resistors, inductors, and capacitors are present, and here the diagonal element is of the from $(U-1)/(U+1)$, with $U = R_i$ or $1/pC_i$ or pL_i] or two diagonal and two symmetrically placed off-diagonal terms such that the four elements are of the form given by the $[S_i]$ noted above (corresponding to the presence of a transformer, gyrator, or coupled coils at this pair of ports). We note that $[Q]$ is a bounded-real matrix, as we may easily demonstrate since $(U-1)/(U+1)$ and $[S_i]$ are bounded-real given that all the resistors, capacitors, and inductors are positive and $L_1 L_2 \geq M^2$, i.e., we may show directly that

$$\mathbf{y}^{*T}\{1_q - [Q]^{*T}[Q]\}\mathbf{y} > 0$$

in Re $p > 0$. With these properties of $[Q]$ established then we see that

$$\mathbf{b} = [S_{11}]\mathbf{a} + [S_{12}][Q]\hat{\mathfrak{b}},$$

$$\hat{\mathfrak{b}} = [S_{21}]\mathbf{a} + [S_{22}][Q]\hat{\mathfrak{b}},$$

and

$$\hat{\mathfrak{b}} = (1_q - [S_{22}][Q])^{-1}[S_{21}]\mathbf{a},$$

and so

$$\mathbf{b} = \{[S_{11}] + [S_{12}][Q](1_q - [S_{22}][Q])^{-1}[S_{21}]\}\mathbf{a} = [S]\mathbf{a},$$

or

$$[S] = [S_{11}] + [S_{12}][Q](1_q - [S_{22}][Q])^{-1}[S_{21}].$$

The inversion of the matrix $1_q - [S_{22}][Q]$ can be justified using the methods employed in the proof of Theorem 2.16; i.e., with $[S_M]$ bounded-real $[S_{22}]$ must be bounded-real, and this fact, together with $\mathbf{y}^{*T}\{1_q - [Q]^{*T}[Q]\}\mathbf{y} > 0$ in Re $p > 0$, then implies $\det(1_q - [S_{22}][Q]) \neq 0$ in Re $p > 0$. Moreover, we see that every element in $[S]$ is a rational function of p, analytic in Re $p > 0$. Finally, since $[S_M]$ is unitary, we have for all p

$$\mathbf{a}_M^{*T}\mathbf{a}_M - \mathbf{b}_M^{*T}\mathbf{b}_M = 0 = \mathbf{a}^{*T}\mathbf{a} - \mathbf{b}^{*T}\mathbf{b} + \hat{\mathbf{a}}^{*T}\hat{\mathbf{a}} - \hat{\mathfrak{b}}^{*T}\hat{\mathfrak{b}},$$

or

$$\mathbf{a}^{*T}\mathbf{a} - \mathbf{b}^{*T}\mathbf{b} = \hat{\mathfrak{b}}^{*T}\hat{\mathfrak{b}} - \hat{\mathbf{a}}^{*T}\hat{\mathbf{a}} = \hat{\mathfrak{b}}^{*T}\{1_q - [Q]^{*T}[Q]\}\hat{\mathfrak{b}} \geq 0,$$

so that $1_n - [S]^{*T}[S]$ is nonnegative-definite in Re $p > 0$, and thus we have established that the scattering matrix of the original n-port is bounded-real. We have thereby proven the following:

Theorem 3.1. Consider an n-port network composed of a finite number of "passive," time-invariant, lumped elements (positive resistors, capacitors, inductors, transformers, gyrators, and coupled coils with $L_1L_2 \geq M^2$) and described by voltages $\mathbf{v}(t)$ and currents $\mathbf{i}(t)$. Then there exists a rational bounded-real matrix $[S(p)]$ such that

$$(\mathbf{V}(p) - \mathbf{I}(p)) = [S(p)](\mathbf{V}(p) + \mathbf{I}(p)),$$

where $V(p) = \mathscr{L}(\mathbf{v}(t))$ and $\mathbf{I}(p) = \mathscr{L}(\mathbf{i}(t))$.

We note that the above theorem is in fact the justification for the identification of $\mathbf{v} + \mathbf{i}$ as inputs and $\mathbf{v} - \mathbf{i}$ as outputs for our system in that this identification, as we have shown, will always guarantee that there exists some operator T such that $\mathbf{v} - \mathbf{i} = T[\mathbf{v} - \mathbf{i}]$ when the system is a network of lumped, constant, and passive elements.

One of the major triumphs of network theory has been the fact that the converse of Theorem 3.1 has been shown to be true, i.e., given a rational $n \times n$ bounded-real matrix, then there exists an n-port network composed of a finite number of the given "passive" time-invariant lumped elements with this matrix as its scattering matrix. We will not give a proof of this important result, but simply refer the reader so [Ne2]. However, for completeness we will formally state:

Theorem 3.2. Given a rational $n \times n$ bounded-real matrix $[S]$, there will always exist some n-port network, composed of a finite number of the passive, time-invariant, lumped elements, which has $[S]$ as its scattering matrix, i.e., such that

$$(\mathbf{V}(p) - \mathbf{I}(p)) = [S(p)](\mathbf{V}(p) + \mathbf{I}(p))$$

We might note at this point that the inclusion of coupled coils in our discussion, which are defined by the equations

$$v_1 = L_1(di_1/dt) + M(di_2/dt)$$

and

$$v_2 = M(di_1/dt) + L_2(di_2/dt),$$

with L_1 and L_2 nonnegative and $L_1L_2 \geq M^2$, is of considerable engineering interest, since there are physical devices whose performance is in good agreement with such a model. The ideal transformer, or the system satisfying

$$v_2 = Kv_1, \qquad i_2 = -(1/K)i_1,$$

on the other hand, is much more difficult to realize with a physical device. However, the circuit in Fig. 3.2 shows that a network composed of two inductors and one ideal transformer can be arranged to have the same terminal performance as two coupled coils. Moreover, since the ideal transformer is simpler to describe when it is combined with other elements, we will, in the future, concentrate on those networks in which the coupled coils are replaced with the equivalent networks shown in Fig. 3.2.

Much more of the detailed internal structure of a network may be deduced from its terminal or n-port performance. We list, again without proof (see, for example, [Ne2]) the following corollaries of Theorem 3.2 which illustrate the type of result we have in mind:

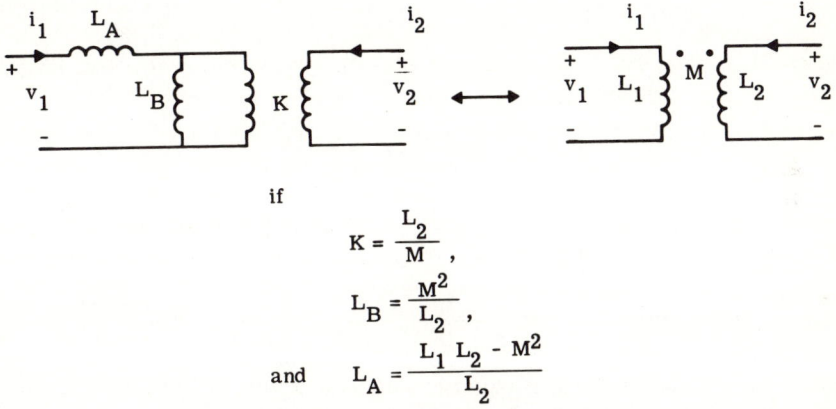

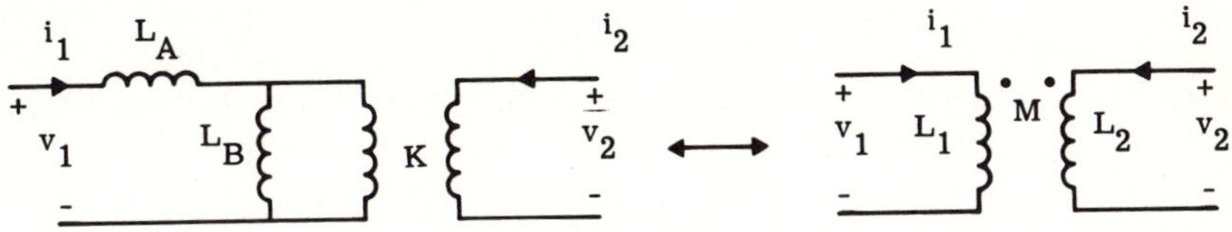

if

$$K = \frac{L_2}{M},$$

$$L_B = \frac{M^2}{L_2},$$

and $$L_A = \frac{L_1 L_2 - M^2}{L_2}$$

FIG. 3.2 Equivalence between coupled coils and ideal transformers.

Corollary 3.1. Assume that $[S]$ is a rational bounded-real matrix and let $r[S]$ denote the order of the largest nonzero cofactor in $1_n - [S(-p)]^T[S(p)]$ (the so-called normal rank of $[S]$); then any network having $[S]$ as its scattering matrix will have at least r resistors in it, but there will exist at least one network, having exactly r resistors, whose scattering matrix is $[S]$.

Corollary 3.2. Assume that $[S]$ is a rational bounded-real matrix and let $\delta(p_i)$ denote the largest order to which a pole, at $p = p_i$, appears in any minor of $[S]$. Now, if $\delta[S] = \sum_{i=1}^{n} \delta(p_i)$ (the so-called Smith–McMillan degree), where the sum is taken over all the poles in $[S]$, then the number of inductors plus the number of capacitors in any network having $[S]$ as its scattering matrix must be greater than or equal to $\delta[S]$, but there will always exist at least one network having exactly this number.

Corollary 3.3. The number of gyrators present in any network must always exceed one-half of the normal rank of $[S] - [S]^T$.

We note that it has not been shown, at least to the author's knowledge, that a minimal network may always be found in the sense that it contain only this number of gyrators, except in the case where $[S]^T = [S]$, i.e., the network is reciprocal, in which case no gyrators are necessary.

We may also summarize the purely mathematical aspects of Theorem 3.2 and its three corollaries as:

Theorem 3.3. Let $[S(p)]$ be an $n \times n$ rational bounded-real matrix of normal rank r and with degree δ; then $[S(p)]$ may always be represented as

$$[S(p)] = [S_{11}] + [S_{12}][Q](1_{r+\delta} - [S_{22}][Q])^{-1}[S_{21}],$$

where $[S_{11}]$ is $n \times n$, $[S_{12}]$ is $n \times (r+\delta)$, $[S_{21}]$ is $(r+\delta)$, $\times\, n$, $[S_{22}]$ is $(r+\delta) \times (r+\delta)$; the matrix

$$[\hat{S}] = \begin{bmatrix} [S_{11}] & [S_{12}] \\ [S_{21}] & [S_{22}] \end{bmatrix}$$

is a unitary $([\hat{S}]^T[\hat{S}] = 1_{n+r+\delta})$ matrix of real constants and $[Q]$ is an $(r+\delta) \times (r+\delta)$ matrix given by

$$[Q] = \operatorname{diag}\left[0, \ldots, 0, \frac{p-1}{p+1}, \ldots, \frac{p-1}{p+1}, \frac{1-p}{p+1}, \ldots, \frac{1-p}{p+1}\right],$$

where the number of zeros is equal to r and the number of terms either of the form $(1-p)/(p+1)$ or $(p-1)/(p+1)$ is equal to δ.†

Proof: Using the arguments of Theorem 3.1, we may remove only the resistors, inductors, and capacitors to form the network M. Moreover, these need only be unit-valued elements, since any other value may always be represented as the input impedance of an ideal transformer terminated in a unit element, i.e., such an input impedance is given as $V_1/I_1 = -(1/K^2)(V_2/I_2) = (Z_2/K^2)$, where Z_2 is either R, pL, or $1/pC$, and the ideal transformers may then be absorbed into the network M. Theorem 3.2 and its corollaries then allow us to conclude that we need only r resistors and a total of δ inductors plus capacitors. The resulting decomposition of the given matrix $[S]$ then follows from the discussion preceding Theorem 3.1. In particular, the form of $[Q]$ obtained is a consequence of the fact that only the resistors, capacitors, and inductors were removed to form M. QED

We note that the representation of $[S]$ given in Theorem 3.3 lumps the gyrators and transformers (the coupled coils, as we noted in our previous

† By the notation $[Q] = \operatorname{diag}[q_1, \ldots, q_n]$ we mean that all the off-diagonal elements of $[Q]$ are zero and that the diagonal elements are q_1, etc.

discussion, have all been replaced by the equivalent inductor-transformer network) into the network M, which, as we see, is now described by the scattering matrix $[\hat{S}]$. One of the outstanding questions in the theory of lumped or rational networks relates to the realization of M with a minimum number of transformers and gyrators, and, in particular, to the case when it can be realized with wires only. One might suspect that we are asking too much if we demand that such be the case using precisely r resistors and δ inductors plus capacitors; however, the answer is not available even if we relax these requirements.

3.1.2. One- and Two-Port Networks

Having described some of the basic results for the general n-port network, we will now turn our attention to the algebraically simpler one- and two-port cases, and in doing so we will obtain many more detailed results.

The first formal realizability theorem was due to Foster [Fo1] and related to lossless one-port networks. Shortly thereafter Cauer [Ca4] obtained alternate realizations of such networks, and we summarize both results in:

Theorem 3.4. Let $S(p)$ be a rational bounded-real function satisfying $S(p)S(-p) = 1$; then $Z(p) = [1 + S(p)]/[1 - S(p)]$ has the two expansions

$$Z(p) = a_0 p + \frac{a_1}{p} + \sum_{i=2}^{n} \frac{a_i p}{p^2 + \omega_i^2},$$

where a_i and ω_i are real and $a_i \geq 0$, and

$$Z(p) = b_0 p + \cfrac{1}{b_1 p + \cfrac{1}{b_2 p + \ddots + \cfrac{1}{b_n p}}},$$

or

$$c_0\left(\frac{1}{p}\right) + \cfrac{1}{c_1\left(\frac{1}{p}\right) + \cfrac{1}{c_2\left(\frac{1}{p}\right) + \ddots + \cfrac{1}{c_n\left(\frac{1}{p}\right)}}},$$

where $b_i \geq 0$, or $c_i \geq 0$.

Proof. With $S(p)S(-p) = 1$ we note that $\lim_{p\to\infty} S(p) = 1$, so that $Z(p) = (1 + S)/(1 - S)$ will be a ratio of polynomials in p such that the numerator and denominator are of different degree. But with $S(p)$ bounded-real $Z(p)$ is positive-real, and Theorem 2.10 implies that its asymptotic behavior be of the form $a_0 p$, where $a_0 \geq 0$. Moreover, from Theorem 2.10 we have the requirement that any poles of $Z(p)$ for $p = j\omega$ be simple and their residues be real and positive. But $S(-p) = 1/S(p)$, and thus $|S(p)| > 1$ in Re $p < 0$, since from the maximum modulus theorem $|S(p)| < 1$ in Re $p > 0$. Thus all the poles of $Z(p)$ lie on the $p = j\omega$ axis, and performing a partial fraction expansion, we get the first representation in the theorem.

The second expansion will be discussed with the assumption that the degree of the numerator of $Z(p)$ is greater than the degree of its denominator. The alternate situation may be treated by letting $p' = 1/p$ and applying the same reasoning. Thus we consider $Z_r(p) = Z(p) - b_0 p$, where $b_0 = \lim_{p\to\infty} Z(p)/p > 0$, again from Theorem 2.10. Using the first expansion described above, we see that $Z_r(p)$ is still positive-real. Moreover, since $S(p)S(-p) = 1$, $Z_r(p) + Z_r(-p) = 0$, so that Z_r must be a ratio of odd to even polynomials in p, and since $\lim_{p\to\infty} Z_r(p)/p = 0$, we see that $1/Z_r(p)$ must be such that the degree of its numerator is larger than the degree of its denominator. However, $1/Z_r(p)$ is also positive-real, and we may therefore repeat the argument just completed to obtain

$$Z(p) = b_0 p + \frac{1}{b_1 p + Z_{r2}(p)}$$

with a new remainder $Z_{r2}(p)$. Continuing the process, we are finally left with the simple function $Z_{rn} = 1/(b_n p)$. QED

The two expansions given in the theorem have simple physical realizations. The first, also referred to as the Foster form, corresponds to a network consisting of a series connection of subnetworks each of which is the parallel combination of an inductor and a capacitor. The second expansion, or Cauer form, is a ladder network whose series branches are all either capacitors or inductors and whose parallel branches are the alternate element. We may easily show, using Corollary 3.2, that both the Foster and Cauer forms result in a minimum number of inductors plus capacitors being used; thus if $Z(p) = a_0(p^m + \cdots)/(p^n + \cdots)$, with, for example, $m > n$, then $S(p) = b_0(p^m + \cdots)/(p^m + \cdots)$ and $\delta(S) = m$, the number of poles of $S(p)$.

The next realizability theorem, historically speaking, was due to Brune [Br3], and it set the standard for most of the subsequent results in the area of rational networks. In particular, he was the first to show that any rational bounded-real function $S(p)$ is the scattering "matrix" (in this case a scalar function) or reflection factor of a lumped network composed of the standard passive elements. In his procedure it is recognized that both $Z(p) = (1 + S)/(1 - S)$ and $Z(p) - R = Z_r(p)$ will be positive-real if $R = \min_\omega\{\text{Re } Z(j\omega)\}$.

This follows from Theorem 2.10 and the maximum modulus theorem, which implies that $|\exp - Z_r(p)| \leq \exp - \text{Re}\ Z_r(j\omega)$, or $\text{Re}\ Z_r(p) \geq \text{Re}\ Z_r(j\omega)$, in $\text{Re}\ p \geq 0$ Let us now assume temporarily that $\min\{\text{Re}\ Z(j\omega)\}$ does not occur at $\omega = 0$ or ∞. Then the function $Z - R - p\{X(\omega_0)/\omega_0\}$, with $Z(j\omega_0) = R + jX(\omega_0)$, will vanish at $p = j\omega_0$, and in the neighborhood of this point

$$Z - R - p\{X(\omega_0)/\omega_0\} \approx a(p - j\omega_0) + \cdots,$$

where, in view of Lemma 2.4,

$$a = \left.\frac{d[Z - R - p\{X(\omega_0)/\omega_0\}]}{dj\omega}\right|_{\omega=\omega_0} = \frac{X'(\omega_0)}{\omega_0} - \frac{X(\omega_0)}{\omega_0} > 0,$$

since the fact that Z has a minimum real part at $p = j\omega_0$ yields

$$\left.\frac{d[\text{Re}\ Z(j\omega)]}{dj\omega}\right|_{\omega=\omega_0} = 0.$$

We will first assume that $X(\omega_0)/\omega_0 < 0$, so that $Z - R - p[X(\omega_0)/\omega_0]$ is positive-real. If we perform a partial fraction expansion, we obtain

$$\frac{1}{Z - R - p[X(\omega_0)/\omega_0]} = \frac{1}{a(p - j\omega_0)} + \frac{1}{a(p + j\omega_0)} + \frac{1}{\hat{Z}_r}$$
$$= \frac{(2/a)p}{p^2 + \omega_0{}^2} + \frac{1}{\hat{Z}_r},$$

where $\hat{Z}_r$, as we see, is also positive-real. With regard to the behavior of $\hat{Z}_r$ at ∞, we observe that

$$\frac{1}{Z - R - p[X(\omega_0)/\omega_0]} \approx \frac{1}{-pX(\omega_0)/\omega_0}$$

as $p \to \infty$, since $\text{Re}[Z(j\omega) - R]$ is nonzero at $\omega = \infty$ [the minimum of $Z(j\omega)$ was assumed to occur at $\omega = \omega_0$]. Thus

$$\frac{1}{\hat{Z}_r} \approx \frac{1}{-pX(\omega_0)/\omega_0} - \frac{2/a}{p}$$

as $p \to \infty$, and we obtain

$$\hat{Z}_r = \frac{-a[X(\omega_0)/\omega_0]}{a + 2[X(\omega_0)/\omega_0]}\, p + Z_2,$$

where $Z_2(p)$ is positive-real. We may interpret the foregoing decomposition of Z, i.e.,

$$Z = R + p\frac{X(\omega_0)}{\omega_0} + \cfrac{1}{\cfrac{2/ap}{p^2+{\omega_0}^2} + \cfrac{1}{\cfrac{-a[X(\omega_0)/\omega_0]}{a+2[X(\omega_0)/\omega_0]}p + Z_2}},$$

as a two-port network composed of a resistor R, a capacitor, and three inductors (not all with positive values of L) arranged in a tee, and this network terminated in Z_2. If we replace the three-terminal network formed by the three inductors by the three-terminal network obtained from two coupled coils, then we find the resultant overall network to be the one shown in Fig. 3.3. In particular, from Lemma 2.4 we have $X'(\omega_0) > |X(\omega_0)/\omega_0|$, so that the coupled coils are realizable, i.e., $L_1 > 0$, $L_2 > 0$, and $L_1L_2 - M^2 = 0$. We may treat the two situations which we have previously excluded—namely, when min Re{Re $Z(j\omega)$} is at $\omega = 0$ or ∞—in essentially the same way, and the resulting two-port networks are very simple: for $\omega = 0$ we obtain a simple shunting inductor, and for $\omega = \infty$ a simple shunting capacitor. Finally, if we encounter a situation where $X(\omega_0)/\omega_0 > 0$, then we may deal with the function $Y(p) = 1/(Z - R)$ and after retracing all the previous steps in this discussion we will find that the resultant Brune network is identical to the one in Fig. 3.3, and the conclusion that $Z_2(p)$ is positive-real still follows. Now, using the network shown in Fig. 3.3, we may show directly that with $Z_{\text{in}}(p)$ specified, assuming $\operatorname{Re} Z_{\text{in}}(j\omega_0) = 0$, the $Z_2(p)$ which must be present at the output of the network to produce this $Z_{\text{in}}(p)$ will be such that the sum of the degree of its numerator and the degree of its denominator will be less than that of $Z_{\text{in}}(p)$. Moreover, as we have shown $Z_2(p)$ will still be positive-real. Thus we may repeat the cycle by finding

$$R_2 = \min\{\operatorname{Re} Z_2(j\omega)\}$$

and considering $Z_2 - R_2$, etc. until we arrive at some Z_2 which will be either R, $1/pC$, or pL. Thus a rational positive-real function may always be represented as the input impedance of a network that has the form of a cascade of networks of the type shown in Fig. 3.3 but with series resistors connecting each pair of networks, the value of the resistors determined by $\min\{\operatorname{Re} Z_k(j\omega)\}$.

One final comment is in order regarding the Brune procedure: the resultant network contains the minimum number of inductors and capacitors. In order to achieve this we replace the two coupled coils by an ideal transformer and *one* inductor. This is possible, as we see from Fig. 3.2, since $L_1L_2 = M^2$. Then, since the Smith–McMillan degree of Z_2, $\delta(Z_2)$, is less than that of the original $Z(p)$ by at least two because the Brune section reduces the degree of both the numerator and denominator of $Z(p)$ by two, we see that this has been achieved with a total of two reactive elements (one capacitor and one inductor).

Brune's result was followed shortly by the discovery that any positive-real function may be realized as the input impedance of a rational, lossless, two-port network terminated in a unit resistor. This result was due to

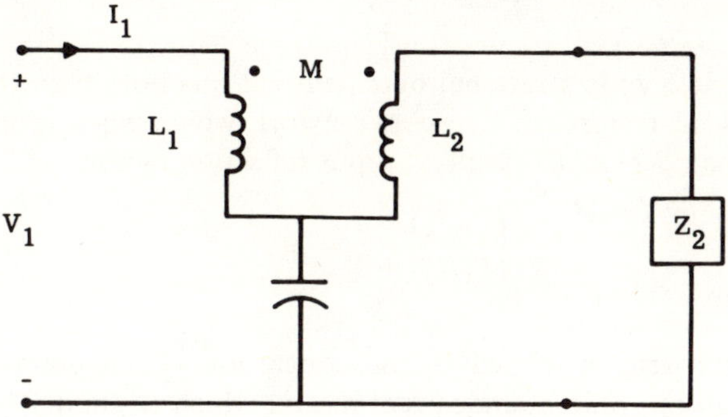

$$\frac{V_1}{I_1} = Z_{in} = R + jX$$

with $R(\omega_0) = 0$

$$L_1 = \frac{\omega_0 X'(\omega_0) + X(\omega_0)}{2\omega_0}$$

$$L_2 = \frac{1}{2\omega_0} \frac{\left[\omega_0 X'(\omega_0) - X(\omega_0)\right]^2}{\omega_0 X'(\omega_0) + X(\omega_0)}$$

$$M = \frac{\omega_0 X'(\omega_0) - X(\omega_0)}{2\omega_0}$$

and $$C = \frac{2}{\omega_0} \frac{1}{\omega_0 X'(\omega_0) - X(\omega_0)}$$

FIG. 3.3 Brune network.

Darlington [Da1], who established it by finding lossless networks (see, for example, [Yo4]), similar to Brune's network which also produced the desired degree reduction. Thus if the Darlington network was formed using the values of $Z(p)$ at some point in Re $p > 0$, where $Z(p) + Z(-p) = 0$, then these lossless Darlington networks required a terminating impedance that was positive-real and of reduced degree. Since the series resistors used in

Brune's procedure were eliminated, although in the Darlington procedure the Brune network itself is used if a zero occurs on $p = j\omega$, the resultant cascade of networks is a lossless network which is finally terminated in a resistor when we arrive at the point where the required $Z_{\text{in}}(p) + Z_{\text{in}}(-p) = K$ [in this case $Z_{\text{in}}(p) = K + Z_{\text{F}}(p)$ where $Z_{\text{F}}(p)$ is a Foster form]. We thus have the following theorem, although the proof we will present is not a constructive one, as was Darlington's proof:

Theorem 3.5. If $S(p)$ is a rational bounded-real function, then there exists a rational, lossless, reciprocal, two-port network which when terminated in a unit resistor has $S(p)$ as its resulting input reflection factor, i.e.,

$$\frac{(V_1/I_1) - 1}{(V_1/I_1) + 1} = S(p).$$

Proof. This theorem is related to the conclusion of Theorem 2.18 that a symmetric meromorphic bounded-real matrix which is unitary for $p = j\omega$ is uniquely determined by $S_{11}(p)$ and the factor $\pm e^{-ap}b(p)$. In particular,

$$S_{12}(p) = S_{21}(p) = \pm S_m(p)e^{-ap}b(p)$$

and

$$S_{22}(p) = -S_{11}(-p)S_{12}(p)/S_{12}(-p),$$

where $a \geq 0$, $b(p)$ is a regular Blaschke product, and $S_m(p)$ is that factorization of

$$S_m(p)S_m(-p) = 1 - S_{11}(p)S_{11}(-p)$$

that is analytic and nonzero in $\text{Re}\, p > 0$.

For the problem at hand we assume that $[S(p)]$ is the scattering matrix of some rational two-port lossless network normalized to 1 at both ports. Then if port two is terminated in a unit resistor, $V_2 = -I_2$ or $a_2 = 0$ and $b_1/a_1 = (V_1 - I_1/)(V_1 + I_1) = S_{11}$, i.e., if a given function $S(p)$ is to be the input reflection factor of some rational, lossless, two-port network terminated in a unit resistor, then $S(p) = S_{11}(p)$, where S_{11} is the 11 element of the scattering matrix of the two-port. Thus with S_{11} determined we consider

$$S_m(p)S_m(-p) = 1 - S_{11}(p)S_{11}(-p)$$

and note that since $S_{11}(p)S_{11}(-p)$ is an even function of p we may always factor the function $1 - S_{11}(p)S_{11}(-p)$ so that $S_m(p)$ is rational and has no poles or zeros in $\text{Re}\, p > 0$. But since $S(p) = S_{11}(p)$ is bounded-real, $|S_m(j\omega)|^2 = 1 - |S_{11}(j\omega)|^2 \leq 1$, and, using the maximum modulus theorem, we conclude that $S_m(p)$ is bounded by 1 in $\text{Re}\, p > 0$.

We then have $S_{12} = S_{21} = \pm S_m(p)b(p)$, where $b(p)$ is a rational Blaschke product because the scattering matrix of the network is to have only rational elements.

Finally, we obtain, using the expressions for S_{12} and S_{21},

$$S_{22}(p) = -S_{11}(-p)S_m(p)b^2(p)/S_m(-p),$$

since $1/b(-p) = b(p)$. Now, the poles of $S_{11}(-p)$, which are not among the poles of $S_m(-p)$ in $\mathrm{Re}\, p > 0$, together with the zeros of $S_m(-p)$ in $\mathrm{Re}\, p > 0$, which are not among those of $S_{11}(-p)$, will become poles of $S_{22}(p)$ unless they are canceled by zeros introduced through $b^2(p)$. But since $b(p)$, in particular its zeros in $\mathrm{Re}\, p > 0$, are at our diposal, we may always cause such a cancellation and thus obtain an $S_{22}(p)$ that is analytic and bounded in $\mathrm{Re}\, p > 0$.

We have shown that a symmetric, "unitary," rational, bounded-real matrix can always be found such that $S_{11}(p)$ is some given arbitrary, rational, bounded-real function. Thus we have established Darlington's theorem, since the normal rank r of $1 - [S(-p)]^T[S(p)]$ will be zero and Corollary 3.1 then shows that a network having this scattering matrix may be found containing no resistors. QED

In the proof of Theorem 3.5 we saw that there was considerable freedom in the selection of the Blaschke product $b(p)$. If we let $b_{\mathrm{m}}(p)$ denote the minimal Blaschke product, i.e., $b_{\mathrm{m}}(p)$ has the smallest number of zeros necessary to render $S_{22}(p)$ analytic in $\mathrm{Re}\, p > 0$, then the most general lossless network satisfying the conditions of the theorem has a scattering matrix with $S_{11}(p) = S(p)$, $S_{12} = S_{21} = S_{12\mathrm{m}}\hat{b}(p)$, and $S_{22} = S_{22\mathrm{m}}\hat{b}^2(p)$, where $S_{12\mathrm{m}} = \pm S_m b_{\mathrm{m}}(p)$ and $S_{22\mathrm{m}} = -S_{11}(-p)S_{12\mathrm{m}}(p)/S_{12\mathrm{m}}(-p)$. In this formulation $\hat{b}$ is an arbitrary, rational, regular Blaschke product that is unconstrained, since $b_{\mathrm{m}}(p)$ was selected to guarantee that $[S(p)]$ is realizable. One can give an interesting physical interpretation of the additional flexibility inherent in $\hat{b}(p)$. We may show directly that the most general lossless two-port network may be realized as the cascade of two lossless networks, the first with a scattering matrix

$$\begin{bmatrix} S_{11} & S_{12\mathrm{m}} \\ S_{12\mathrm{m}} & S_{22\mathrm{m}} \end{bmatrix}$$

and the second with a scattering matrix

$$\begin{bmatrix} 0 & \hat{b} \\ \hat{b} & 0 \end{bmatrix}.$$

In both cases it is assumed that the normalization used at all ports is unity, and we see that terminating the second lossless network, which is sometimes referred to as an all-pass network, in a unit resistor yields an input impedance equal to unity for that network. Thus the performance of the first network, which is called the minimal network, is the same as if a unit resistor was used to terminate it instead of the all-pass network. The additional flexibility in the selection of $S_{21}(p) = S_{12}(p)$ provided by the all-pass network has no direct engineering interest if attention is focused solely on the input reflection factor $S_{11}(p) = S(p)$. But as we will see in the following discussion of the realization of networks having prescribed $S_{21}(p)$ or $S_{12}(p)$ coefficients, this flexibility is of considerable utility for many other applications.

Before terminating our discussion of Darlington's theorem we note that one is prompted by this result to question whether or not it holds if the input reflection factor is an arbitrary bounded-real function and the corresponding two-port network is allowed to be nonrational, although still lossless. The following counterexample demonstrates that the answer to this question is no. Thus consider the positive-real function

$$Z(p) = \tan^{-1} p = \frac{j}{2} \log \frac{1 - jp}{1 + jp},$$

where we have rendered the log single-valued by selecting the branch $-\pi < \theta \leq \pi$, i.e., $\log f(p) = \log Ae^{j\theta} = \log A + j\theta$, where θ is restricted to that range. For $p = \sigma + j\omega$ we note that

$$\operatorname{Re} Z(p) = -\theta/2,$$

where θ is the phase angle of the function

$$\frac{1 - \omega^2 - \sigma^2 - 2j\sigma}{(1 - \omega)^2 + \sigma^2}.$$

Thus

$$\begin{aligned} \lim_{\sigma \to 0} \operatorname{Re} Z(p) &= 0 \qquad \text{for} \quad |\omega| < 1 \\ &= \frac{\pi}{2} \qquad \text{for} \quad |\omega| > 1, \end{aligned}$$

since the corresponding limits for θ are 0 and $-\pi$. Now, consider the bounded-real function

$$S(p) = \frac{Z(p) - 1}{Z(p) + 1}$$

and note that the behavior of $\operatorname{Re} Z(j\omega)$ implies that

$$\begin{aligned} |S(j\omega)| &= 1 \qquad \text{for} \quad |\omega| < 1 \\ &\neq 1 \qquad \text{for} \quad |\omega| > 1. \end{aligned}$$

If we assume that a lossless two-port network exists which has $S(p)$ as its input reflection factor when terminated in a unit resistor, then $S_{11}(p) = S(p)$ and the S_{21} element of the scattering matrix of the network must satisfy

$$\begin{aligned} |S_{21}(j\omega)|^2 = 1 - |S_{11}(j\omega)|^2 &= 0 \qquad \text{for} \quad |\omega| < 1 \quad \text{a.e.} \\ &\neq 0 \qquad \text{for} \quad |\omega| > 1 \quad \text{a.e.} \end{aligned}$$

But $S_{21}(j\omega)$ must also be the boundary value, as Re $p \to 0$, of a bounded analytic function, and we conclude on the basis of Theorem 1.7 that S_{21} cannot be zero pointwise for almost all $|\omega| < 1$ unless it is identically zero for all ω. Since this would not be the case for the given function, i.e.,

$$S(p) = \frac{(\tan^{-1}p - 1)}{(\tan^{-1}p + 1)},$$

we have arrived at a contradiction and we may state that *if $S(p)$ is an arbitrary bounded-real function, then it is not always possible to find some lossless two-port network such that $S(p)$ will be its input reflection factor when the network is terminated in a unit resistor.*

The lossless two-port network has also been used as a filter network, i.e., as a means of controlling the voltage transfer function between some source and some load. Specifically, we will consider the system shown in Fig. 2.2, where the purpose of the network N, assumed to be lossless, is to produce some desired voltage transfer function between E_S and V_2. We will further assume that both Z_g and Z_L are real nonnegative constants, i.e., they are resistors. If we consider the scattering matrix of the network N normalized to R_g at port one and R_L at port two, then we find that

$$\frac{V_2(p)}{E_S(p)} = \left(\frac{R_L}{R_g}\right)^{1/2} \frac{S_{21}(p)}{2},$$

where S_{21} is the 21 element of the scattering matrix. Thus the problem of determining the network given the voltage transfer function may then be viewed as the problem of finding a rational "unitary" scattering matrix such that $2\{V_2(p)/E_S(p)\}(R_g/R_L)^{1/2}$ is its S_{21} element. The fact that the two-port network is assumed to be lossless imposes certain limitations on the voltage transfer function obtainable from reciprocal networks over and above the requirement that $S_{21}(p)$ be bounded-real. However, there is no additional limitation on the magnitude of $S_{21}(j\omega)$. The following theorem states these conclusions explicitly:

Theorem 3.6. Let $[S(p)]$ be a 2×2 unitary, rational, bounded-real matrix, with $S_{21}(p)$ its 21 element. If $[S(p)]$ is symmetric ($S_{21} = S_{12}$), then $|S_{21}(p)| \leq 1$ in Re $p > 0$ and any zero of $S_{21}(-p)$ in Re $p > 0$ must also be a zero of $S_{21}(p)$ in Re $p > 0$ of at least the same multiplicity. However, the only constraint on $|S_{21}(j\omega)|^2$ is that it be an even rational function of ω bounded by 1 for all ω. If on the other hand $[S(p)]$ is not symmetric, then $S_{21}(p)$ may be an arbitrary, rational, bounded-real function.

Proof. The proof proceeds along the same lines as the one employed for Theorem 3.5. In particular, given $S_{21}(p)$ we may always represent it as $S_{21}(p) = S_{21\text{m}}(p)b(p)$, where $S_{21\text{m}}$ and $b(p)$ are uniquely determined by the fact that $S_{21\text{m}}$ is to be nonzero in $\text{Re}\, p > 0$. Thus from Theorem 2.18 we see that $S_{12}(p) = S_{21\text{m}}(p)\hat{b}(p)$, where $\hat{b}(p)$ is an arbitrary, regular, Blaschke product unless the matrix is symmetric, in which case $\hat{b} = b(p)$. Now, since

$$S_{11}(p)S_{11}(-p) = 1 - S_{21}(p)S_{21}(-p),$$

we obtain the most general $S_{11}(p)$ as

$$S_{11}(p) = C(p)b_1(p),$$

with $C(p)$ that factorization of $C(p)C(-p) = 1 - S_{21}(p)S_{21}(-p)$ that is analytic and nonzero in $\text{Re}\, p > 0$, and $b_1(p)$ an arbitrary regular Blaschke product. We thus conclude that the unitary property of the scattering matrix reduces to the constraint that

$$S_{22}(p) = -\frac{S_{11}(-p)S_{12}(p)}{S_{21}(-p)} = -\frac{C(-p)b_1(-p)S_{21\text{m}}(p)\hat{b}(p)b(p)}{S_{21\text{m}}(-p)}$$

be analytic in $\text{Re}\, p > 0$. If the matrix is nonsymmetric, then both $b_1(-p)$ and $\hat{b}(p)$ are at our disposal, so that we may set $b_1(-p) = 1$ for example, and then utilize $\hat{b}(p)$ to cancel the potential poles of $S_{22}(p)$ introduced by $C(-p)$ and $S_{21\text{m}}(-p)$. However, if the matrix is to be symmetric, then $\hat{b} = b$, and the only freedom left is in $b_1(-p)$. But the zeros of $b_1(-p)$ are in $\text{Re}\, p < 0$, so that it cannot be used to cancel the potential poles introduced through the zeros of $S_{21\text{m}}(-p)$ in $\text{Re}\, p > 0$. Now, we will assume that the denominator of $S_{21\text{m}}(p)$ is identical to that of $C(p)$; since $C(p)C(-p) = 1 - S_{21}(p)S_{21}(-p)$, we see that this will be the case unless cancellations take place between the numerator and denominator of $1 - S_{21}(p)S_{21}(-p)$; however, these cancellations may always be compensated by augmenting both the numerator and denominator of the resulting $C(p)$ with these factors. Thus we see that if $S_{22}(p)$ is to be analytic in $\text{Re}\, p > 0$, then the zeros of $S_{21}(-p)$ in $\text{Re}\, p > 0$ must be among the zeros of $S_{12}(p) = S_{21}(p)$ in $\text{Re}\, p > 0$. QED

In Section 3.3.2 we will consider the more general filter problem, also referred to as the broadband matching problem, where Z_L and Z_g need not be real, positive constants, but may in fact be arbitrary, rational, positive-real functions. As we will note, no generally applicable solution exists to this more general problem, and the results contained in Theorem 3.6 are in fact unique because of their completeness

3.2. INTERPOLATION WITH BOUNDED-REAL FUNCTIONS

We noted in Section 2.4 that a positive-real function $Z(p)$ may not assume arbitrary values at a point p_0 if Re $p_0 > 0$. We certainly expect that Re $Z(p_0)$ must be nonnegative at such a point, but the limitations on Im $Z(p_0)$ expressed in Lemma 2.4 came as a surprise. Using the transformation $S(p) = \{Z(p) - 1\}/\{Z(p) + 1\}$, we note that $S(p)$ will be bounded-real given that $Z(p)$ is positive-real. The conditions of Lemma 2.4 may then be applied to conclude that at a point $p_0 = \sigma_0 + j\omega_0$ ($\sigma_0 \geq 0$), where $S(p_0) = A_0 \exp(j\theta_0)$, we must have $A_0 \leq 1$,

$$\frac{1 - A_0{}^2}{\sigma_0} > \left|\frac{2A_0 \sin \theta_0}{\omega_0}\right| \qquad \text{if} \quad \sigma_0 \neq 0 \quad \text{and} \quad \omega_0 \neq 0$$

and

$$-\frac{d\theta_0}{d\omega}(\omega_0) > \left|\frac{\sin \theta_0}{\omega_0}\right| \qquad \text{if} \quad \sigma_0 = 0 \quad \text{and} \quad A_0 = 1.$$

We have assumed that $S(p)$ is not of the form $\pm(p - K)/(p + K)$ and that $S(p)$ is analytic at the point where $\sigma_0 = 0$ and $A_0 = 1$, these being the corresponding conditions under which Lemma 2.4 is applicable. One might then inquire as to the conditions which must be met if the same bounded-real function is to take on prescribed values at two or more points in Re $p \geq 0$. This problem, which is referred to as the problem of interpolation, will be discussed in this section using a result due to Youla [Yo4] and a constructive procedure will be presented to determine if such an interpolation is possible in any given instance. Youla's result is based on the realization that the synthesis procedures developed by Brune and Darlington and described in the previous section may be viewed as interpolation procedures. Using the form of the networks obtained by Brune and Darlington, Youla was then led to the conclusion that the following theorem was in fact the origin of both synthesis techniques. Moreover, he showed that the theorem could be used to generate an algorithm which in turn solves the interpolation problem.

Theorem 3.7 [Yo4]. Let $S_0(p)$ be an arbitrary, rational, bounded-real function not of the form $\pm(p - K)/(p + K)$; then the function

$$S_1(p) = \frac{[(I_1 - I_4)p^2 + p(I_3 - I_2)] + S_0(p)[(I_1 + I_4)p^2 + (I_2 + I_3)p + 2|p_0|^2]}{[(I_1 + I_4)p^2 - (I_2 + I_3)p + 2|p_0|^2] + S_0(p)[(I_1 - I_4)p^2 + p(I_2 - I_3)]} \tag{3.1}$$

is a rational bounded-real function if I_1, I_2, I_3, and I_4 are selected in accordance with Table 3.1 and $S_0(p_0) = Ae^{j\theta}$, Re $p_0 \geq 0$. Moreover, $S_1(p_0)$ is independent of $S_0(p_0)$. If in addition p_0 is a root of $1 - S_0(-p)S_0(p)$, then the sum of the degree of the numerator plus the degree of the denominator of $S_1(p)$ is at least four less than the comparable sum for $S_0(p)$ when $\omega_0 \neq 0$ and at least two less when $\omega_0 = 0$. Finally, given an arbitrary bounded-real function $S_1(p)$, the corresponding $S_0(p)$ obtained by inverting the above expression is also bounded-real.

TABLE 3.1

Coefficients for Eq. (3.1)

	$p_0 = \sigma_0 + j\omega_0$, σ_0 and $\omega_0 \neq 0$	$p_0 = j\omega_0$, $A = 1$	$p_0 = \sigma_0 \neq 0$
$I_1 \geq 0$	$\dfrac{\left(\dfrac{1-A^2}{\sigma_0}\right) - \left(\dfrac{2A\sin\theta}{\omega_0}\right)}{\left(\dfrac{1-A^2}{\sigma_0}\right) + \left(\dfrac{2A\sin\theta}{\omega_0}\right)}$	$\dfrac{-\dfrac{d\theta}{d\omega} - \dfrac{\sin\theta}{\omega_0}}{-\dfrac{d\theta}{d\omega} + \dfrac{\sin\theta}{\omega_0}}$	0
$I_2 \geq 0$	$2\dfrac{1 + 2A\cos\theta + A^2}{\left(\dfrac{1-A^2}{\sigma_0}\right) + \left(\dfrac{2A\sin\theta}{\omega_0}\right)}$	$\dfrac{2(1+\cos\theta)}{-\dfrac{d\theta}{d\omega} + \dfrac{\sin\theta}{\omega_0}}$	$\sigma_0\left(\dfrac{1 + A\cos\theta}{1 - A\cos\theta}\right)$
$I_3 \geq 0$	$2\dfrac{1 - 2A\cos\theta + A^2}{\left(\dfrac{1-A^2}{\sigma_0}\right) - \left(\dfrac{2A\sin\theta}{\omega_0}\right)}$	$\dfrac{2(1-\cos\theta)}{-\dfrac{d\theta}{d\omega} - \dfrac{\sin\theta}{\omega_0}}$	$\sigma_0\left(\dfrac{1 - A\cos\theta}{1 + A\cos\theta}\right)$
$I_4 \geq 0$	$1/I_1$	I/I_4	0

Proof.† With $S_0(p)$ bounded-real and not of the form $\pm(p - K)/(p + K)$ we see that the inequality constraints on the values which S_0 may assume at some point p_0 in Re $p \geq 0$ guarantee that the constants I_1, I_2, I_3, and I_4 are all nonnegative. Now, if we define $\alpha(p) = (I_1 - I_4)p^2 + p(I_3 - I_2)$ and $\beta(p) = (I_1 + I_4)p^2 + (I_2 + I_3)p + 2|p_0|^2$, then we may rewrite the expression defining $S_1(p)$ as

$$S_1(p) = \frac{\alpha(p) + \beta(p)S_0(p)}{\beta(-p) + \alpha(-p)S_0(p)}.$$

† We will explicitly discuss the case σ_0 and $\omega_0 \neq 0$; similar reasoning applies to the other two cases.

In particular, we have

$$|S_1(j\omega)|^2 = \frac{|(\alpha/\beta)(j\omega)|^2 + 2\,\mathrm{Re}(\alpha/\beta)(j\omega)S_0(-j\omega) + |S_0(j\omega)|^2}{1 + 2\,\mathrm{Re}(\alpha/\beta)(j\omega)S_0(-j\omega) + |(\alpha/\beta)(j\omega)|^2|S_0(j\omega)|^2}$$

$$= 1 + \frac{(|S_0(j\omega)|^2 - 1)(1 - |(\alpha/\beta)(j\omega)|^2)}{1 + 2\,\mathrm{Re}(\alpha/\beta)(j\omega)S_0(-j\omega) + |(\alpha/\beta)(j\omega)|^2|S_0(j\omega)|^2},$$

so that if we can demonstrate that $|(\alpha/\beta)(j\omega)| \leq 1$, then we may conclude that $|S_1(j\omega)| \leq 1$, since with $S_0(p)$ a bounded-real function $|S_0(j\omega)|^2 \leq 1$. But

$$\left|\frac{\alpha}{\beta}(j\omega)\right|^2 = \frac{(I_1 - I_4)^2\omega^4 + \omega^2(I_2 - I_3)^2}{(I_1 + I_4)^2\omega^4 + [(I_2 + I_3)^2 - 4\,|p_0|^2(I_1 + I_4)]\omega^2 + 4\,|p_0|^4} = \frac{N}{D},$$

where N and D are both nonnegative. Then, using the fact that $I_1 I_4 = 1$, we find

$$D - N = 4\{\omega^4 + [I_2 I_3 - |p_0|^2(I_1 + I_4)]\omega^2 + |p_0|^4\}.$$

Now, the inequality constraints of Lemma 2.4 also imply, since $|(X_0/Z_0) \times (p_0/\omega_0)| < 1$, that

$$\left|\frac{2A_0 \sin\theta_0}{1 - A_0{}^2 + j2A_0 \sin\theta_0}\,\frac{p_0}{\omega_0}\right| < 1,$$

so that we see, using the definitions from Table 3.1, that

$$I_2 I_3 - |p_0|^2(I_1 + I_4) > -2\,|p_0|^2.$$

Therefore $D - N > 0$, or $|(\alpha/\beta)(j\omega)| < 1$.

We also observe using Table 3.1 that the numerator and denominator of the expression for S_1 both vanish at $p = p_0$ and $p = p_0{}^*$. Therefore if we can establish that the number of roots, including their multiplicity, of the denominator may not exceed two then we may conclude that $S_1(p)$ is analytic in $\mathrm{Re}\, p > 0$, since these roots also appear, and are thus cancelled, in the numerator polynomial. To this end consider the function

$$F(p) = \frac{\beta(-p) + \alpha(-p)S_0(p)}{\beta(-p)} = 1 + \frac{\alpha(-p)}{\beta(-p)}\,S_0(p)$$

and note that $|\{\alpha(-j\omega)/\beta(-j\omega)\}S_0(j\omega)| < 1$. Moreover, observe that $\lim_{|p|\to\infty} \{\alpha(-p)/\beta(-p)\}S_0(p) = \text{const}$ independent of the angle of p. Thus the number of zeros of $F(p)$ is equal to the number of its poles in $\mathrm{Re}\, p > 0$, since the net change in the phase angle of $F(j\omega)$ as ω varies from $-\infty$ to $+\infty$ will be zero given that $|\{\alpha(-j\omega)/\beta(-j\omega)\}S_0(j\omega)| < 1$, and so the net change in the phase of $F(p)$ as p moves around the contour formed by the

$j\omega$ axis and a semicircle in Re $p > 0$ which has infinite radius is zero. However, $\beta(-p)$ has at most two roots (it is only a second-order polynomial) in Re $p > 0$, so the maximum number of roots of $\beta(-p) + \alpha(-p)S_0(p)$ in Re $p > 0$ is two. Thus we have established that $S_1(p)$ will be analytic in Re $p > 0$, and the combination of this with the conclusion that $|S_1(j\omega)| < 1$ implies via the maximum modulus theorem that $S_1(p)$ is bounded-real.

Now, by definition, the total degree (sum of the numerator degree and the denominator degree) of $S_1(p)$ is at most four greater than $S_0(p)$, and we may then conclude from the above arguments concerning the cancellation of roots at p_0 and $p_0{}^*$ that the total degree of $S_1(p)$ is at least no greater than that of $S_0(p)$. But if p_0 is a root of $1 - S_0(-p)S_0(p)$, we may show by direct algebraic arguments that there will be additional cancellations of degree two in both numerator and denominator when $\omega_0 \neq 0$ and of degree 1 in both when $\omega_0 = 0$, so that the total degree of $S_1(p)$ will be reduced in this situation by the amounts stated in the theorem.

Finally, we observe that since both the numerator and denominator in the expression for $S_1(p)$ vanish at $p = p_0$, then $S_1(p_0)$ is not determined from $S_0(p_0)$ alone.

The fact that the transformation defined by $I_1, \ldots, I_4$ yields a bounded-real $S_0(p)$ given that $S_1(p)$ is bounded-real may be shown directly, since

$$S_0(p) = \frac{S_1(p)[(I_1 + I_4)p^2 - (I_2 + I_3)p + 2|p_0|^2] - [(I_1 - I_4)p^2 + p(I_3 - I_2)]}{[(I_1 + I_4)p^2 + (I_2 + I_3)p + 2|p_0|^2] - [(I_1 - I_4)p^2 - p(I_3 - I_2)]S_1(p)}$$

$$= \frac{\beta(-p)S_1(p) - \alpha(p)}{\beta(p) - \alpha(-p)S_1(p)}.$$

But $1 - \{\alpha(-p)/\beta(p)\}S_1(p)$ is nonzero in Re $p > 0$ since $\alpha(-p)/\beta(p)$, being analytic in Re $p > 0$ and bounded by 1 for $p = j\omega$, must be bounded by 1 in all Re $p > 0$. Finally, we show, as above, that $|S_0(j\omega)| \leq 1$ given that $|S_1(j\omega)| \leq 1$, and again using the maximum modulus theorem, we obtain the conclusion that $S_0(p)$ is bounded-real. QED

We note that the above theorem can be given the following physical or network interpretation: $S_0(p)$ will be the input reflection factor of a lossless network terminated in an impedance $Z_2 = \{1 + S_1(p)\}/\{1 - S_1(p)\}$, where these networks are of three types—the Brune network if $\sigma_0 = 0$ and $A_0 = 1$ and the Darlington networks when $\sigma_0 \neq 0$. We will not specifically describe these two Darlington networks, corresponding to $\omega_0 = 0$ and $\omega_0 \neq 0$; the interested reader will find the details in either [Da1] or [Yo4].

However, we will apply Theorem 3.7 to the question of interpolation. Specifically, let us assume that we are given n points p_i such that Re $p_i \geq 0$, and n corresponding complex numbers $M(p_i)$, with $|M(p_i)| \leq 1$ and $M(p_i{}^*) =$

$M^*(p_i)$, and we wish to find a single bounded-real function $S(p)$ such that

$$S(p_i) = M(p_i)$$

at these n points. First let us assume that all the points p_i are on the $j\omega$ axis and note that we will have satisfied our aim if we can find a positive-real impedance $Z(p)$ such that $Z(j\omega_i) = \{1 + M(j\omega_i)\}/\{1 - M(j\omega_i)\}$. We may now employ a technique based on the Brune synthesis procedure suggested by Smilen [Sm1]:

1. Note that Re $Z(j\omega_i) \geq 0$, since $|M(j\omega_i)| \leq 1$, and select that point ω_{m} having the minimum value of Re $Z(j\omega_i)$, say R_{m}.
2. Using $X(\omega_{\mathrm{m}}) = \mathrm{Im}\ Z(j\omega_{\mathrm{m}})$, select any real number q satisfying $q > |X(\omega_{\mathrm{m}})/\omega_{\mathrm{m}}|$ and identify it as $X'(\omega_{\mathrm{m}})$.
3. Use the $X(\omega_{\mathrm{m}})$ and the q selected in 2 to generate a Brune section as in Fig. 3.3.
4. Calculate the values of the terminating impedance (Z_2 in Fig. 3.3) of the Brune section that is required to produce an input impedance Z_{in} such that $Z_{\mathrm{in}}(j\omega_i) + R_{\mathrm{m}} = Z(j\omega_i)$.
5. From our previous discussion of the Brune section we note that at $\omega_i = \omega_{\mathrm{m}}$ any value of Z_2 will suffice and that at the remaining points the values of Re $Z_2(j\omega_i)$ will be greater than or equal to zero. Therefore the process may be repeated until all the points are considered.
6. Arbitrarily setting $Z_2(p)$ equal to a positive-real function in the last section, we may compute the resultant input impedance of the entire cascade of Brune networks. This impedance will be positive-real and have the required values at the specified points.

From this discussion we note that as long as each Re $Z(j\omega_i) \geq 0$ it is always possible to find a positive-real function to perform the interpolation when $p_i = j\omega_i$, and thus it is always possible to find a bounded-real function to interpolate at points on the $j\omega$ axis if $|M(j\omega_i)| \leq 1$. Now we may return to the general problem. The following algorithm is then justified on the basis of Theorem 3.7 [we include the possibility that the derivate of $S(p)$ is also prescribed at certain points, since we will have occasion to use this result in some subsequent discussions]:

1. Given n points at which the value of $S(p)$ (and possibly its derivatives) are prescribed, select one of these points such that Re $p_i > 0$, and using the prescribed value of $S(p)$, compute the appropriate coefficients from Table 3.1.
2. Using the coefficients obtained in 1, form the function $S_1(p)$ using the expression given in Theorem 3.7 and then calculate the values that $S_1(p)$ must assume at the remaining $n - 1$ points in order that $S(p)$ will be equal to the prescribed values at these points. Note that if only the value of $S(p)$ is

prescribed at the first point, $p = p_0$, as opposed to its value together with the values of its first k derivatives, then $S_1(p)$ may be an arbitrary number at $p = p_0$ and we have in fact only $n - 1$ remaining values which $S_1(p)$ must assume. However, if the first k derivatives of $S(p)$ were also specified at $p = p_0$, then one must determine the required value of $S_1(p_0)$ together with its first $k - 1$ derivatives using L'Hospital's rule to evaluate the indeterminate form of the expression at $p = p_0$. But in every case there will be at least one less specification imposed on the values of $S_1(p)$.

3. Repeat the above cycle using any one of the remaining $n - 1$ values in $\operatorname{Re} p > 0$.

4. When eventually the only points remaining are on the $j\omega$ axis employ the Brune-type procedure described above to accommodate these points. At the last section we may terminate with an arbitrary positive-real function or, equivalently, an arbitrary bounded-real function and then compute the resulting $S(p)$ by retracing all of the transformations defined by each of the n cycles, computing at each step $S_0(p)$ in terms of $S_1(p)$ using either

$$S_0(p) = \frac{S_1(p)[(I_1 + I_4)p^2 - (I_2 + I_3)p + 2|p_0|^2] - [(I_1 - I_4)p^2 + p(I_3 - I_2)]}{[(I_1 + I_4)p^2 + (I_2 + I_3)p + 2|p_0|^2] - [(I_1 - I_4)p^2 - p(I_3 - I_2)]S_1(p)}$$

or in the case of a Brune section

$$S_0(p) = (R_m + \hat{Z} - 1)/(R_m + \hat{Z} + 1),$$

where

$$\hat{Z} = p(L_1 - M) + \frac{[pM + (1/pC)][Z_2 + p(L_2 - M)]}{Z_2 + pL_2 + (1/pC)},$$

with $Z_2 = (1 + S_1)/(1 - S_1)$ and L_1, L_2, M, and C as computed in Fig. 3.2 from X', X, and ω_0.

Theorem 3.7 guarantees that if the algorithm can be completed, i.e., at each cycle the coefficients $I_1, \ldots, I_4$ are all nonnegative, then $S(p)$ will have the prescribed values and, moreover, it will be bounded-real. But if the process should terminate prematurely, then one may conclude that no bounded-real function can be found which will have the prescribed values.

3.3. SOME APPLICATIONS

Much of the effort in the area of lumped or rational network theory has been devoted to the question of synthesis. There are, however, many problems in which the structure of a network is of secondary interest, the basic question being one of feasibility. Thus we classify our discussion of Darlington's theorem, Theorem 3.5, as a feasibility argument in that we simply demonstrated that there will exist some rational, lossless, two-port network which when terminated in a unit resistor yields an input reflection factor that is an arbitrary, rational, bounded-real function. In this section we will consider two problems of engineering interest, and in both we will be concerned with demonstrating the feasibility of a particular solution. The second problem, that of broadband matching, if of greater practical significance than the first problem, that of compatible impedances. However, the mode of analysis used in the discussion of the first problem is of considerable theoretical interest.

3.3.1. Compatible Impedances

Two positive-real impedances $Z_1(p)$ and $Z_2(p)$ are said to be compatible if Z_1 is the input impedance of some rational, reciprocal, lossless, two-port network that is terminated in $Z_2(p)$. This problem was first considered by Schoeffler [Sc1], who obtained a restricted solution to it. The discussion we will present was subsequently developed [Wo3] using the techniques of complex normalization, and as we will see this approach is closely related to the interpolation problem described in the previous section.† In particular, if the scattering matrix of the two-port network is defined with normalization 1 at port one and Z_2 at port two [assuming $Z_2(p) + Z_2(-p)$ is not identically zero, so that this normalization is possible—see Definition 10], then since Z_2 introduces the boundary condition

$$I_2 = -Z_2 V_2,$$

we find $a_2 = 0$, and thus

$$\frac{b_1}{a_1} = \frac{V_1 - I_1}{V_1 + I_1} = \frac{Z_1 - 1}{Z_1 + 1} = S_{11},$$

where S_{11} is the 11 element of the scattering matrix, and since by assumption the input impedance of the network, when it is terminated at port two by $Z_2(p)$, is $Z_1(p)$, we have $Z_1 = V_1/I_1$.

The problem is now analogous to the one encountered in Darlington's theorem, since $S_{11}(p)$ is specified and we must find the conditions under which we may determine $S_{12} = S_{21}$, and S_{22} so that the matrix

$$[S] = \begin{bmatrix} S_{11} & S_{12} \\ S_{12} & S_{22} \end{bmatrix}$$

is the scattering matrix of a rational, two-port, lossless network normalized

† More recently Ho and Balabanian considered the general problem in which Z_1 and Z_2 are not positive-real [Ho 1].

to 1 and Z_2. From Theorem 2.16 we note that if the network is to be lossless, then $[S(-p)]^T[S(p)] = 1_n$, or

$$S_{11}(p)S_{11}(-p) + S_{12}(p)S_{12}(-p) = 1,$$

and

$$S_{22}(p) = -S_{11}(-p)S_{12}(p)/S_{12}(-p).$$

Since S_{11} has already been uniquely determined as $(Z_1 - 1)/(Z_1 + 1)$, we find that

$$S_{12}(p)S_{12}(-p) = \frac{2(Z_1 + Z_1(-p)}{(Z_1 + 1)(Z_1(-p) + 1)} = \frac{4K(p)K(-p)}{(Z_1 + 1)(Z_1(-p) + 1)},$$

where $K(p)$ is identified so that it is analytic together with its inverse in $\operatorname{Re} p > 0$ [we thus also assume that $Z_1(p) + Z_1(-p)$ does not vanish identically]. That this is always possible follows from the evenness of $Z_1 + Z_1(-p)$. We then identify the most general S_{12} as

$$S_{12}(p) = \frac{2K(p)}{Z_1 + 1}\, b(p),$$

where $b(p)$ is a regular Blaschke product.

The final scattering coefficient is then obtained as

$$S_{22}(p) = \left[\frac{1 - Z_1(-p)}{1 + Z_1}\right] \frac{K(p)}{K(-p)}\, b^2(p).$$

Having defined the most general scattering coefficients (normalized in the way indicated previously) that are consistent with the lossless and reciprocal character of the network and the fact that the input impedance at port one is Z_1 when port two is terminated in Z_2, what remains is to determine the conditions that $b(p)$ must satisfy so that these three scattering coefficients characterize a physical system. For this we resort to Theorem 2.17. Applying the theorem to the three scattering coefficients, we see that the first two statements of the theorem have been satisfied. From the third statement we require that

$$y_{11} = 1/(Z_1 + 1),$$

$$y_{12} = -\frac{K(p)}{h_2}\frac{b(p)}{Z_1 + 1},$$

and

$$y_{22} = \frac{1}{2h_2{}^2}\left[\frac{h_2}{h_2(-p)} - \frac{1 - Z_1(-p)}{1 + Z_1}\frac{K(p)}{K(-p)}\, b^2\right]$$

$$= \frac{1}{2h_2{}^2}\left[\frac{h_2}{h_2(-p)} - \frac{K(p)}{K(-p)}\, b^2\right] + \frac{K^2(p)b^2}{h_2{}^2(Z_1 + 1)}$$

be analytic in Re $p > 0$. The last statement of the theorem requires that

$$[1-(Z-1)Y](Z+1) = \begin{bmatrix} 1 & 0 \\ (1-Z_2)y_{12} & (Z_2+1)(1-(Z_2-1)y_{22}) \end{bmatrix}$$

have simple poles on the real frequency axis and a positive residue matrix. Since the residue matrix must satisfy $A^{*T} = A$ (see Theorem 2.10), we conclude that $(1-Z_2)y_{12}$ must be analytic, and $(Z_2+1)(1-(Z_2-1)y_{22})$ must have simple poles with positive residues on the real frequency axis. The analyticity of $(1-Z_2)y_{12}$ for $p = j\omega$ implies that $K(p)$ must include all the real frequency zeros of h_2 to at least the same multiplicity, except when Z_2 has a pole but Z_1 does not. In the latter case $K(p)$ must have a zero of at least one order greater than h_2 (assuming h_2 is zero). But since

$$K(p)K(-p) = \tfrac{1}{2}\{Z_1(p) + Z_1(-p)\} = E_v Z_1 \qquad \text{(the even part of } Z_1\text{)},$$

and

$$h_2 h_2(-p) = \tfrac{1}{2}\{Z_2(p) + Z_2(-p)\} = E_v Z_2,$$

this can be rephrased as: "the even part of Z_1 must contain all the real frequency zeros of $E_v Z_2$ to at least the same multiplicity, except when Z_2 is singular but Z_1 is not." If $E_v Z_2$ is zero in this case, $E_v Z_1$ must have a zero of order two or more greater (the zeros of the even part of Z are always of even order on the real frequency axis if Z is a positive real function).

Returning to the expression for y_{12}, we note that $b(p)$ must contain all the right-half-plane (r.h.p.) zeros of h_2 since $K(p)$ has no zeros in the r.h.p. Finally, in the expression for y_{22}, we observe:

1. $h_2/h_2(-p)$ is analytic in the r.h.p., since h_2 is selected to have its zeros in the r.h.p.

2. $h_2/h_2(-p)$ contains the r.h.p. zeros of h_2 to exactly the same order, a consequence of the fact that a zero of the even part of a function cannot occur at a pole of the function except on the real frequency axis, so that the denominator of $h_2(-p)$ cannot contain zeros of the numerator of $h_2(p)$.

3. $(\{1 - Z_1(-p)\}/(1 + Z_1))K(p)/K(-p)$ is analytic in the r.h.p. except at the zeros of $E_v Z_1$, and the order of these poles is equal to the order of the zeros of $E_v Z_1$. This follows since $K(-p)$ contains all the r.h.p. poles of $Z_1(-p)$ to the same order, and at a zero of $E_v Z_1$ in the r.h.p. $Z_1(-p)$ cannot equal 1, since this would imply that $Z_1(p)$ would equal -1.

From the three previous observations we conclude that if y_{22} is to be analytic in Re $p > 0$, then the order of the zero of $E_v Z_1$ plus the order of $E_v Z_2$ at a zero of $E_v Z_2$ in the r.h.p. must be an even integer and the all-pass, must be of the form

$$b^2(p) = \frac{n_1(-p)}{n_1(p)} \frac{n_2(-p)}{n_2(p)} \hat{b}^2(p),$$

where $n_2(-p)$ contains all the r.h.p. zeros of $E_v Z_2$, $n_1(-p)$ all the r.h.p. zeros of $E_v Z_1$ which coincide with the zeros of $E_v Z_2$, and $\hat{b}^2(p)$ is not zero at the zeros of $E_v Z_2$ in the r.h.p. As a consequence of the required form of b^2, we note that y_{12} will be analytic in $\operatorname{Re} p > 0$ only if the order of the zeros of $E_v Z_1$ is at least equal to the order of the zeros of $E_v Z_2$ (when $E_v Z_2$ is zero in the r.h.p.).

We can summarize the preceding statements as:

Theorem 3.8. The necessary and sufficient conditions that two rational positive-real impedances Z_1 and Z_2 chosen such that $Z_{1,2}(p) + Z_{1,2}(-p)$ are not identically zero be compatible are:

1. $E_v Z_1 = \frac{1}{2}\{Z_1 + Z_1(-p)\}$ must contain all the zeros of $E_v Z_2$ to at least the same multiplicity except when Z_2 is singular on the real frequency axis but Z_1 is not. In the latter case the order of the zero must exceed that of $E_v Z_2$ (if $E_v Z_2 = 0$) by at least two. In addition, the order of the zeros of $E_v Z_1$ plus the order of the zeros of $E_v Z_2$ must be an even integer at the zeros of $E_v Z_2$ in $\operatorname{Re} p > 0$;

2. A regular all-pass $\hat{b}$ must exist so that with $K(p)$ defined as a factorization of $K(p)K(-p) = Z_1(p) + Z_1(-p)$ that is analytic together with its inverse in $\operatorname{Re} p > 0$,

$$\frac{1}{2h_2{}^2}\left[\frac{h_2}{h_2(-p)} - \frac{K(p)}{K(-p)}\,\frac{n_1(-p)n_2(-p)}{n_1(p)n_2(p)}\,\hat{b}^2\right]$$

is analytic in $\operatorname{Re} p > 0$, and $(Z_2 + 1)\{1 - (Z_2 - 1)y_{22}\}$ have simple poles on the real frequency axis, with positive residues (i.e., if Z_2 is analytic, then y_{22} must have simple poles with positive residues, and if Z_2 is singular at $p = j\omega_0$, then $\lim_{p\to j\omega_0} Z_2 y_{22}$ must be a real number less than or equal to 1).

The statements in the theorem follow from the previous discussion, except for the second one, which is justified since $K^2(p)b^2/[h_2{}^2(Z_1 + 1)]$ will be analytic in $\operatorname{Re} p > 0$ given the restrictions imposed on b^2 and the zeros of $E_v Z_1$, so that y_{22} will be analytic if and only if condition 2 is satisfied.

We note that one can show by direct calculation that y_{22} will have simple poles with positive residues, and the $\lim_{p\to j\omega_0} Z_2 y_{22}$ will be a real number less than 1 (when Z_2 is singular) if and only if the following holds:

$$\theta_d^{(k)} = \theta_{\hat{s}}^{(k)}, \qquad 0 \le k < 2n - 1,$$

and $\theta_{\hat{s}}^{(2n-1)} \ge \theta_d^{(2n-1)}$ if Z_2 is analytic or

$$\theta_{\hat{s}}^{(2n+1)} \ge \theta_d^{(2n+1)} + \frac{|a|^2 \lim_{p\to 0}(K(p)b(p)/h_2(p))^2}{\operatorname{Res} Z_1} - \frac{|a|^2}{\operatorname{Res} Z_2}$$

when Z_2 is singular (if Z_1 is analytic, the next to the last term is zero). In these expressions $\theta_d^{(k)}$ is the kth derivative (with respect to ω) of the phase of $h_2(p)/h_2(-p)$ evaluated at $p = j\omega_0$, $\theta_{\hat{b}}^{(k)}$ is the kth derivative (with respect to ω) of the phase of $[K(p)/K(-p)]b^2$ evaluated at $p = j\omega_0$, n is the order of the zero of $h(p)$ at $p = j\omega_0$,

$$a = \frac{1}{n!} \frac{d^n h_2(p)}{d(j\omega)^n}\bigg|_{p=j\omega_0},$$

and Res Z is the residue at $p = j\omega_0$.

Thus $\hat{b}$ must satisfy certain conditions when $p = j\omega_0$, and these can be obtained in the general form $\theta_{\hat{b}}^{(k)}$, specified for all k less than a certain integer, and $\theta_{\hat{b}}^{(k+1)}$, restricted to be greater than or equal to a known constant. In addition,

$$\frac{1}{2h_2{}^2}\left[\frac{h_2(p)}{h_2(-p)} - \frac{K(p)}{K(-p)} \frac{n_1(-p)}{n_1(p)} \frac{n_2(-p)}{n_2(p)} \hat{b}^2\right]$$

must be analytic in Re $p > 0$, so that $\hat{b}$ will be required to assume specific values, possibly together with its derivatives, at points in Re $p > 0$ where $h_2(p)$, i.e., $E_v\, Z_2(p)$, vanishes.

Thus we may interpret the bulk of Theorem 3.8 as a reformulation of the problem of compatible impedances, i.e., it reduces the problem to a question of interpolation, in particular the interpolation with a Blaschke product. Fortunately, the algorithm described in Section 3.2 may be directly applied to answer this question. We note with regard to the algorithm that if we select $S_{\text{in}}(p)$ as ± 1, or in fact any arbitrary Blaschke product, then the resulting $S(p)$ will also be a Blaschke product, assuming of course that for the given problem it is possible to find the interpolating function at all, i.e., that the algorithm does not terminate prematurely. We also note that in applying the interpolation algorithm to the problem of compatible impedances there may be a requirement, noted above, that the highest order derivative of $\hat{b}$ satisfy an inequality rather than an equality at certain points where $p = j\omega_0$. Fortunately, the interpolation procedure will accommodate this situation, and we refer the interested reader to [Wo3] for a discussion of the pertinent details.

3.3.2. Broadband Matching

We will now investigate a generalization of the filtering problem which was described in Section 3.2, particularly Theorem 3.6. Thus we consider a rational, two-port, lossless network terminated in some rational positive-real function Z_L and excited at its other port by a voltage source E_S that has a series output impedance $Z_g(p)$ (see Fig. 2.2). If we define the scattering matrix of the two-port network by using as normalizing functions Z_g at port one and Z_L at port two [we are thus assuming that $Z_L(p)+Z_L(-p)$ and $Z_g(p)+Z_g(-p)$ are not identically zero; see Definition 10], then as we saw in Section 2.4.1, the voltage transfer function of the system, $T(p)=V_2(p)/E_S(p)$, is related to the 21 element of the scattering matrix, and in fact we have

$$S_{21}(p)=T(p)2h_L(p)h_g(p)/Z_L(p)$$

where $h_L(p)$ and $h_g(p)$ are the factorizations of the functions $Z_L(p)+Z_L(-p)$ and $Z_g(p)+Z_g(-p)$. We will now assume that the problem presented to us is to find a lossless network given Z_L and Z_g, such that $|T(j\omega)|$ is some prescribed function of ω (this is what we will mean by broadband matching, although a more practical problem, in many instances, would be one in which we restrict our attention to some finite band of frequencies rather than consider all ω). We see that this requires that

$$S_{21}(p)=S_{21\mathrm{m}}b_1(p),$$

where $b_1(p)$ is an arbitrary, regular, Blaschke product and $S_{21\mathrm{m}}$ is that unique factorization of

$$S_{21m}(p)S_{21m}(-p)=\frac{T(p)T(-p)2h_L(p)h_L(-p)h_g(p)h_g(-p)}{Z_L(p)Z_L(-p)}$$

which is analytic and nonzero in Re $p>0$. In particular, the function $T(p)T(-p)$ may be found from $|T(j\omega)|^2$ by substituting p for $j\omega$. The first limitation on the magnitudes of the allowable voltage transfer functions that can be realized using such a system is apparent, since we must have $|S_{21\mathrm{m}}(j\omega)|\le 1$ for all ω. Assuming that the given $|T(j\omega)|$ and the given Z_L and Z_g are such that this condition is satisfied, we may then proceed as in our discussion of Theorem 3.6 to find the remaining three elements of the scattering matrix as

$$S_{12}(p)=S_{21\mathrm{m}}(p)b_2(p),$$
$$S_{11}(p)=K(p)b_3(p),$$

and

$$S_{22}(p)=-\frac{S_{11}(-p)S_{12}(p)}{S_{21}(-p)}=-\frac{K(-p)b_3(-p)S_{21\mathrm{m}}(p)b_1(p)b_2(p)}{S_{21\mathrm{m}}(-p)},$$

where $K(p)$ is the factorization of $K(p)K(-p)=1-S_{21}(p)S_{21}(-p)$ that is analytic and nonzero in Re $p>0$, and the $b(p)$ are arbitrary, regular, Blaschke products. If $|T(j\omega)|$ is such that $|S_{21\mathrm{m}}(j\omega)|\le 1$, we see from the form of the parameters, so identified, that they define a rational, unitary, bounded-real matrix if the zeros of b_1b_2 cancel the zeros of $S_{21\mathrm{m}}(-p)$ and the poles of $K(-p)b_3(-p)$ in Re $p>0$ so that $S_{22}(p)$ will be analytic in

$\operatorname{Re} p > 0$. We may now appeal to Theorem 2.17 in order to find the appropriate conditions which will guarantee that the matrix so identified is in fact the scattering matrix of some lossless network normalized to Z_g and Z_L. However, in order to simplify our discussion and thus bring out the essential features of the problem without being encumbered by the details, we will assume that both Z_g and Z_L are analytic on $p = j\omega$ and, moreover, that their real parts, i.e., h_L and h_g, are nonzero for $p = j\omega$. With these assumptions we see that the conditions of Theorem 2.17 reduce to the requirement that

$$\begin{bmatrix} h_g & 0 \\ 0 & h_L \end{bmatrix}^{-1} \left\{ \begin{bmatrix} h_g & 0 \\ 0 & h_L \end{bmatrix} \begin{bmatrix} h_g(-p) & 0 \\ 0 & h_L(-p) \end{bmatrix}^{-1} - [S] \right\} \begin{bmatrix} h_g & 0 \\ 0 & h_L \end{bmatrix}^{-1}$$

have elements that are analytic in $\operatorname{Re} p > 0$. Performing the indicated algebraic operations, we see that this will be the case if and only if

$$\frac{h_g - h_g(-p)S_{11}}{h_g^2 h_g(-p)}, \qquad \frac{S_{12}}{h_L h_g}, \qquad \frac{S_{21}}{h_L h_g}, \qquad \text{and} \qquad \frac{h_L - h_L(-p)S_{22}}{h_L^2 h_L(-p)}$$

are all analytic in $\operatorname{Re} p > 0$. By substituting the expressions previously obtained for S_{11}, S_{12}, S_{21}, and S_{22} and noting that both $h_g(-p)$ and $h_L(-p)$, by their definitions, are nonzero in $\operatorname{Re} p > 0$, we then arrive at the requirement that $h_g - h_g(-p)K(p)b_3$ contain all the zeros of h_g^2, that b_2 and b_1 both contain all the zeros of h_L and h_g, and finally that

$$\frac{1}{h_L^2}\left[\frac{h_L}{h_L(-p)} + \frac{K(-p)b_3(-p)S_{21\mathrm{m}}b_1 b_2}{S_{21\mathrm{m}}(-p)}\right]$$

be analytic in $\operatorname{Re} p > 0$. We may now delineate two special cases in which these requirements may be further simplified:

1. Assume that Z_g is a constant, so that h_g is a constant. Then with $b_1 = b_0 \hat{b}_1$ and $b_2 = b_0 \hat{b}_2$, where b_0 contains only the zeros of h_L, so that S_{12}/h_L and S_{21}/h_L are then analytic in $\operatorname{Re} p > 0$, we see that the conditions reduce to the requirement that

$$\frac{1}{h_L^2}\left[\frac{h_L}{h_L(-p)} + \frac{K(-p)b_3(-p)S_{21m}b_0^2}{S_{21m}(-p)}\hat{b}_1\hat{b}_2\right]$$

be analytic in $\operatorname{Re} p > 0$. Since $h_L/h_L(-p)$ has precisely the same zeros (including their multiplicity) as h_L in $\operatorname{Re} p > 0$, we see that the zeros of h_L^2 can only be absorbed if $b_3(-p)$ is selected, so that

$$\frac{K(-p)b_3(-p)S_{21m}b_0^2}{S_{21\mathrm{m}}(-p)}\hat{b}_1\hat{b}_2$$

does not have the double order zeros that b_0^2 can introduce. Thus we will assume that $b_3(-p)$ is selected so that this expression has precisely the zeros of the same order as h_L. Note that in some cases we may have $b_3(-p) = 1$, since $S_{21m}(-p)$ may cancel some of the zeros in b_0^2. In any event, the problem then reduces to the requirement that a regular Blaschke product $\hat{b}_1\hat{b}_2$ (in the reciprocal case $\hat{b}_1 = \hat{b}_2$, so that we have $\hat{b}^2$) must exist so that

$$\frac{K(-p)b_3(-p)S_{21m}\,b_0^2}{S_{21m}(-p)}\,\hat{b}_1\hat{b}_2$$

is analytic in Re $p > 0$ and

$$\frac{h_L}{h_L(-p)} + \frac{K(-p)b_3(-p)S_{21m}\,b_0^2}{S_{21m}(-p)}\,(\hat{b}_1\hat{b}_2)$$

have at least zeros of the same multiplicity as h_L^2.

2. Assume that Z_L is a constant, so that h_L is a constant. Here the problem reduces to a requirement that a regular Blaschke product b_3 exist so that $h_g - h_g(-p)Kb_3$ has at least all the zeros of h_g in Re $p > 0$. This follows from our previous discussion, since if such a b_3 exists, then b_1 and b_2 may always be selected to cancel the zeros of $S_{21m}(-p)$ and the poles of $K(-p)b_3(-p)$ in Re $p > 0$ and, in particular, to have the zeros of $h_g(p)$.

The two cases discussed above may then be reduced to a problem of interpolation, as was the case in our discussion of the problem of compatible impedances. Since the interpolation question may be answered using the algorithm of Section 3.2, we conclude that the problem of broadband matching is solved for these two special cases, at least when Z_L and Z_g are analytic and h_L and h_g are nonzero for $p = j\omega$. To remove the latter restriction we can resort to detailed discussion of the additional conditions required by Theorem 2.17 using the same techniques as were employed in the discussion of compatible impedances (see, for example, [Yo5]).

If we now return to the general problem, in which neither Z_L nor Z_g are constant, we may reason as follows. Since b_3 must be such as to force $h_g - h_g(-p)Kb_3$ to have at least the same zeros as h_g^2 in Re $p > 0$, we see that $b_3(p)$ must assume, perhaps together with a certain number of its derivatives, specific values at these points in Re $p > 0$. If we employ the interpolation algorithm of Section 3.2, we can find the most general b_3 having these specific values (assuming it exists at all) by terminating the algorithm at its last cycle and selecting $S_{1n}(p) = \hat{b}_3$ where $\hat{b}_3$ is some arbitrary regular Blaschke product. We may then find, using the algorithm, a representation for $b_3(p)$ expressed as $b_3 = f(p, \hat{b}_3)$, where the function f as indicated depends both on p and $\hat{b}_3$, but is well defined once the interpolation algorithm is completed. But $b_1(p)$ and $b_2(p)$ are partially defined by the requirement that they both contain all the zeros of h_L and h_g and, moreover, that b_1b_2

cancel the zeros of $S_{21m}(-p)$ and the poles of $K(-p)$ in Re $p > 0$. We thus arrive at the point where we must determine if a $\hat{b}_3$ and a $\hat{b}$ can be found so that

$$h_L + \frac{K(-p)f(-p, \hat{b}_3(-p))S_{21m}\, b_0\, h_L(-p)\hat{b}}{S_{21m}(-p)}$$

has at least the zeros of h_L^2 in Re $p > 0$, and $\hat{b}$ cancels the poles of $f(-p, b_3(-p))$ in Re $p > 0$. The term b_0 is uniquely determined from those portions of $b_1 b_2$ that have already been described above. Unfortunately, there is no known procedure which will enable one to determine if such a set of functions, $\hat{b}_3$ and $\hat{b}$, can be found. The virtue of our discussion is that it points out the basic algebraic problem involved.

3.4. ACTIVE LUMPED NETWORKS

If we say that a system is active if and only if it is not passive, then the study of such active systems is equivalent to the study of passive systems in the trivial sense that all theorems concerning passive systems may be applied to active systems by simply prefacing them with the statement "a system is active if and only if the following conditions are *not* satisfied." However, there is considerable interest in the structure of active systems, and, as we will demonstrate, considerable insight into this structure may be derived from the study of passive systems. We will consider two specific illustrations of active systems using the negative resistor as the basic active element. We will also describe a very useful approach to the study of active systems in which the activity is confined to one specific element in the system.

Thus we consider as an element a device whose input-output relationship is of the form $v(t)/i(t) = -R$, where R is a positive-real constant, i.e., the negative resistor. We see immediately that $\int_{-\infty}^{t} v(\eta)i(\eta)\, d\eta < 0$ for all t, so that this device is certainly not passive. Although the model is an idealization, such devices as the tunnel diode are representable, at least to first order, by such a model if v and i are assumed to be deviations from some quiescent or bias values of voltage and current; in other words, the negative resistor corresponds to the incremental model for such a device. One may now inquire as to how to use such an element to achieve, for example, voltage gain or amplification. An approach to such a study is to consider the negative resistor as being imbedded in some otherwise passive network, for example, the one shown in Fig. 3.4, and then determining how the voltage transfer ratio $(V_2/E_S)(p)$ is affected by the presence of the negative resistor. To this end, assume that we define the scattering matrix of the three-port network shown in Fig. 3.4 by normalizing to it R_g at port one, R_L at port two, and $|R|$

at port three. Then with the boundary condition $V_3 = RI_3$ we find $b_3 = 0$, and thus

$$\frac{V_2}{E_S}(p)2\left(\frac{R_g}{R_L}\right)^{1/2} = \frac{b_2}{a_1}(p) = \frac{S_{21}S_{33} - S_{31}S_{23}}{S_{33}}.$$

If we further assume that the three-port network is lossless, then by evaluating $[S(p)]^{-1}$, we find that

$$S_{21}S_{33} - S_{31}S_{23} = -(\det[S])S_{12}(-p),$$

so that

$$\left|\frac{V_2}{E_S}(p)2\left(\frac{R_g}{R_L}\right)^{1/2}\right| = \left|\frac{S_{12}}{S_{33}}(j\omega)\right|,$$

since $|\det[S(j\omega)]| = 1$ given that the three-port network is lossless. We see

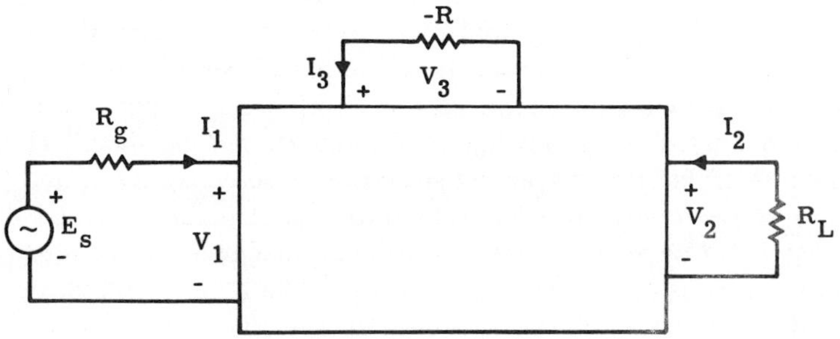

FIG. 3.4. Active network.

immediately that the magnitude of the voltage transfer ratio is no longer constrained, and in fact we will prove:

Theorem 3.9. There will exist a rational, lossless, three-port network which when used in the circuit of Fig. 3.4 yields a voltage transfer ratio $(V_2/E_S)(p)$ that is an arbitrary, real, rational function of p.

Proof. First we note that given $(V_2/E_S)(p)$ we may always find two rational bounded-real functions $A(p)$ and $B(p)$ such that

$$\frac{V_2}{E_S}(p)2\left(\frac{R_g}{R_L}\right)^{1/2} = \frac{A(p)}{B(p)}.$$

Now consider the network shown in Fig. 3.5, in which the second network has a scattering matrix satisfying

$$S_{12} = 1, \quad S_{22} = S_{32} = S_{13} = S_{11} = 0,$$
$$S_{21}(-p)S_{23} + S_{31}(-p)S_{33} = 0,$$
$$S_{21}S_{21}(-p) + S_{31}S_{31}(-p) = 1,$$

and

$$S_{23}S_{23}(-p) + S_{33}S_{33}(-p) = 1.$$

We may show directly that if S_{21}, S_{23}, and S_{33} are rational and analytic in Re $p > 0$, then such a matrix is bounded-real and $[S(-p)]^T[S(p)] = 1_3$,

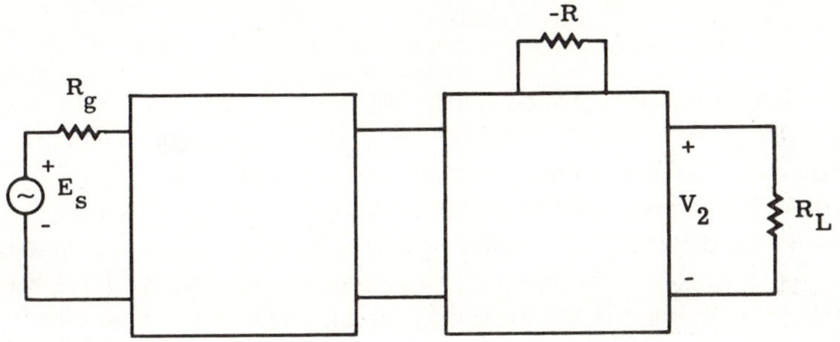

FIG. 3.5. Active amplifier.

i.e., it represents a rational lossless network. Moreover, we may also show that since $[S(-p)]^T = [S(p)]^{-1}$,

$$\left|\frac{b_1}{a_1}\right| = \left|\frac{S_{33}S_{11} - S_{13}S_{31}}{S_{33}}\right| = \left|\frac{S_{22}}{S_{23}}\right| = 0,$$

or that the input reflection factor at port one of this network is zero when ports two and three are terminated in R_L and $-R$. We also find that

$$b_2/a_1 = 1/S_{33}(p),$$

and we then conclude that the ratio b_2/a_1 may be the inverse of an arbitrary, rational, bounded-real function, since with S_{33} so specified we may always satisfy the conditions

$$S_{21}(-p)S_{23} + S_{31}(-p)S_{33} = 0,$$
$$S_{21}S_{21}(-p) + S_{31}S_{31}(-p) = S_{23}S_{23}(-p) + S_{33}S_{33}(-p) = 1,$$

using the same technique employed in the proof of Darlington's theorem, Theorem 3.5.

Now consider the first network of Fig. 3.5 and observe that since $b_1/a_1 = 0$, where b_1 and a_1 correspond to the input of the second network, then

$$\hat{b}_2/\hat{a}_2 = \hat{S}_{21},$$

where $\hat{b}_2$ and $\hat{a}_2$ correspond to the parameters of the first network and $\hat{S}_{21}$ is the 21 element of its scattering matrix. But $\hat{b}_2 = a_1$, and we then obtain

$$b_2/\hat{a}_1 = \hat{S}_{21}/S_{33},$$

so that the voltage transfer function of the composite network becomes

$$\frac{V_2}{E_S}(p)2\left(\frac{R_g}{R_L}\right)^{1/2} = \frac{\hat{S}_{21}}{S_{33}}$$

and the theorem is established, since we may always find some rational, lossless, two-port network having an arbitrary, rational, bounded-real function as its 21 element (see Theorem 3.6). QED

As we see, the proof of the previous theorem rests heavily on the results obtained from the study of passive systems. Moreover, we see that by the addition of a single active element, the negative resistor, we have removed all restrictions on our ability to realize voltage transfer ratios. Further, we see that by employing the results of passive network theory we have found networks that will utilize the "activity" of the negative resistor in order to produce gain on amplification.

As one final result we will also show that the restriction on the input reflection factor of a network, that it be bounded-real, may also be removed when active networks are considered. Thus we have:

Theorem 3.10. It is always possible to find a passive, rational, two-port network such that its input reflection factor at port one is an arbitrary, real, rational function when the other port is terminated in a negative resistor.

Proof. Consider the network shown in Fig. 3.6, where the network labeled N has the scattering matrix

$$\begin{bmatrix} 0 & 0 & 1 \\ 1 & 0 & 0 \\ 0 & 1 & 0 \end{bmatrix},$$

so that it is certainly lossless and rational (in fact, such a network is called a circulator). If the other two networks are characterized by $(V_3 - I_3)/(V_3 + I_3) = S_A$ and $(V_2 - I_2)/(V_2 + I_2) = S_B$, where S_A and S_B are both bounded-real functions, then we may easily show that

$$\frac{b_1}{a_1} = -\frac{S_B}{S_A} = \frac{V_1 - I_1}{V_1 + I_1},$$

But as we noted in the proof of Theorem 3.9, any real rational function can be written as the ratio of two rational bounded-real functions, and the present theorem is then established, since S_A and S_B represent rational passive networks, and we have used only one negative resistor in the network of Fig. 3.6, which may be considered as terminating a port of the network. QED

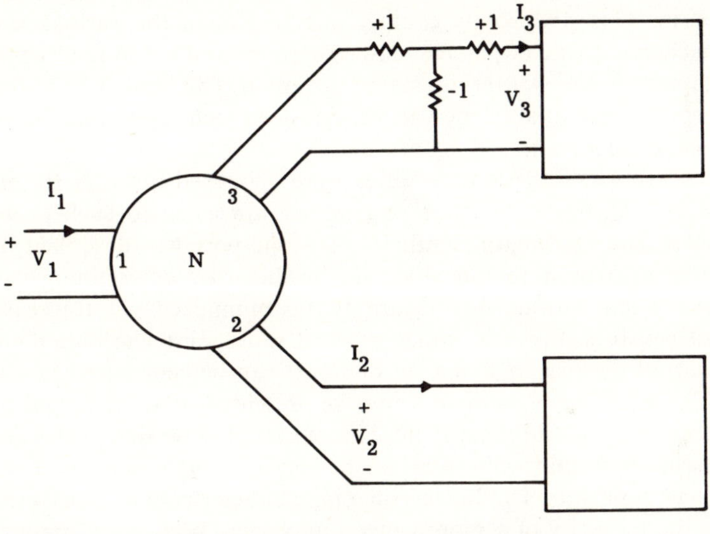

FIG. 3.6. Input reflection factor realization.

We have thus briefly described some of the properties of active rational networks and, in particular, a method for studying means for the utilization of activity to create gain. We should note that the first theorem employed a network developed by Youla and Smilen [Yo9] in their study of tunnel diode circuits, and the second theorem may be refined (see [Ca3]) to show that the two-port network may always be realized with only one positive resistor, i.e., its normal rank need not exceed 1. In fact, it was shown in [Ca3] that an arbitrary real rational $n \times n$ matrix may always be realized as the scattering matrix of an n-port containing no more than n positive resistors and n negative resistors together with lossless elements.

3.5. SUMMARY

In this chapter we have applied the techniques and results of our generalized discussion of passive systems to the study of lumped or rational networks. We have focused our attention primarily on those aspects of the problem that would be termed analysis, as opposed to synthesis, although we have illustrated the techniques applied in the n-port synthesis problem by discussing the algebraically simpler one-port problem.

We discussed the structure of lumped n-ports in terms of the scattering matrix of those systems. The approach entailed the partitioning of the network so that the frequency-dependent elements, the inductors and capacitors, were separated from the interconnecting elements, the wires, transformers, and gyrators. This philosophy enables one to isolate the various features which are characteristic of passive lumped systems and led to the "canonic" representation of the scattering matrix given in Theorem 3.3. Moreover, this viewpoint is essential in the various synthesis techniques that have been developed for the n-port.

Some one-port synthesis procedures were described in order to indicate technique and illustrate the scattering matrix approach to such problems. In particular, the Darlington synthesis of a one-port was described in this way. It was also used to introduce the problem of determining canonic realizations when various factors are to be minimized—in this case the number of resistors. From a purely physical standpoint Darlington's result implies that all the loss in a lumped one-port can be centralized in a single resistor. The subsequent example which we presented of an irrational system then demonstrates that in general one cannot expect to be able to characterize the system in a way that will centralize the loss in a single element. Thus this example has a profound influence on our characterization of nature, and illustrates the necessity of a more general approach, both mathematical and physical, to the study of passive systems.

We illustrated the central role that interpolation plays in the study of many physical problems. Youla was the first to appreciate the importance of interpolation questions in passive network theory, and he applied this viewpoint to the study of such problems as broadband matching. We presented a slightly different, although equivalent, discussion of this problem, and also described the problem of compatible impedances. Although we avoided some of the algebraic detail that is present in the general statement of these problems, our discussion points out the basic algebraic question which underlies both, i.e., the ability to find interpolating functions that are bounded-real or positive-real.

We also illustrated the utility of the complex normalization technique for scattering matrices by applying it to the two problems mentioned above.

The viewpoint associated with complex normalization has only recently been developed and its immediate success in the simplification of many problems is certainly an indication of the central role that it will play in the study of future problems.

The theory of passive networks was also applied to the more basic study of active systems. In particular, it was shown that an analogous Darlington theorem was true, i.e., any rational input impedance can be realized with a lossless three-port, one positive resistor, and one negative resistor. Thus for lumped systems we can also isolate all the activity or gain potential in a single element, the negative resistor. Moreover, we showed that the techniques which are needed in the study of active networks rest heavily on the studies of that subset of active networks which we call passive systems.

There are many problems still remaining in the study of lumped passive systems. Some of these are described in Appendix III, with emphasis on those problems which are suggested by the discussions which we have presented.

IV

Distributed Networks

INTRODUCTION

We saw in the last chapter that lumped passive networks may be completely characterized by rational bounded-real matrices. We noted, in particular, that the inverse or synthesis question for such networks has been resolved, since one may show that any rational bounded-real matrix is the scattering matrix of some network composed of a finite number of passive building blocks: resistors, capacitors, inductors, gyrators, and ideal transformers. The present state of the theory of passive distributed systems is not nearly as complete, since we must admit networks whose terminal description is no longer in the form of rational matrices. The prototype for such a network is one whose internal behavior is characterized by partial differential equations, as, for example, the networks representing the propagation of electromagnetic waves in materials, which we discussed in Chapter I. One of the major difficulties in the study of distributed systems, at least to date, is the fact that one cannot find a finite number of building blocks or basic elements such that a relatively large class of distributed systems may be realized by networks consisting of an interconnection of these basic elements. In this chapter we will discuss one possible building block—a lossless transmission line of finite length. In fact, we will see that at least one class of distributed systems may be delineated by considering networks composed of such elements.

We will first consider some general results concerning distributed networks. Here we will for concreteness use the problem of electromagnetic wave propagation in materials. Specifically, we will be led to a definition of a passive material and its corresponding passive distributed network. We will then consider those lossless distributed networks that are described by a simple set of wave equations. Although we will be able to obtain a considerable amount of information concerning the terminal behavior of such systems, the inherent mathematical difficulty of the corresponding analysis limits us severely. Since such lossless distributed networks are only a small subset of passive distributed networks, we may conclude that the general subject area of this chapter is not only in its infancy but that in its future development one can expect to face similar mathematical complexities.

4.1. MODAL ANALYSIS OF PASSIVE DISTRIBUTED SYSTEMS

We will assume that the internal behavior of the systems under consideration is described by partial differential equations—specifically, those that are generally referred to as "wave" equations. In order to state our results in a more concrete fashion we will consider in particular the problem of electromagnetic wave propagation in materials. However, studies in rigid body motion, thermodynamics, quantum mechanics, to name a few areas of physics, lead to a consideration of similar mathematical models, so our results have wide physical applicability.

We have in mind the physical situation described in Section 1.3.2, in which electromagnetic fields are caused to exist in some region of space surrounded by a perfectly conducting surface or envelope. We will consider those envelopes that are in the form of cylinders whose cross-sectional dimensions do not vary in a direction (taken to be the z axis) perpendicular to these cross sections. Moreover, we will assume that this cylinder is filled with a material whose electrical properties are more general than those of the simple lossy dielectric materials considered in Section 1.3.2. Specifically, we will assume that Maxwell's equations in the materials we are considering may be written as

$$\nabla \times \mathbf{E} = -[\hat{\xi}_0(z, t) * \mathbf{H}],$$

$$\nabla \times \mathbf{H} = -[\hat{\psi}_0(z, t) * \mathbf{E}],$$

where $\hat{\xi}_0(z, t)$ and $\hat{\psi}_0(z, t)$ are the microscopic scalar representations of the material (we note that allowance has been made for a variation of these parameters, but only along the axis of symmetry of the system, the z axis), and $*$ denotes the convolutions of these scalar functions with the corresponding field vectors $\mathbf{E}$ or $\mathbf{H}$. We note that the system described by these equations is both linear and time-invariant in the sense that if $(\mathbf{E}_1, \mathbf{H}_1)$ and $(\mathbf{E}_2, \mathbf{H}_2)$ are solutions of the system equations, then $(\alpha\mathbf{E}_1 + \beta\mathbf{E}_2, \alpha\mathbf{H}_1 + \beta\mathbf{H}_2)$ is also a solution for any two constants α and β, and $(\mathbf{E}_1(t + \tau), \mathbf{H}_1(t + \tau))$ is also a solution to the equations. Now, in order to identify input–output variables to characterize the system we consider the modal decomposition of the electromagnetic fields described in Section 1.3.2. Thus we consider the various TEM, TE, and TM modes in which the transverse (x, y) dependence of the fields is obtained as an eigenfunction of the appropriate system of equations noted in Table 1.1 with eigenvalues k_m or k_e, and the z dependence is governed by the equations

$$\partial\mathscr{E}(z, p)/dz = -\hat{Z}_0(z, p)\mathscr{H}(z, p),$$
$$\partial\mathscr{H}(z, p)/\partial z = -\hat{Y}_0(z, p)\mathscr{E}(z, p),$$

where

$$\hat{Z}_0(z, p) = \mathscr{L}[\hat{\xi}_0(z, t)], \qquad \text{TEM and TE}$$
$$= \mathscr{L}[\hat{\xi}_0(z, t)] + \frac{k_m}{\mathscr{L}[\hat{\psi}_0(z, t)]}, \qquad \text{TM}$$
$$\hat{Y}_0(z, p) = \mathscr{L}[\hat{\psi}_0(z, t)], \qquad \text{TEM and TM}$$
$$= \mathscr{L}[\hat{\psi}_0(z, t)] + \frac{k_e}{\mathscr{L}[\hat{\xi}_0(z, t)]}, \qquad \text{TE}$$

and both $\mathscr{E}(z, p)$ and $\mathscr{H}(z, p)$ are Laplace transforms of the appropriate scalar time functions evaluated for each fixed value of z.

For our subsequent discussion we will assume that only one mode, either TEM, TE, or TM, is present in the material. Therefore we must consider the two coupled partial differential equations describing $\mathscr{E}$ and $\mathscr{H}$, and our first observation is that in general one cannot solve these equations in closed form, i.e., find solutions in terms of the elementary transcendental functions, since $\hat{Z}_0$ and $\hat{Y}_0$ may have arbitrary functional dependences on the independent variable z. However, certain specific information can be obtained about these solutions, and in particular a direct iterative technique can be found to obtain these solutions. To this end we convert the equations into an equivalent integral equation. Thus consider the solution $\mathscr{E}(z, p)$ of the integral equation

$$\mathscr{E}(z, p) = A(p)\int_z^l \hat{Z}_0(\eta, p)\, d\eta + B(p) + \int_z^l \left[\int_z^\tau \hat{Z}_0(\eta, p)\, d\eta\right] \hat{Y}_0(\tau, p)\mathscr{E}(\tau, p)\, d\tau, \qquad z \le l, \tag{4.1}$$

with A and B arbitrary functions of p, and note that if

$$\mathscr{H}(z, p) = A(p) + \int_z^l \hat{Y}_0(\tau, p)\mathscr{E}(\tau, p)\, d\tau, \qquad z \le l \tag{4.2}$$

then $\mathscr{E}$ and $\mathscr{H}$ so obtained are solutions of the original differential equations, as we can verify by direct substitution. But Eq. (4.1) is a Volterra integral equation, and it is known [Ko2] that solutions to such equations exist when relatively weak assumptions are made concerning the functions $\hat{Z}_0$ and $\hat{Y}_0$. For our future purposes we will state one such set of conditions as [Ko2]:

Lemma 4.1. Let $\mathscr{M}(p)$ denote the set of all points p_0 such that $\hat{Z}_0(z, p_0)$ and $\hat{Y}_0(z, p_0)$ are bounded and piecewise continuous functions of z on some interval, say $0 \leq z \leq l$. Then the integral equation (4.1) has a continuous solution $\mathscr{E}(z, p)$, with respect to z, for all fixed $p \in \mathscr{M}$.

We may also obtain conditions guaranteeing that these solutions are unique by appealing again to the classic theory of the Volterra equation. For our purposes we will need the following:

Lemma 4.2. Assume the boundary conditions $\mathscr{E}(l, p) = V(p)$ and $\mathscr{H}(l, p) = I(p)$ are prescribed. Then the integral equation possesses a unique continuous solution $\mathscr{E}(z, p)$ for all fixed $p_0 \in \mathscr{M}$ such that $I(p_0)$ and $V(p_0)$ are bounded. Moreover, the corresponding function $\mathscr{H}(z,p)$ is also uniquely determined.

Proof. The functions $A(p)$ and $B(p)$ are determined by the two boundary conditions as $B(p) = \mathscr{E}(l, p) = V(p)$ and $A(p) = \mathscr{H}(l, p) = I(p)$, so the integral equation for $\mathscr{E}(z, p)$ is then completely specified. The Volterra theory (see, for example, [Ko2]) then guarantees the existence of a unique solution $\mathscr{E}(z, p_0)$ for each $p_0 \in \mathscr{M}$ such that $I(p_0)$ and $V(p_0)$, or, equivalently, $A(p_0)$ and $B(p_0)$, are finite. Finally, having found $\mathscr{E}(z, p)$, we compute a unique $\mathscr{H}(z, p)$ from Eq. (4.2). QED

We now turn our attention to the situation illustrated in Fig. 4.1, where the problem we have been discussing is shown schematically and further constrained by the presence of boundary conditions of the form

$$\mathscr{E}(l, p)/\mathscr{H}(l, p) = Z_l(p)$$

and

$$E_s(p) = Z_g(p)H(0, p) + \mathscr{E}(0, p),$$

where $Z_l(p)$, $Z_g(p)$, and $E_s(p)$ are all known functions of p. The physical significance of these boundary conditions is that they correspond to the case where no sources of electromagnetic fields are present to the right of $z = l$. Thus that region of space simply imposes some constraint on the ratio of fields at $z = l$, i.e., $Z_l(p)$. Moreover, the sources of electromagnetic fields existing to the left of $z = 0$ are such that they may be represented by a lumped impedance $Z_g(p)$ and a fixed source $E_s(p)$. The network model of this situation, depicted in Fig. 4.1, then follows from the boundary conditions and

Kirchhoff's laws assuming we identify $\mathscr{E}(z, p) = V(z, p)$ as a voltage and $\mathscr{H}(z, p) = I(z, p)$ as a current, with their interrelationship given by the corresponding partial differential equations. If we return to the integral equation formulation of the problem, we observe that now

$$\frac{B(p)}{A(p)} = \frac{V(l, p)}{I(l, p)} = Z_l(p),$$

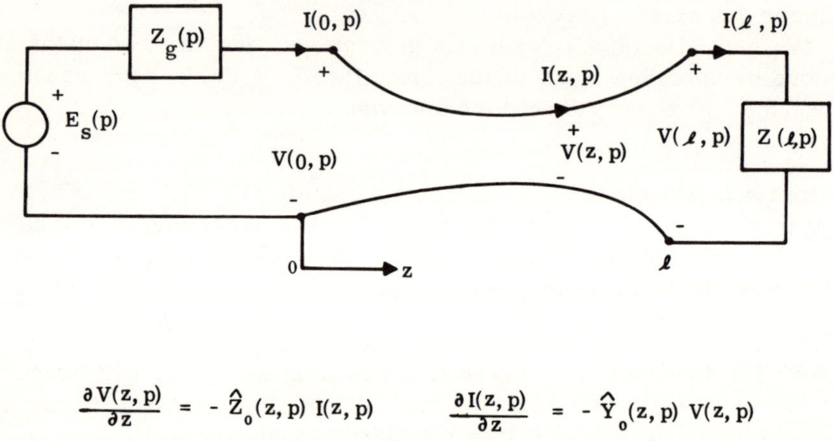

FIG. 4.1 A distributed network.

so that by defining $V(z, p) = A(p)r(z, p)$ we see that $r(z, p)$ must satisfy the integral equation

$$r(z, p) = \int_z^l \hat{Z}_0(\eta, p)\, d\eta + Z_l(p) + \int_z^l \left[\int_z^\tau \hat{Z}_0(\eta, p)\, d\eta\right] \hat{Y}_0(\tau, p) r(\tau, p)\, d\tau. \quad (4.3)$$

On the basis of Lemma 4.2 we note that unique solutions to the network problem are then obtained for all $p_0 \in \mathscr{M}$ such that $Z_l(p_0)$ is finite. Moreover, for all such p_0, $V(z, p) = A(p)r(z, p)$ and

$$I(z, p) = A(p)\left[1 + \int_z^l \hat{Y}_0(\tau, p) r(\tau, p)\, d\tau\right]$$

for all $z \leq l$. In particular, the ratio V/I is uniquely determined by $Z_l(p)$ as

$$\frac{V(z, p)}{I(z, p)} = \frac{r(z, p)}{1 + \int_z^l \hat{Y}_0(\tau, p) r(\tau, p)\, d\tau}.$$

In addition, we may evaluate $A(p)$ from the boundary condition

$$E_s(p) = Z_g(p) I(0, p) + V(0, p)$$

and obtain

$$\frac{V(z, p)}{E_s(p)} = \frac{r(z, p)}{Z_g(p)\left[\int_0^l \hat{Y}_0(\tau, p) r(\tau, p)\, d\tau + 1\right] + r(0, p)}$$

or, since $r(l, p) = Z_l(p)$, we have for $z = l$

$$\frac{V(l, p)}{E_s(p)} = \frac{Z_l(p)}{Z_g(p)\left[\int_0^l \hat{Y}_0(\tau, p) r(\tau, p)\, d\tau + 1\right] + r(0, p)}$$

Thus the study of a single mode of an electromagnetic field in some finite region of space subject to boundary conditions of the type $\mathscr{E}(l, p)/\mathscr{H}(l, p) = f_1(p)$ and $f_2(p)\mathscr{H}(0, p) + \mathscr{E}(0, p) = f_3(p)$ has been reduced to the study of the solutions of a single Volterra integral equation, Eq. (4.3). Fortunately, we know that unique solutions exist to such equations, and in addition the Volterra theory also indicates a method of solution. Thus if we define

$$r_1(z, p) = \int_z^l \hat{Z}_0(\eta, p)\, d\eta + Z_l(p)$$

and

$$r_{n+1}(z, p) = r_1(z, p) + \int_z^l \left[\int_z^\tau \hat{Z}_0(\eta, p)\, d\eta\right] \hat{Y}_0(\tau, p) r_n(\tau, p)\, d\tau,$$

then one may show that for all $z \leq l$

$$r(z, p) = \lim_{n \to \infty} r_n(z, p)$$

for each fixed $p \in \mathscr{M}$ such that $Z_l(p)$ is bounded. Moreover, the convergence of r_n to r is uniform over any finite interval—say, $0 \leq z \leq l$.

We may note at this point that the solution we obtained for, say, $V(l, p)$, was in terms of Laplace transforms of certain quantities. Since our basic problem is phrased in such a way that the independent variables are functions of time, we must still invert these transforms to obtain the actual solutions. Thus if we view the system as a mapping from $E_s(t)$ to $V(l, t)$, i.e., $E_s(t)$ is the input and $V(l, t)$ the corresponding output, then we see, as we noted earlier, that such a mapping is linear, continuous, and time-invariant, and we will have succeeded in characterizing this system if we can show that $V(l, p)/E_s(p)$ is in fact the transform of some well-defined function or distribution. In particular, if one would like to determine if this operator is causal, then, as we know from Chapter I, one is led to the problem of determining if the expression for $V(l, p)/E_s(p)$ is analytic for all p in some half plane $\operatorname{Re} p > \sigma_0$. In other words, if the system is to behave in a causal manner, then certain constraints must be imposed on the functions $\hat{Z}_0(z, p)$, $\hat{Y}_0(z, p)$, and $Z_l(p)$ so that the ratio $V(l, p)/E_s(p)$, which is uniquely determined by solutions of Eq. (4.3), is analytic for $\operatorname{Re} p > \sigma_0$ for some σ_0. The resolution of such a question has not yet been obtained; however, certain specific statements can be made relative to the problem. For example, we have:

Theorem 4.1. Let $\mathcal{N}$ denote that subset of the set $\mathcal{M}$ of Lemma 4.1 defined by those points p such that (1) $\partial\hat{Z}_0(z, p)/\partial p$ and $\partial\hat{Y}_0(z, p)/\partial p$ exist and are piecewise continuous functions with respect to both p and z for all $0 \leq z \leq l$, and (2) $dZ_l(p)/dp$ exists. Then the solution $r(z, p)$ of Eq. (4.3) is an analytic function of p for all $p \in \mathcal{N}$, i.e., $\{\partial r(z, p)/\partial p\}_{p=p_0}$ exists for each fixed $0 \leq z_0 \leq l$ and any $p_0 \in \mathcal{N}$.

Proof. As previously noted, a Volterra integral equation, such as Eq. (4.3), may be solved in an iterative manner using the functions $r_n(z, p)$. We can easily demonstrate that $\partial r_n(z, p)/\partial p$ exists and is continuous with respect to p for any fixed $0 \leq z_0 \leq l$ given the existence of the partial derivatives of $\hat{Z}_0$, $\hat{Y}_0$, and Z_l. Moreover, the sequence generated by $\partial r_n/\partial p$ converges uniformly. These facts are then sufficient, together with the fact that $\lim_{n\to\infty} r_n = r$ uniformly, to guarantee that

$$\frac{\partial r(z, p)}{\partial p} = \lim_{n\to\infty} \frac{\partial r_n(z, p)}{\partial p}$$

exists for all $0 \leq z \leq l$. QED

The following are simple consequences of the theorem, but they are of independent interest, so we will state them:

Corollary 4.1. The function $V(z, p)/I(z, p)$ is, for each fixed z, an analytic function of p in a domain defined by the intersection of $\mathcal{N}$ with the set of points for which

$$\int_z^l \hat{Y}_0(\tau, p) r(\tau, p)\, d\tau \neq -1.$$

Moreover, if the equality holds, then $V(z, p)/I(z, p)$ will have a pole at that value of p.

Corollary 4.2. The transfer function $V(l, p)/E_s(p)$ can have zeros in the domain defined by the set $\mathcal{N}$ only at those points where either $Z_l(p)$ vanishes or $Z_g(p)$ is unbounded.

The next corollary is also relatively simple to establish once Theorem 4.1 is available, but since we shall make extensive use of it in the subsequent sections, we will present its proof.

Corollary 4.3. Assume that $\hat{Z}_0(z, p) = F(p)f(z)$ and $\hat{Y}_0(z, p) = G(p)g(z)$ where $F(p)$ and $G(p)$ are entire functions of p and $f(z)$ and $g(z)$ are continuous functions of z. Then if $Z_l(p)$ is a meromorphic function of p, we will obtain solutions of the integral (4.3) for $r(z, p)$ such that for each fixed z they are meromorphic functions of p. Moreover, the poles of $r(z, p)$, including multiplicity, are among the poles of $Z_l(p)$.

Proof. Consider the two Volterra integral equations

$$r_1(z, p) = F(p) \int_z^l f(\eta)\, d\eta + F(p)G(p) \int_z^l \left[\int_z^\tau f(\eta)\, d\eta \right] g(\tau) r_1(\tau, p)\, d\tau$$

and

$$\frac{r_2(z, p)}{Z_l(p)} = 1 + F(p)G(p) \int_z^l \left[\int_z^\tau f(\eta)\, d\eta \right] g(\tau) \frac{r_2(\tau, p)}{Z_l(p)}\, d\tau.$$

Since $F(p)$ and $G(p)$ are entire functions of p, then, using the arguments employed in the proof of Theorem 4.1, we see that both $r_1(z, p)$ and $r_2(z, p)/Z_l(p) = m(z, p)$ are analytic functions of p for all finite p and all fixed z. Thus $r_2(z, p) = m(z, p)Z_l(p)$ is a meromorphic function of p whose poles are the poles of $Z_l(p)$ unless $m(z, p)$ vanishes at these points. Thus the poles of $r_2(z, p)$, including multiplicity, are among the poles of $Z_l(p)$. But from the uniqueness of the solutions of a Volterra integral equation we see that $r = r_1 + r_2$ is the solution to Eq. (4.3), and we conclude that $r(z, p)$ has the properties described in the theorem. QED

Although we have not resolved the question of the analyticity of the various transfer functions, and thus the causality of the corresponding operators, appearing in the study of the distributed network of Fig. 4.1, we do have certain results concerning the behavior of the solutions of the basic integral equation describing such networks Eq. (4.3). We will see, however, in the following discussion that Corollary 4.3 will allow us to conclude that the model assumed for such networks is in fact a causal one if in addition we impose a requirement of passivity.

We will now turn our attention to those materials in which the propagation of electromagnetic fields takes place in a "passive" manner, i.e., the fields do not gain net energy from the material. We noted in Section 1.3.2 that a meaningful measure of such an energy exchange may be expressed by means of the Poynting theorem. In particular, such considerations prompt us to state:

Definition 11. A material supporting an electromagnetic field [defined by $\mathbf{E}(\mathbf{r}, t)$ and $\mathbf{H}(\mathbf{r}, t)$] is said to be passive if for all t

$$-\int_{-\infty}^{t}\left[\int_{V} \nabla \cdot [\mathbf{E}(\mathbf{r}, \eta) \times \mathbf{H}(\mathbf{r}, \eta)]\, dv\right] d\eta \geq 0,$$

where V is some arbitrary finite volume of the material and $\mathbf{E}$ and $\mathbf{H}$ are arbitrary real solutions of Maxwell's equations continuous in both $\mathbf{r}$ and t.

Now consider the situation in which a *single mode* of the type discussed previously is present inside the cylindrical envelope or waveguide. By applying Definition 11 we find that if the material filling the waveguide is passive, then it is necessary that

$$-\int_{-\infty}^{t}\int_{a}^{b} \frac{\partial[v(z, t)i(z, t)]}{\partial z}\, dz\, d\eta \geq 0,$$

where $\mathscr{L}[v(z, t)] = \mathscr{E}(z, p)$, $\mathscr{L}[i(z, t)] = \mathscr{H}(z, p)$, and the volume considered is that defined by the waveguide and the two surfaces located at $z = a$ and $z = b$. The fact that the waveguide is a perfect conductor has been used in arriving at this conclusion, since it imposes as a boundary condition the vanishing of the tangential component of the electric field, so that the integral in the definition need only be taken over the two parallel surfaces. The resultant integrals may in turn be simplified to the stated form using the transverse properties of the electromagnetic fields as described in Section 1.3.2. Then using the fact that

$$\partial v(z, t)/\partial z = -[\hat{z}_0(z, t) * i(z, t)]$$

and

$$\partial i(z, t)/\partial z = -[\hat{y}_0(z, t) * v(z, t)],$$

which is the time domain version of the equations $\partial\mathscr{E}/\partial z = -\hat{Z}_0\mathscr{H}$ and $\partial\mathscr{H}/\partial z = -\hat{Y}_0\mathscr{E}$ which we described at the beginning of this section, with $\mathscr{L}(\hat{z}_0(z, t)) = \hat{Z}_0(z, p)$ and $\mathscr{L}(\hat{y}_0(z, t)) = \hat{Y}_0(z, p)$, allows us to rephrase the integral requirement as

$$\int_{-\infty}^{t}\int_{a}^{b} [v(v * \hat{y}_0) + i(i * \hat{z}_0)]\, dz\, d\eta \geq 0.$$

Now, since this integral inequality must be satisfied for all electromagnetic fields that are continuous functions of position, we may apply the mean value theorem and rewrite the integral as

$$\int_{a}^{b} [v(v * \hat{y}_0) + i(i * \hat{z}_0)]\, dz = (b - a)[v(v * \hat{y}_0) + i(i * \hat{z}_0)]_{z=k}$$

for some $a \leq k \leq b$. Thus we find that a necessary condition for passivity of the material is that

$$\int_{-\infty}^{t} [v(v * \hat{y}_0) + i(i * \hat{z}_0)]_{z=k}\, d\eta \geq 0.$$

However, we saw in Lemma 4.2 that unique solutions to the basic Maxwell equation for a single mode are obtained when the electric and magnetic fields are simultaneously specified at some cross-sectional plane, i.e., if we specify $v(z, t)$ and $i(z, t)$ as functions of time at some point $z = z_0$. But by the definition of a passive material the inequalities discussed above are to be nonnegative for arbitrary volumes, and thus in the single-mode case we have been considering this implies that the inequalities must be satisfied for arbitrary b and a, and, in particular, regardless of how small the difference $b - a$ becomes. This then implies that we may consider the point $z = k$ to be an arbitrary point at which we specify the boundary conditions on $v(z, t)$ and $i(z, t)$. Thus we may require that $v(k, t) = 0$, and then passivity would demand that $\int_{-\infty}^{t} i(i * \hat{z}_0) d\eta \geq 0$ when $i = i(k, t)$ is an arbitrary function contained, for example, in C_0^{∞}. Similarly, we may select $i(k, t) = 0$, and then passivity demands that $\int_{-\infty}^{t} v(v * \hat{y}_0)\, d\eta \geq 0$.

From our discussion of passive immittance operators in Chapter II we conclude that $\mathscr{L}(\hat{z}_0) = \hat{Z}_0(z, p)$ and $\mathscr{L}(\hat{y}_0) = \hat{Y}_0(z, p)$ must then be positive-real functions of p for each fixed z (see Theorem 2.5). We state this result as:

Theorem 4.2. Assume that the Maxwell equations describing the z dependence of each mode inside a cylindrical envelope are of the form

$$\partial \mathscr{E}(z, p)/\partial z = -\hat{Z}_0(z, p)\mathscr{H}(z, p),$$

$$\partial \mathscr{H}(z, p)/\partial z = -\hat{Y}_0(z, p)\mathscr{E}(z, p),$$

where $\hat{Z}_0$ and $\hat{Y}_0$ depend upon the specific mode in question. Then if the material filling the cylindrical envelope or waveguide is passive, it is necessary that $\hat{Z}_0(z, p)$ and $\hat{Y}_0(z, p)$ both be positive-real functions of p for each fixed z. Conversely, if each $\hat{Z}_0$ and $\hat{Y}_0$ are positive-real functions of p for each fixed z, then the material is passive.

Proof. The necessity of the theorem follows from our preceding discussion, whereas the sufficiency is established using the fact that the modes are orthogonal in the sense that $\int (\mathbf{e}_i \times \mathbf{h}_j) \cdot d\mathbf{S} = 0\ (i \neq j)$ (see the discussion in Section 1.3.2), so that if each mode $(\mathbf{E}_i, \mathbf{H}_i)$ yields

$$-\int_{-\infty}^{t} \left[\int_V \nabla \cdot [\mathbf{E}_i(\mathbf{r}, \eta) \times \mathbf{H}_i(\mathbf{r}, \eta)]\, dv \right] d\eta \geq 0,$$

then the integral with $\mathbf{E}$ and $\mathbf{H}$ is in fact the sum of the integrals for each mode. QED

We will define a scattering matrix for the two-port network of Fig. 4.1 corresponding to a section of waveguide containing a passive material that is supporting electromagnetic fields in the form of a single mode. If we select as normalizing functions $Z_l(p)$ at port two ($z = l$) and $Z_g(p)$ at port one ($z = 0$), then we have, using the techniques of Section 2.3.3,

$$\begin{aligned} 2h_g(p)a_1 &= V(0) + Z_g I(0), \\ 2h_g(-p)b_1 &= V(0) - Z_g(-p)I(0), \\ 2h_l(p)a_2 &= V(l) - Z_l I(l), \\ 2h_l(-p)b_2 &= V(l) + Z_l(-p)I(l), \end{aligned}$$

with h_g and h_l the factorizations of $Z_g(p) + Z_g(-p) = 2h_g(p)h_g(-p)$ and $Z_l(p) + Z_l(-p) = 2h_l(p)h_l(-p)$ as described in Definition 10 of Section 2.3.3, and

$$\mathbf{b} = [S]\mathbf{a}.$$

In particular, we may express the quantities $V(0)$, $I(0)$, $V(l)$, and $I(l)$ in terms of the solution to the integral equation describing the system, Eq. (4.3), to obtain

$$\begin{aligned} S_{11}(p) &\equiv \left.\frac{b_1}{a_1}\right|_{a_2=0 \text{ or } V(p)=Z_l I(l)} = \frac{h_g(p)}{h_g(-p)} \frac{\alpha - Z_g(-p)\beta}{\alpha + Z_g(p)\beta} \\ S_{21}(p) &\equiv \left.\frac{b_2}{a_1}\right|_{a_2=0 \text{ or } V(p)=Z_l I(l)} = \frac{2h_g(p)h_l(p)}{\alpha + Z_g(p)\beta}, \end{aligned} \tag{4.4}$$

where

$$\alpha(p) = \int_0^l \hat{Z}_0(\eta, p)\, d\eta + Z_l(p) + \int_0^l \left[\int_0^\tau \hat{Z}_0(\eta, p)\, d\eta\right] \hat{Y}_0(\tau, p) r(\tau, p)\, d\tau = r(0, p)$$

$$\beta(p) = 1 + \int_0^l \hat{Y}_0(\eta, p) r(\eta, p)\, d\eta, \tag{4.5}$$

with $r(z, p)$ a solution of Eq. (4.3). A central theorem to our discussions in the following section may now be stated:

Theorem 4.3. Assume that $\hat{Z}_0(z, p) = Z_0(p)f(z)$ and $\hat{Y}_0(z, p) = Y_0(p)g(z)$, where $f(z)$ and $g(z)$ are real continuous nonnegative functions of z, and $Z_0(p)$ and $Y_0(p)$ are entire positive-real functions. Then the scattering matrix representation $[S(p)]$ of a single mode in a waveguide filled with a passive material characterized by such functions $\hat{Z}_0$ and $\hat{Y}_0$ is a meromorphic, symmetric, bounded-real matrix. Moreover, if $\text{Re}\,[Z_0(j\omega)] = \text{Re}\,[Y_0(j\omega)] = 0$, then

$$[S(-p)][S(p)] = 1_n.$$

Proof. Consider first the scattering matrix obtained using $Z_l = Z_g = 1$ as normalizing functions, i.e., $[S_{1_n}]$. From Corollary 4.3 we see that the $r(z, p)$ corresponding to $Z_l = 1$ is an entire function of p, since its only poles are those of $Z_l(p)$ for each fixed z, so that the α and β of Eq. (4.5) are entire functions. Thus, as we see from Eq. (4.4), S_{11} and S_{21} will be meromorphic functions of p. But a similar set of expressions can be obtained for the S_{22} and $S_{12}(p)$ elements of the scattering matrix using an integral equation formulation with $z = 0$ as its fixed limit, and so we may also conclude that S_{22} and S_{12} will be meromorphic functions of p. In addition, we see from the basic partial differential equations for $V(z, p) = \mathscr{E}(z, p)$ and $I(z, p) = \mathscr{H}(z, p)$ that

$$
\begin{aligned}
-\int_0^l \frac{\partial}{\partial z}[VI^* + V^*I]\,dz &= V(0, p)I^*(0, p) + V^*(0, p)I(0, p) \\
&\quad - [V(l, p)I^*(l, p) + V^*(l, p)I(l, p)] \\
&= \mathbf{a}_1^{*T}(1_n - [S_{1n}]^{*T}[S_{1n}])\mathbf{a}_1,
\end{aligned}
$$

where $\mathbf{a}_1$ correspond to the normalizations $Z_l = Z_g = 1$. We may also obtain from these partial differential equations the inequality

$$
\begin{aligned}
-\int_0^l \frac{\partial}{\partial z}[VI^* + V^*I]\,dz &= \int_0^l \{|V(z, p)|^2[\hat{Y}_0(z, p) + \hat{Y}_0^*(z, p)] + |I(z\ p)|^2 \\
&\qquad \times [\hat{Z}_0(z, p) + \hat{Z}_0^*(z, p)]\}\,dz \\
&\geq 0
\end{aligned}
$$

for $\operatorname{Re} p > 0$, since by assumption $\hat{Z}_0(z, p)$ and $\hat{Y}_0 = Z_0(z, p)$ are positive-real functions of p for each fixed z. Thus $[S_{1_n}]$ is a bounded-real matrix: the nonnegativity has been established, so that, in particular, $|S_{jk}(p)| \leq 1$ for $\operatorname{Re} p > 0$, and thus $S_{kj}(p)$, being meromorphic, must be analytic in $\operatorname{Re} p > 0$.

Now we may appeal to Theorem 2.16 to conclude that $[S_Z]$ will also be a meromorphic bounded-real matrix when $Z_l(p)$ and $Z_g(p)$ are regular normalizing functions. In particular, when $\operatorname{Re}[Z_0(j\omega)] = \operatorname{Re}[Y_0(j\omega)] = 0$ for all ω we see that

$$1_n - [S_1(j\omega)]^{*T}[S_1(j\omega)] = 0_n,$$

and we may appeal again to Theorem 2.16 to conclude that

$$[S_Z(-p)]^T[S_Z(p)] = 1_n \text{ for all } p.$$

Finally, with regard to the symmetry of the scattering matrix we see that if $V_c(z, p)$, $I_c(z, p)$ and $V_d(z, p)$, $I_d(z, p)$ are two pairs of solutions to the partial differential equations describing the system, then multiplication and subsequent subtraction yields

$$\frac{\partial V_c I_d}{\partial z} = \frac{\partial V_d I_c}{\partial z}$$

By integrating from zero to l we then obtain

$$V_c(l, p)I_d(l, p) - I_c(l, p)V_d(l, p) = V_c(0, p)I_d(0, p) - I_c(0, p)V_d(0, p)$$

which may be written in terms of the scattering matrix, normalized to arbitrary Z_l and Z_g, as

$$[S_{21} - S_{12}](a_{2c}a_{1d} - a_{1c}a_{2d}) = 0.$$

But a_1 and a_2 are independent, and we conclude that $S_{21} = S_{12}$. QED

We have seen that a class of distributed networks may be studied by means of the passive network formalism—e.g., if $\hat{Z}_0 = pL(z)$ and $\hat{Y}_0 = pC(z)$, where L and C are continuous functions of z, we meet the conditions of Theorem 4.3. Similar results can be obtained for larger classes of distributed networks. For example, we might want to consider the propagation of single-mode electromagnetic waves in materials where $\hat{Z}_0$ and $\hat{Y}_0$ are not necessarily entire functions of p. We note that the elements of the scattering matrices of such networks may no longer be meromorphic functions, so that we might have some difficulty establishing, for example, that $[S_Z]$ exists. However, if we select $Z_l = Z_g = 1$, then on the basis of Theorem 4.2 we can still establish:

Theorem 4.4. The scattering matrix representation (normalized to 1 at both ports) of a single mode in a waveguide filled with passive material exists and is bounded-real.

Proof. First we note from Theorem 4.1 that $r(z, p)$ and thus $\alpha(p)$ and $\beta(p)$ will be analytic functions of p in $\operatorname{Re} p > 0$ given that $\hat{Z}_0(z, p)$ and $\hat{Y}_0(z, p)$ are positive-real functions and thus analytic in $\operatorname{Re} p > 0$. Therefore the elements of the scattering matrix $[S_1(p)]$ can have at most isolated singularities or poles in $\operatorname{Re} p > 0$. But one can show directly that $1_n - [S_1]^{*T}[S_1]$ is nonnegative-definite in $\operatorname{Re} p > 0$, so that $|S_{jk}(p)| \leq 1$, and thus such poles cannot exist. We then conclude that $S_{jk}(p)$ is analytic in $\operatorname{Re} p > 0$, and $[S_1(p)]$ is therefore a bounded-real matrix. QED

4.2. NONUNIFORM LOSSLESS TRANSMISSION LINES

In this section we will further restrict our study by considering only those materials that are characterized by functions $\hat{Z}_0(z, p) = pL(z)$, $L(z) > 0$, and $\hat{Y}_0(z, p) = pC(z)$, $C(z) > 0$. The corresponding two-port network determined from the partial differential equations by defining the currents $I(z, p)$ and voltages $V(z, p)$ is referred to as a nonuniform lossless transmission line. We will see that a rather complete theory can be obtained for this class of distributed networks by considering its terminal properties as reflected in the corresponding scattering matrix of the network. We will divide our discussion into three basic parts. In the first we will show that the $S_{11}(p)$ element of the scattering matrix uniquely determines the remaining three elements of the matrix. In the second part we will show how the terminal properties of the network determine the behavior of the functions $L(z)$ and $C(z)$. Here we will proceed by using the formalism of the Sturm–Liouville operator, and, in particular, some results concerning the inversion of such operators. The final section discusses some additional topics of a related or illustrative nature.

4.2.1. The Scattering Matrix of a Lossless Nonuniform Transmission Line

In the present case the partial differential equations defining the network become

$$\begin{aligned} \partial V(z,p)/\partial z &= -pL(z)I(z,p) \\ \partial I(z,p)/\partial z &= -pC(z)V(z,p), \end{aligned} \tag{4.6}$$

where z is the physical position of the line, $V(z, p)$ and $I(z, p)$ are the Laplace transforms (with respect to time) of the voltage and current, respectively, and $L(z) > 0$ and $C(z) > 0$ are the distributed inductance and capacitance, respectively, per unit length of the line. If the electrical position along the line and the local characteristic impedance are defined by

$$y(z) = \int_0^z \{L(\eta)C(\eta)\}^{1/2}\, d\eta$$

and

$$Z_0(y) = \{L[z(y)]/C[z(y)]\}^{1/2},$$

respectively, then by changing the independent variable z to y, Eqs. (4.6) are transformed to

$$\begin{aligned} \partial V(y,p)/\partial y &= -pZ_0(y)I(y,p) \\ \partial I(y,p)/\partial y &= -[p/Z_0(y)]V(y,p). \end{aligned} \tag{4.7}$$

On the basis of our discussion in Section 4.1 we see that the scattering matrix of such a network, when it is normalized to 1 at $y = 0$ and some regular normalizing function $Z_\Delta(p)$ at $\Delta = \int_0^l \{L(\eta)C(\eta)\}^{1/2}\, d\eta$, is a meromorphic bounded-real matrix satisfying $[S_Z]^T = [S_Z]$, or $S_{12}(p) = S_{21}(p)$, and

$[S_Z(-p)][S_Z(p)] = 1$ (see Theorem 4.3). Moreover, we have

$$S_{11}(p) = \left.\frac{b_1}{a_1}\right|_{a_2=0} = \frac{\alpha(p) - \beta(p)}{\alpha(p) + \beta(p)}$$

$$S_{21}(p) = \left.\frac{b_2}{a_1}\right|_{a_2=0} = \frac{2h_\Delta(p)}{\alpha(p) + \beta(p)}, \tag{4.8}$$

where $\alpha(p) = r(0, p)$,

$$\beta(p) = 1 + p\int_0^\Delta \frac{r(\tau, p)}{Z_0(\tau)}\, d\tau = \frac{(dr/dy)|_{y=0}}{[-pZ_0(0)]}, \tag{4.9}$$

and $r(y, p)$ is a solution of the integral equation

$$r(y, p) = p\int_y^\Delta Z_0(\eta)\, d\eta + Z_\Delta(p) + p^2 \int_y^\Delta \frac{\int_y^\tau Z_0(\eta)\, d\eta}{Z_0(\tau)}\, r(\tau, p)\, d\tau. \tag{4.10}$$

In particular, from Corollary 4.3 we have the fact that $r(y, p)$ is a meromorphic function of p with poles that are among the poles of $Z_\Delta(p)$.

In the preceding sections certain properties of the scattering matrix were established by assuming that the local characteristic impedance $Z_0(y)$ of the line was a continuous function of y. In order to proceed with our discussion it is necessary at this point to strengthen this assumption considerably. Specificially, it will be assumed that $Z_0(y)$ is at least twice continuously differentiable. Now it can be demonstrated directly that the solution $r(y, p)$ of Eq. (4.10) satisfies, if $Z_0(y)$ is continuously differentiable,

$$\frac{d^2 r}{dy^2} - \frac{d \ln Z_0}{dy}\frac{dr}{dy} - p^2 r = 0, \tag{4.11}$$

with $r(\Delta) = Z_\Delta(p)$ and $(dr/dy)|_\Delta = -pZ_0(\Delta)$. Defining $m(y, p) = r(y, p)/\{Z_0(y)\}^{1/2}$ and substituting into Eq. (4.11) shows that $m(y, p)$ must satisfy

$$\frac{d^2 m}{dy^2} - Qm - p^2 m = 0,$$

with $m(\Delta) = Z_\Delta(p)/\{Z_0(\Delta)\}^{1/2}$

$$\left.\frac{dm}{dy}\right|_\Delta = -p[Z_0(\Delta)]^{1/2} + Z_\Delta(p)\left.\frac{d(1/\{Z_0(y)\}^{1/2})}{dy}\right|_\Delta,$$

and

$$Q = \{Z_0(y)\}^{1/2}\frac{d^2(1/\{Z_0(y)\}^{1/2})}{dy^2}.$$

To investigate the asymptotic behavior of $m(y, p)$ with respect to p it is convenient to define $m(y, p) = e^{yp}\hat{m}(y, p)$. Substitution in the differential equation for $m(y, p)$ then leads to the requirement that

$$\frac{d^2\hat{m}}{dy^2} + 2p\frac{d\hat{m}}{dy} - Q\hat{m} = 0. \tag{4.12}$$

A particular solution of Eq. (4.12)—specifically, one that satisfies $\hat{m}(0) = 1$, $(d\hat{m}/dy)|_0 = 0$—is obtained as a solution to the integral equation

$$(\hat{m} - 1) - \frac{1}{2p}\int_0^y [1 - e^{-2p(y-\eta)}]Q(\eta)(\hat{m} - 1)\, d\eta$$

$$= \frac{1}{2p}\int_0^y [1 - e^{-2p(y-\eta)}]Q(\eta)\, d\eta. \tag{4.13}$$

In this equation the right-hand term is $\mathrm{O}(1/p)$ for $\mathrm{Re}\, p \geq 0$ and $|p|$ sufficiently large, and the second term is $\mathrm{O}(1/p)(\hat{m} - 1)$ in the same region.† Thus

$$\hat{m}(y, p) = 1 + \mathrm{O}(1/p)$$

in $\mathrm{Re}\, p \geq 0$ for $|p|$ sufficiently large. A similar argument with $m = e^{-yp}\hat{\hat{m}}$ and

$$\hat{\hat{m}} = 1 + \int_y^\Delta \frac{1 - e^{2p(y-\eta)}}{2p} Q(\eta)\hat{\hat{m}}\, d\eta$$

establishes that there exists another independent solution to the differential equation for $m(y, p)$ that is of the form

$$m(y, p) = e^{-yp}[1 + \mathrm{O}(1/p)]$$

in $\mathrm{Re}\, p \geq 0$ for $|p|$ sufficiently large. Thus a fundamental set of solutions for $m(y, p)$ have been found whose asymptotic behavior is known. The above argument may be repeated for $\mathrm{Re}\, p \leq 0$ with the same conclusions. Thus, since $r = m\sqrt{Z_0}$, the asymptotic behavior of $r(y, p)$ is of the form

$$r(y, p) = \{Z_0(y)\}^{1/2}\left\{A(p)e^{yp}\left[1 + \mathrm{O}\left(\frac{1}{p}\right)\right] + B(p)e^{-yp}\left[1 + \mathrm{O}\left(\frac{1}{p}\right)\right]\right\}$$

for $|p|$ sufficiently large. The coefficients A and B may be evaluated from the boundary conditions prescribed at $y = \Delta$ by $Z_\Delta(p)$. One then obtains the

† The symbol O is defined as follows: $f(p) = \mathrm{O}(1/p^n)$ if $|f(p)| < K/|p^n|$ for some fixed K when $|p|$ is sufficiently close to $+\infty$.

following asymptotic expressions for $\alpha(p)$ and $\beta(p)$ [see Eqs. (4.9)]:

$$\alpha(p) = \frac{\{Z_0(0)\}^{1/2}}{2}\left\{\left[\frac{Z_\Delta(p)}{\{Z_0(\Delta)\}^{1/2}} - \{Z_0(\Delta)\}^{1/2}\right]e^{-p\Delta} + \left[\frac{Z_\Delta(p)}{\{Z_0(\Delta)\}^{1/2}} + \{Z_0(\Delta)\}^{1/2}\right]e^{p\Delta}\right\}$$

and

$$\beta(p) = \frac{1}{2\{Z_0(0)\}^{1/2}}\left\{-\left[\frac{Z_\Delta(p)}{\{Z_0(\Delta)\}^{1/2}} - \{Z_0(\Delta)\}^{1/2}\right]e^{-p\Delta} + \left[\frac{Z_\Delta(p)}{\{Z_0(\Delta)\}^{1/2}} + \{Z_0(\Delta)\}^{1/2}\right]e^{p\Delta}\right\}$$

where we have retained only the dominant terms and the specific dependence on $Z_\Delta(p)$. For future use we will also need the following:

Lemma 4.3. If $Z_0(y)$ is an infinitely differentiable function, then it is necessary that the following asymptotic expansion hold as $|\omega| \to \infty$:

$$\beta(j\omega) = e^{j\omega\Delta}\left[c_0 + \frac{c_1}{\omega} + \cdots + \frac{c_n}{\omega^n} + \cdots\right] + e^{-j\omega\Delta}\left[d_0 + \frac{d_1}{\omega} + \cdots + \frac{d_n}{\omega^n} + \cdots\right],$$

where for any n c_n and d_n are finite constants, and, in particular, $c_0 > 0$ and $|d_0/c_0| < 1$.

Proof. It may be shown directly, assuming $Q(y)$ exists or that $Z_0(y)$ is twice continuously differentiable, that

$$m(y, p) = \frac{r(y, p)}{\{Z_0(y)\}^{1/2}} = a \cosh p(y - \Delta) + \frac{bp + c}{p} \sinh p(y - \Delta) - \frac{1}{p}\int_y^\Delta \sinh p(y - t)Q(t)m(t)\, dt, \tag{4.14}$$

where $a = 1/\{Z_0(\Delta)\}^{1/2} > 0$, $b = -\{Z_0(\Delta)\}^{1/2} < 0$, and

$$c = (d\{Z_0(y)\}^{-1/2}/dy)\,|_{y=\Delta}.$$

Thus

$$\beta(j\omega) = \frac{1}{\{Z_0(0)\}^{1/2}}\,[a \sinh j\omega\Delta - b \cosh j\omega\Delta]$$
$$- \frac{1}{j\omega Z_0(0)}\left\{\cosh j\omega\Delta\left[a\,\frac{d(Z_0)^{1/2}}{dy}(0) + c\{Z_0(0)\}^{1/2}\right] - b \sinh j\omega\Delta\right.$$
$$\left. - \int_0^\Delta \cosh j\omega t Q(t) m(t)\, dt\right\}$$
$$+ \frac{1}{\omega^2 Z_0(0)}\left[-c \sinh j\omega\Delta + \int_0^\Delta \sinh j\omega t Q(t) m(t)\, dt\right]. \qquad (4.15)$$

But the integral equation defining $m(y, p)$, Eq. (4.14), is of the Volterra type, and thus one may iterate it to obtain

$$m(y, j\omega) = a \cosh j\omega(y - \Delta) + \frac{bj\omega + c}{j\omega} \sinh j\omega(y - \Delta)$$
$$- \frac{1}{j\omega}\int_y^\Delta \sinh j\omega(y - t) Q(t)$$
$$\times \left[a \cosh j\omega(t - \Delta) + \frac{bj\omega + c}{j\omega} \sinh j\omega(t - \Delta)\right] dt$$
$$- \frac{1}{\omega^2}\int_y^\Delta \sinh j\omega(y - t) Q(t)$$
$$\times \left[\int_t^\Delta \sinh j\omega(t - \eta) Q(\eta)\left(a \cosh j\omega(\eta - \Delta)\right.\right.$$
$$\left.\left. + \frac{bj\omega + c}{j\omega} \sinh j\omega(\eta - \Delta)\right) d\eta\right] dt + O\left(\frac{1}{\omega^3}\right).$$

Now, substituting this expression into Eq. (4.15) yields an asymptotic expansion for $\beta(j\omega)$ containing various integrals of $Q(y)$. However, since $Z_0(y)$ is infinitely differentiable, we may integrate these various integrals by parts an unlimited number of times. For example, the integral in the $1/j\omega$ term of Eq. (4.15) may be represented, after substitution of the iterated form of $m(y, j\omega)$, as

$$\int_0^\Delta \cosh j\omega t Q(t) m(t)\, dt = k_1 e^{j\omega\Delta} + k_2 e^{-j\omega\Delta} + k_3 e^{-j\omega\Delta}\int_0^\Delta e^{-2j\omega t} Q(t)\, dt$$
$$+ k_4 e^{-j\omega\Delta}\int_0^\Delta e^{2j\omega t} Q(t)\, dt + O(1/\omega),$$

and successive integration by parts yields

$$\int_0^{\Delta} e^{\pm 2j\omega t} Q(t)\, dt = \frac{Q(\Delta)e^{\pm 2j\omega\Delta} - Q(0)}{\pm 2j\omega} - \frac{[Q'(\Delta)e^{\pm 2j\omega\Delta} - Q'(0)]}{(\pm 2j\omega)^2} + \cdots .$$

By this process we may conclude in an inductive fashion that the stated expansion in the lemma is a consequence of the fact that $Z_0(y)$ is infinitely differentiable.

Finally, we note that

$$c_0 = \frac{a-b}{2} = \frac{1}{2}\left\{\frac{1}{[Z_0(\Delta)]^{1/2}} + [Z_0(\Delta)]^{1/2}\right\}$$

$$d_0 = -\left(\frac{a+b}{2}\right) = -\frac{1}{2}\left\{\frac{1}{[Z_0(\Delta)]^{1/2}} - [Z_0(\Delta)]^{1/2}\right\}$$

so that $c_0 > 0$ and $|d_0/c_0| < 1$. QED

On the basis of our previous discussion we are now in a position to prove that the scattering matrix of a smooth line is uniquely determined by $S_{11}(p)$. It is convenient to separate this proof into two segments, one of which is contained in:

Lemma 4.4. The zeros (including order) of $S_{21}(p) = 2h_\Delta(p)/\{\alpha(p) + \beta(p)\}$ in Re $p \geq 0$ are precisely those of $h_\Delta(p)$.

Proof. Since $[S(-p)][S(p)] = 1_n$, we have in particular that $1 = S_{11}(p)S_{11}(-p) + S_{21}(p)S_{21}(-p)$. Substitution from Eqs. (4.8) then yields

$$\frac{2h_\Delta(p)h_\Delta(-p) - [\alpha(p)\beta(-p) + \alpha(-p)\beta(p)]}{[\alpha(p) + \beta(p)][\alpha(-p) + \beta(-p)]} = 0. \tag{4.16}$$

This statement, together with the meromorphic nature of h_Δ, α, and β, implies that

$$2h_\Delta(p)h_\Delta(-p) = \alpha(p)\beta(-p) + \alpha(-p)\beta(p) \tag{4.17}$$

except possibly at the poles of the denominator of Eq. (4.16), i.e., the poles of α and β. We are thus led to the conclusion that the numerator of Eq. (4.16) vanishes identically in any finite domain of the complex p plane except possibly at a finite number of points of the domain. However, since Eq. (4.17) is meromorphic, it cannot behave in this manner unless it vanishes everywhere, i.e., Eq. (4.17) holds for all p.

But as we may prove, $r(y, p) + r(y, -p)$ must have at least the same zeros (including order) as $Z_\Delta(p) + Z_\Delta(-p) = 2h_\Delta(p)h_\Delta(-p)$.† Thus $\alpha(p) + \alpha(-p) = r(0, p) + r(0, -p)$ must have the same zeros as $h_\Delta(p)h_\Delta(-p)$. Note also that $\beta(p) - \beta(-p)$ has at least the same zeros (including order) as $h_\Delta(p)h_\Delta(-p)$. Let us assume now that $h_\Delta(p)h_\Delta(-p)$ has a zero of order n at $p = p_1$. Then in some neighborhood of p_1

$$h_\Delta(p)h_\Delta(-p) \approx a(p - p_1)^n, \qquad a \neq 0,$$

and from the previous discussion

$$\alpha(p) + \alpha(-p) \approx b(p - p_1)^n$$

$$\beta(p) - \beta(-p) \approx c(p - p_1)^n$$

in this neighborhood. Solving for $\alpha(-p)$ and $\beta(-p)$ and substituting into Eq. (4.17), we find

$$-\alpha(p)c(p - p_1)^n + b(p - p_1)^n\beta(p) \approx 2a(p - p_1)^n,$$

or, since $a \neq 0$, we conclude that $\alpha(p)$ and $\beta(p)$ cannot simultaneously vanish at $p = p_1$ or, in general, at a zero of $h_\Delta(p)h_\Delta(-p)$.

Finally, let us assume that $S_{21}(p) = 2h_\Delta/(\alpha + \beta)$ does not have the same zeros as $h_\Delta(p)$, i.e., that $\alpha + \beta$ has a zero at a zero of $h_\Delta(p)$ in $\mathrm{Re}\, p \geq 0$. Then since $S_{11}(p) = (\alpha - \beta)/(\alpha + \beta)$ must be analytic in this region, it follows that $\alpha - \beta$ must also be zero or that α and β must simultaneously be zero. Since this was shown above to be impossible, it may be concluded that $S_{21}(p)$ has the same zeros as $h_\Delta(p)$ in $\mathrm{Re}\, p \geq 0$. QED

The remaining portion of our discussion is contained in:

Theorem 4.5 [Wo6]. Let $Z_\Delta(p)$ be a regular normalizing function (see Definition 10 of Section 2.3.3). If $S(p)$ is the input reflection factor of a lossless line whose characteristic impedance $Z_0(y)$ is at least twice continuously differentiable and which is of length Δ and is terminated in $Z_\Delta(p)$, then $S(p)$ uniquely determines that scattering matrix of the line which is normalized to 1 at port one and $Z_\Delta(p)$ at port two. In particular, then, the input reflection factor for an arbitrary terminating impedance can be found.

† We have from Eq. (4.10) that

$$r(y, p) + r(y, -p) = Z_\Delta(p) + Z_\Delta(-p) + p^2 \int_y^\Delta \frac{\int_y^\tau Z_0(\eta)\, d\eta}{Z_0(\tau)} [r(\tau, p) + r(\tau, -p)]\, d\tau,$$

and from the theory of Volterra integral equations we know that the only solutions of this equation when $Z_\Delta(p_0) + Z_\Delta(-p_0) = 0$ are themselves zero, i.e., $r(y, p_0) + r(y, -p_0) = 0$ for all $y \leq \Delta$.

Proof. It is known that $S_{21}(p)$ is a bounded analytic function in $\operatorname{Re} p \geq 0$. But such a function may be represented, for $\operatorname{Re} p > 0$ (see Theorem 2.14), as

$$S_{21}(p) = \pm e^{-cp} B(p) \exp\left\{\frac{p}{\pi}\int_{-\infty}^{\infty} \frac{\log|S_{21}(j\eta)|}{p^2+\eta^2}\, d\eta\right\}, \tag{4.18}$$

where $B(p)$ is a regular Blaschke product formed with the zeros of $S_{21}(p)$ in $\operatorname{Re} p > 0$, i.e.,

$$B(p) = \prod_j \frac{p - a_j}{p + a_j^*} \prod_k \frac{a_k - p}{a_k^* + p}\frac{a_k^*}{a_k},$$

with a_j those zeros of $S_{21}(p)$ in $\operatorname{Re} p > 0$ such that $|a_j| < 1$; a_k the remaining zeros in $\operatorname{Re} p > 0$; and $-c = \lim_{|p|\to\infty}(\{\log S_{21}(p)\}/p)$, $\operatorname{Re} p > 0$. Now, from Theorem 4.3 $1 - |S_{11}(j\omega)|^2 = |S_{21}(j\omega)|^2$, so that the exponential term of the representation is determined by $S_{11}(j\omega)$. Lemma 4.4 implies that the zeros of $S_{21}(p)$ are precisely those of $h_\Delta(p)$, so that $B(p)$ is determined completely by $h_\Delta(p)$. From the asymptotic behavior of the scattering parameters described above we have

$$\alpha(p) + \beta(p) = \frac{e^{p\Delta}}{2}\left[\frac{Z_\Delta(p)}{\{Z_0(\Delta)\}^{1/2}} + \{Z_0(\Delta)\}^{1/2}\right]\left[\{Z_0(0)\}^{1/2} + \frac{1}{\{Z_0(0)\}^{1/2}}\right]$$

in $\operatorname{Re} p > 0$ for $|p|$ sufficiently large. Thus $\lim_{|p|\to\infty}\{(\log S_{21})/p\} = -\Delta$, in $\operatorname{Re} p > 0$, since with Z_Δ a regular normalizing function $\lim_{|p|\to\infty}\{(\log h_\Delta)/p\} = 0$ in $\operatorname{Re} p > 0$. Thus by direct comparison $c = \Delta$. Finally, the $(\pm)$ ambiguity can be removed by comparing the asymptotic behavior of $S_{21}(p)$, which is given, using the asymptotic behavior of α and β previously obtained, as

$$S_{21}(p) \approx \frac{2h_\Delta}{(\frac{1}{2}\{Z_0(0)\}^{1/2} + \frac{1}{2}\{Z_0(0)\}^{-1/2})(Z_\Delta(p)\{Z_0(\Delta)\}^{-1/2} + \{Z_0(\Delta)\}^{1/2})e^{p\Delta}},$$

to the behavior of the function constructed as per Eq. (4.18) using $B(p)$, $S_{11}(j\omega)$, and Δ. Note that $\operatorname{Re} Z_\Delta(p) \geq 0$ in $\operatorname{Re} p > 0$, and thus the sign of $S_{21}(p)$ can be uniquely determined from the above asymptotic expansion. By our convention we will assume that a unique $b(p) = (\pm)B(p)$ is determined which incorporates the proper sign. Thus $S_{21}(p)$ has been uniquely determined from $S_{11}(p)$, $Z_\Delta(p)$, and Δ. However, with the normalization used to define the scattering matrix, $S_{11}(p) = S(p)$ [the input reflection factor with $Z_\Delta(p)$ termination], and if $A(p)$ is as defined in the corollary, then

$$S_{21}(p) = A(p)b(p)e^{-\Delta p}$$

where $b(p)$, as described above, has been altered to remove the $(\pm)$ ambiguity.

Finally, a particular consequence of $[S(-p)][S(p)] = 1_n$ is that

$$S_{22}(p) = -S_{11}(-p)S_{12}(p)/S_{12}(-p),$$

and we see that $S_{22}(p)$ is uniquely determined. QED

We have also established by means of the previous proof the following:

Corollary 4.4. Assume the conditions of Theorem 4.5 apply and define $A(p)$ to be that unique factorization of

$$1 - S(p)S(-p) = A(p)A(-p)$$

which is analytic and nonzero in $\operatorname{Re} p > 0$, satisfying $\lim_{|p|\to\infty} (\{\log A(p)\}/p) = 0$ and $A(\sigma) \geq 0$ in $\sigma = \operatorname{Re} p > 0$. Then the scattering parameters of the line are given by $S_{11}(p) = S(p)$, $S_{12}(p) = S_{21}(p) = A(p)b(p)e^{-\Delta p}$, and $S_{22}(p) = -S_{11}(-p)S_{12}(p)/S_{12}(-p)$, where $b(p)$ is the regular Blaschke or all-pass function defined by the zeros of $h_\Delta(p)$ in $\operatorname{Re} p > 0$.

The following discussion justifies the assumptions of this corollary in those cases when the function $S(p)$ is known to be the input reflection factor of some line terminated in some regular Z_Δ. Thus for such a line

$$S_{21}(p) = 2h_\Delta(p)/(\alpha + \beta),$$

where from Corollary 4.3 we know that α and β are meromorphic functions (from Definition 10 of Section 2.3.3 Z_Δ is meromorphic). Moreover, we also have from Definition 10 that $h_\Delta(p)$ is meromorphic, analytic, and bounded in $\operatorname{Re} p > 0$, and satisfies $\lim_{|p|\to\infty} (\{\log h_\Delta(p)\}/p) = 0$ in $\operatorname{Re} p > 0$. Thus, appealing to Theorem 2.14 of Section 2.3.2, we know that $h_\Delta(p)$ may be represented as $h_\Delta(p) = \hat{h}_\Delta(p)b(p)$, where $b(p)$ is a Blaschke product formed from the zeros of $h_\Delta(p)$ in $\operatorname{Re} p > 0$, $\hat{h}_\Delta(p)$ is analytic and nonzero in $\operatorname{Re} p > 0$, and $\lim_{|p|\to\infty} (\{\log \hat{h}_\Delta(p)\}/p) = 0$ also in $\operatorname{Re} p > 0$. Since Lemma 4.4 established that the only zeros of $S_{21}(p)$ in $\operatorname{Re} p > 0$ are those of $h_\Delta(p)$, we have

$$S_{21}(p) = \left[\frac{2\hat{h}_\Delta(p)}{\alpha + \beta}\,\frac{1}{e^{-\Delta p}}\right] b(p)e^{-\Delta p} = \hat{S}_{21}(p)b(p)e^{-\Delta p},$$

where $\hat{S}_{21}(p)$ is analytic and nonzero in $\operatorname{Re} p > 0$ and satisfies

$$\lim_{|p|\to\infty} (\{\log \hat{S}_{21}(p)\}/p) = 0$$

in $\operatorname{Re} p > 0$. Note that the factor $e^{-\Delta p}$ accounts for the asymptotic behavior of the denominator $(\alpha + \beta)$ of S_{21}. But since the scattering matrix of the line satisfies $[S(-p)][S(p)] = 1_n$ and $S_{11}(p) = S(p)$ we have

$$1 - S(p)S(-p) = 1 - S_{11}(p)S_{11}(-p) = S_{21}(p)S_{21}(-p) = \hat{S}_{21}(p)\hat{S}_{21}(-p),$$

since $b(p)b(-p) = 1$. Thus we may identify the desired factorization for $A(p)$ as $\hat{S}_{21}(p) = A(p)$, and from our preceding discussion it is clear that such a factorization will exist. Finally, on the basis of Theorem 2.14 and the fact that $|A(j\omega)|^2 = 1 - |S(j\omega)|^2$ we may obtain an explicit expression for $A(p)$ as

$$A(p) = \exp\left\{\frac{p}{\pi}\int_{-\infty}^{\infty} \frac{\log(1-|S(j\eta)|^2)^{1/2}}{p^2+\eta^2}\,d\eta\right\}$$

in $\operatorname{Re} p > 0$ and $A(-p) = \{1 - S(p)S(-p)\}/A(p)$ defines $A(p)$ for $\operatorname{Re} p < 0$. We also note that this expression for $A(p)$ is the unique factorization for the properties delineated in the corollary, since the arbitrary multiplying product $[\pm b(p)e^{-ap}]$ possible in the factorization must be unity, i.e., the $(\pm)$ ambiguity is resolved, since, as we see, $A(\sigma) \geq 0$ in $\operatorname{Re} p = \sigma > 0$ and the $b(p)e^{-\Delta p}$ factor must be one since $A(p) \neq 0$ in $\operatorname{Re} p > 0$ and $\lim_{|p|\to\infty}(\{\log A(p)\}/p) = 0$ in $\operatorname{Re} p > 0$.

4.2.2. A Realizability Theory for Smooth Lines

In the previous sections certain properties of the scattering matrix of a section of a smooth lossless line were described. In particular, it was demonstrated that this matrix is uniquely determined by knowledge of the input reflection factor of the line when it is terminated in a known load. But this implies that we can then find the input impedance of the line under any termination. In this section it will be shown that when the termination is an inductor, the poles of the input impedance uniquely determine the local characteristic impedance as a function of position along the line, and we are led to a realizability theory based on this fact.

Combining the two partial differential equations defining $V(y, p)$ and $I(y, p)$ leads to the following differential equation for $V(y, p)$:

$$\frac{d^2V(y,p)}{dy^2} - \frac{d\ln Z_0(y)}{dy}\frac{dV(y,p)}{dy} - p^2V(y,p) = 0. \tag{4.19}$$

If the line is terminated in an impedance Z_Δ, the corresponding boundary conditions on this equation may be expressed as

$$\left.\frac{V(y,p)}{I(y,p)}\right|_{y=\Delta} = \left.\frac{V(y,p)}{-(dV/dy)/pZ_0(y)}\right|_{y=\Delta} = Z_\Delta(p)$$

and

$$\left.\frac{V(y,p)}{-(dV/dy)/pZ_0(y)}\right|_0 = Z_{\text{in}}(p),$$

where $Z_{\text{in}}(p)$ is the input impedance of the line terminated in $Z_\Delta(p)$. Defining $V(y, p) = \{Z_0(y)\}^{1/2}q(y, p)$ and employing Eq. (4.19) leads to a differential equation for q, assuming d^2Z_0/dy^2 exists, expressed as

$$\frac{d^2q(y, p)}{dy^2} - Qq(y, p) - p^2q(y, p) = 0 \tag{4.20}$$

with the boundary conditions

$$\begin{aligned} \frac{dq}{dy}(0) + q(0)\left[\frac{\frac{1}{2}(dZ_0/dy)(0)}{Z_0(0)} + \frac{pZ_0(0)}{Z_{in}(p)}\right] &= 0 \\ \frac{dq}{dy}(\Delta) + q(\Delta)\left[\frac{\frac{1}{2}\, dZ_0/dy)(\Delta)}{Z_0(\Delta)} + \frac{pZ_0(\Delta)}{Z_\Delta(p)}\right] &= 0 \end{aligned} \tag{4.21}$$

and where $Q(y) = [Z_0(y)]^{1/2}\, d^2\{Z_0(y)\}^{-1/2}/dy^2$. Now, assume that the terminating impedance is an inductor, i.e., $Z_\Delta(p) = pK$. The following lemma shows the correspondence between the input impedance with this inductive load and the eigenvalues of a Sturm–Liouville problem:

Lemma 4.5. Every p for which a nonzero solution exists to

$$\frac{d^2u}{dy^2} - Qu - p^2u = 0$$

with the boundary conditions

$$\frac{du}{dy}(0) + u(0)\left[\frac{\frac{1}{2}(dZ_0/dy)(0)}{Z_0(0)}\right] = 0,$$

$$\frac{du}{dy}(\Delta) + u(\Delta)\left[\frac{\frac{1}{2}(dZ_0/dy)(\Delta)}{Z_0(\Delta)} + \frac{Z_0(\Delta)}{K}\right] = 0,$$

where $Q = \{Z_0(y)\}^{1/2}\, d^2\{Z_0(y)\}^{-1/2}/dy^2$ and $Z_0(y) > 0$, is a pole of the input impedance $Z_{in}(p)$ of a smooth lossless line with length Δ and local characteristic impedance $Z_0(y)$ and which is terminated in $Z_\Delta = pK$, and conversely.

Proof. In the preceding discussion it was shown that to each bounded, piecewise continuous $Z_0(y) > 0$ there corresponds a lossless line and further a $q(y, p)$ which satisfies Eq. (4.20) and the boundary conditions of Eqs. (4.21), for all values of p. However, all possible solutions to $d^2x/dy^2 - Qx - p^2x = 0$, $(dx/dy)(\Delta) + ax(\Delta) = 0$ differ at most by a constant (with respect to y) nonzero scale factor, i.e., if q and u satisfy the differential equation and the boundary condition at $y = \Delta$ stated in the lemma, then $q = \alpha(p)u$, with

$\alpha(p)$ a bounded, nonzero function of p. Therefore if in addition

$$\frac{du}{dy}(0) + u(0)\left[\frac{\frac{1}{2}(dZ_0/dy)(0)}{Z_0(0)}\right] = 0$$

for some value of p, then for the same value of p

$$\frac{d\alpha u}{dy}(0) + \alpha u(0)\left[\frac{\frac{1}{2}(dZ_0/dy)(0)}{Z_0(0)} + \frac{pZ_0(0)}{Z_{\text{in}}(p)}\right] = 0.$$

Factoring the nonzero α and subtracting, we get

$$\frac{u(0)pZ_0(0)}{Z_{\text{in}}(p)} = 0$$

for this value of p. However, $u(0)$ cannot be zero, since that would imply $(du/dy)(0) = 0$ and, since u is a solution of $d^2u/dy^2 - Qu - p^2u = 0$, it would be necessary for $u(y) = 0$, the trivial case. Therefore $p/Z_{\text{in}}(p)$ must vanish for this value of p, since $Z_0(0) > 0$. We conclude that $Z_{\text{in}}(p)$, being a meromorphic function, must have a pole at this value of p [it can be shown from Eq. (4.10) with $Z_\Delta = pK$ that $Z_{\text{in}}(p)$ has a zero at $p = 0$, so that this point is not to be included].

The converse follows immediately from Eqs. (4.21) with $Z_\Delta = pK$. QED

It should be noted that the problem posed in Lemma 4.5 is the classic Sturm–Liouville problem in which the p are the eigenvalues for the two boundary conditions. There exists an extensive literature on this subject, but the most pertinent for our discussion is the work of Marčenko [Ma1] (see also [Le1]). We have summarized some of the discussion leading to Marčenko's results in Appendix II. In particular, it is known that the most general solution of the differential equation

$$(d^2\Theta/dy^2) - Q(y)\Theta + \lambda\Theta = 0, \tag{4.22}$$

where λ is some constant, subject to the boundary conditions

$$\Theta(0) = 1 \qquad \text{and} \qquad (d\Theta/dy)(0) = h, \tag{4.23}$$

may be represented in terms of a fixed kernel $K(y, x)$ by

$$\Theta(\lambda, y) = \cos(y\sqrt{\lambda}) + \int_0^y K(y, t)\cos(t\sqrt{\lambda})\,dt.$$

If we impose the additional boundary condition

$$(d\Theta/dy)(\Delta) + H_n\Theta(\Delta) = 0, \tag{4.24}$$

then only certain λ exist for which a solution may be found; denote this set by $S_Q[h, H_n]$ (the so-called spectra of the system). We note that these λ

are eigenvalues of the equation $d^2\Theta/dy^2 - Q(y)\Theta = -\lambda\Theta$ subject to the boundary conditions of Eqs. (4.23) and (4.24). Marčenko has shown [Ma1, Lemma 2, p. 388], that for all λ and u

$$\int_0^\Delta \Theta(\lambda^2, t)\Theta(u^2, t)\,dt = \frac{L(u)M(\lambda) - L(\lambda)M(u)}{(H_1 - H_2)(u^2 - \lambda^2)}, \tag{4.25}$$

where

$$L(z) = \Delta(\lambda_0 - z^2)\prod_{n=1}^{\infty} \frac{\lambda_n - z^2}{(\pi/\Delta)^2 n^2}$$

and

$$M(z) = \Delta(\beta_0 - z^2)\prod_{n=1} \frac{\beta_n - z^2}{(\pi/\Delta)^2 n^2}$$

are uniquely determined from the two spectra,

$$S_Q[h, H_1] = \{\lambda_i\}$$

and

$$S_Q[h, H_2] = \{\beta_i\},$$

i.e., two sets of eigenvalues of the equation $d^2\Theta/dy^2 - Q(y)\Theta = -\lambda\Theta$ obtained by imposing two different boundary conditions at $y = \Delta$ [H_1 and H_2 in Eq. (4.24)] but the same boundary conditions at $y = 0$ [Eq. (4.23)]. Moreover, $L(z)$ and $M(z)$ are known to be entire functions of z with asymptotic behavior that satisfies

$$\lim_{t\to\infty} [L(jt) \quad \text{or} \quad Mj(t)] = -jt \sin j\,\Delta t.$$

Finally, it may be shown that the two spectra have asymptotic expansions of the form

$$\sqrt{\lambda_n} = \frac{\pi n}{\Delta} + \frac{A_1}{\pi n} + O\left(\frac{1}{n^2}\right), \qquad \sqrt{\beta_n} = \frac{\pi n}{\Delta} + \frac{B_1}{\pi n} + O\left(\frac{1}{n^2}\right),$$

where, in particular, $H_1 - H_2 = A_1 - B_1$.

Marčenko was then able to show that two spectra uniquely characterize the function $Q(y)$ which defined the eigenvalue problem.

Theorem 4.6 [Ma1]. Denote by $S_Q[h, H_n]$ (the so-called spectra) the set of all λ such that a nonzero solution of $d^2u/dy^2 - Q(y)u + \lambda u = 0$ satisfies the boundary conditions

$$(du/dy)(0) - hu(0) = 0 \qquad \text{and} \qquad (du/dy)(\Delta) + H_n u(\Delta) = 0$$

[it is assumed that $\int_0^\Delta |Q|\,dy < \infty$, and that h, H_n, and $Q(y)$ are real]. Then two spectra $S_Q[h, H_1]$ and $S_Q[h, H_2]$, with $H_1 - H_2 \neq 0$, uniquely determine h, H_1, H_2, and $Q(y)$.†

Not only is this uniqueness theorem available, but the following algorithm provides a means of determining the four quantities described in the theorem:

Marčenko's Algorithm. Assume the elements of the two spectra obtained from some $Q(y)$ are ordered as

$$[\lambda_0 < \lambda_1 < \cdots] = S_Q[h, H_1]$$

and

$$[\beta_0 < \beta_1 < \cdots] = S_Q[h, H_2].$$

Now if we define

$$G(\lambda, \mu) = \frac{L(\mu)M(\lambda) - L(\lambda)M(\mu)}{(A_1 - B_1)(\mu^2 - \lambda^2)} - \frac{\mu \sin \mu\Delta \cos \lambda\Delta - \lambda \sin \lambda\Delta \cos \mu\Delta}{\mu^2 - \lambda^2}$$

and

$$F(y, x) = \left(\frac{1}{\Delta}\right)^2 \sum_{m=-\infty}^{\infty} \sum_{n=-\infty}^{\infty} G\left(\frac{n\pi}{\Delta}, \frac{m\pi}{\Delta}\right) \cos\left(\frac{n\pi}{\Delta} y\right) \cos\left(\frac{m\pi}{\Delta} x\right),$$

where A_1 and B_1 are obtained from the asymptotic expansions of the two spectra, then there exists a solution $K(y, x)$ to the nonlinear integral equation

$$F(y, x) = K(y, x) + \int_y^\Delta K(z, y)K(z, x)\, dz$$

(which may be obtained in an iterative manner by successive approximations), and one may determine

$$Q(y) = \frac{d^2\Theta(0, y)/dy^2}{\Theta(0, y)},$$

$$h = \frac{\partial}{\partial y} \Theta(0, y)\,|_{y=0},$$

and

$$H_1 = -\left.\frac{(\partial/\partial y)\Theta(\lambda_0, y)}{\Theta(\lambda_0, y)}\right|_{y=\Delta}, \qquad H_2 = -\left.\frac{(\partial/\partial y)\Theta(\beta_0, y)}{\Theta(\beta_0, y)}\right|_{y=\Delta},$$

† In other words, given two spectra (two sets of eigenvalues) of a Sturm–Liouville system one can find the function $Q(y)$ which defines the system and in addition the boundary conditions which produced the two sets of eigenvalues.

with

$$\Theta(\lambda, y) = \cos(y\sqrt{\lambda}) + \int_0^y K(y, t)\cos(t\sqrt{\lambda})\, dt.$$

Now, by evaluating the indeterminate form of Eq. (4.25) we may also obtain

$$\int_0^\Delta \Theta^2(z^2, t)\, dt = \frac{(dL/dz)M(z) - L(z)(dM/dz)}{2(H_1 - H_2)z}.$$

Thus $(dL/dz)M(z) - (dM/dz)L(z)$ is either positive or negative for all positive values of z depending on the sign of $H_1 - H_2$, and we conclude that the zeros of $L(z)$ and $M(z)$ (or, equivalently, the two sets of eigenvalues $S_Q[h, H_1]$ and $S_Q[h, H_2]$) interlace. In addition, if we set $z = \sqrt{\lambda_i}$, then because $L(\sqrt{\lambda_i}) = 0$ we obtain

$$\int_0^\Delta \Theta^2(\lambda_i, t)\, dt = \frac{(dL/dz)(\sqrt{\lambda_i})M(\sqrt{\lambda_i})}{2(H_1 - H_2)\sqrt{\lambda_i}} = \rho_i,$$

and conclude that the two spectra $S_Q[h, H_1] = \{\lambda_i\}$ and $S_Q[h, H_2] = \{\beta_i\}$ uniquely determine the set $\{\rho_i\}$, which are the integrals of the squares of the eigenfunctions corresponding to the set of eigenvalues $\{\lambda_i\}$. Conversely, we may show that if given $S_Q[h, H_1]$ and $\{\rho_i\}$, then $S_Q[h, H_2] = \{\beta_i\}$ is uniquely determined: from Eq. (4.25) we see that up to a constant, i.e., independent of i, $M(\sqrt{\lambda_i})$ is determined from ρ_i and $(dL/dz)(\sqrt{\lambda_i})$ or, since $L(z)$ is determined by $\{\lambda_i\}$, $M(z)$ is determined from the two sets $\{\lambda_i\}$ and $\{\rho_i\}$ at all the points $z = \sqrt{\lambda_i}$; but the interlacing of the zeros of M and L, together with the known asymptotic behavior of the entire function $M(z)$, implies that there can exist only one $M(z)$ which takes on these values at all the points. Therefore we may place in one-to-one correspondence two sets of eigenvalues of the system with one set of eigenvalues and its corresponding set of ρ_i.

The motivation for the above remarks is the following result of Gel'fand and Levitan [Ge2, Theorem 2, p. 302]: if two sets $\{\lambda_i \geq 0\}$ and $\{\rho_i > 0\}$ have the asymptotic behavior

$$\begin{aligned} \sqrt{\lambda_n} &= \frac{\pi n}{\Delta} + \frac{b_1}{n} + \frac{b_3}{n^3} + O\left(\frac{1}{n^4}\right) \\ \rho_n &= \frac{\Delta}{2} + \frac{a_1}{n^2} + O\left(\frac{1}{n^4}\right), \end{aligned} \tag{4.26}$$

where a_1, b_1, and b_3 are constants, then there exists a continuous $Q(y)$ such that the set $\{\lambda_i\}$ are eigenvalues of Eq. (4.22) subject to the boundary condition of Eqs. (4.23) and (4.24), and $\{\rho_i\}$ are the integrals of the corresponding eigenfunctions. These conditions are not known to be necessary, since in fact one can only show that a necessary condition is the existence of expansions where b_1 is a constant, but a_1 and b_3 need not exist. However, a single necessary and sufficient condition is available if we demand that $Q(y)$ be infinitely differentiable, namely, that the expansions of Eq. (4.26) contain all terms, i.e., that constants a_k and b_k exist for all k. In both cases it is also known [Ge2, Theorem 1, p. 301] that the kernel $K(y, x)$ exists and has continuous partial derivatives of the first and second order if $Q(y)$ is continuous and of all orders if $Q(y)$ is infinitely differentiable.

Before applying these results of Marčenko, and Gel'fand and Levitan to the theory of nonuniform lines it will prove advantageous to express the quantities $L(z)$ and $M(z)$ in terms of $S(p)$. To this end we first note from Lemma 4.5 that if $p_n = j\omega_n$ is a pole of the input impedance $Z_{\text{in}}(p)$ of the line terminated in $Z_\Delta = pK$, then $\lambda_n = \omega_n{}^2$ must be an eigenvalue of Eq. (4.20). Moreover, this input impedance may be expressed using Eq. (4.8) and Corollary 4.4 as

$$Z_{\text{in}}(p) = \frac{pK[\alpha(p) + \alpha(-p)] + [\alpha(p) - \alpha(-p)]}{pK[\beta(p) - \beta(-p)] + [\beta(p) + \beta(-p)]}, \tag{4.27}$$

where α and β are given by

$$\alpha(p) = \frac{1 + S(p)}{A(p)e^{-\Delta p}}, \qquad \beta(p) = \frac{1 - S(p)}{A(p)e^{-\Delta p}}, \tag{4.28}$$

$S(p)$ is the input reflection factor corresponding to a 1 Ω terminating load [$Z_\Delta(p) = 1$ Ω] and $A(p)$ is its factorization, i.e., $1 - S(p)S(-p) = A(p)A(-p)$. Note that when $Z_\Delta = 1$ Ω, $h_\Delta = 1$, and from Corollary 4.3 $r(y, p)$, and thus $\alpha(p)$ and $\beta(p)$, will be entire functions. It was also show in Section 4.2.1 that $\beta(p) + \beta(-p)$ and $\beta(p) - \beta(-p)$ have the asymptotic behavior given by

$$[Z_0(\Delta)/Z_0(0)]^{1/2}[e^{p\Delta} + e^{-p\Delta}] \qquad \text{and} \qquad \frac{1}{[Z_0(\Delta)Z_0(0)]^{1/2}}[e^{p\Delta} - e^{-p\Delta}],$$

respectively. Moreover the roots of the denominator of Eq. (4.27) [i.e., the poles of $Z_{\text{in}}(p)$] are simple and restricted to lie on the $j\omega$ axis. Therefore, in view of the asymptotic behavior of $L(z)$ and $M(z)$ noted above we conclude that

$$L(z) = k_1\{jzK_1[\beta(jz) - \beta(-jz)] + \beta(jz) + \beta(-jz)\} \tag{4.29}$$

and

$$M(z) = k_2\{jzK_2[\beta(jz) - \beta(-jz)] + \beta(jz) + \beta(-jz)\}, \tag{4.30}$$

where k_1 and k_2 are determined such that this asymptotic behavior is satisfied. In particular, if $K_1 = \infty$, then

$$L(z) = k_1\{jz[\beta(iz) - \beta(-jz)\}, \tag{4.31}$$

since in this case

$$Z_{\text{in}}(p) = \frac{\alpha(p) + \alpha(-p)}{\beta(p) - \beta(-p)}.$$

By combining Eqs. (4.28) with Eq. (4.30) and either Eq. (4.29) or (4.31) we obtain an explicit formula for both $L(z)$ and $M(z)$ expressed in terms of $S(p)$ and its factorization $A(p)$. We are now in a position to establish the following existence theorem, which is the major result of this section. First we note from Theorem 4.5 and Corollary 4.4 of Section 4.2.1 that if the input reflection factor of the line is known for some regular terminating impedance Z_Δ, then the scattering matrix of the line is uniquely determined. Therefore the input reflection factor for an arbitrary terminating impedance can be computed. For simplicity we will assume that this has been done, and we state our theorem explicitly for the case where $Z_\Delta = 1\ \Omega$.

Theorem 4.7 [Wo8]. The following conditions are both necessary and sufficient if $S(p)$ is to be the input reflection factor (normalized to 1) of a nonuniform lossless transmission line of length Δ terminated in a unit resistor and the local characteristic impedance $Z_0(y)$ of the line is positive and infinitely differentiable:

1. $S(p)$ is a meromorphic function, analytic and bounded by 1 in $\operatorname{Re} p \geq 0$, satisfying $S(p^*) = S^*(p)$.
2. $S(0) = 0$.
3. $1 - S(p)S(-p)$ is nonzero in the finite p-plane.
4. The poles of $1 - S(p)S(-p)$ in $\operatorname{Re} p < 0$ are identical (including order) to the poles of $S(p)$ in $\operatorname{Re} p < 0$.
5. If $A(p)$ is that factorization† of $1 - S(p)S(-p) = A(p)A(-p)$ such that $A(p)$ is analytic and nonzero in $\operatorname{Re} p \geq 0$, $\lim_{|p| \to \infty} \{(\log A)/p\} = 0$ in $\operatorname{Re} p > 0$, and $A(\sigma) \geq 0$ in $\sigma = \operatorname{Re} \rho > 0$, then (a) for $|p| \to \infty$

$$\frac{1 \pm S(p)}{A(p)e^{-\Delta p}} = e^{\Delta p}\left[c_\pm + O\left(\frac{1}{p}\right)\right] + e^{-\Delta p}\left[d_\pm + O\left(\frac{1}{p}\right)\right],$$

† From the discussion following Corollary 4.4 in Section 4.2.1 we know that such a factorization must exist if $S(p)$ is a realizable input reflection factor.

where $c_{\pm} > 0$ and $|d_-/c_-| < 1$, and (b) for $|\omega| \to \infty$

$$\frac{1 - S(j\omega)}{A(j\omega)e^{-j\omega\Delta}} = e^{j\Delta\omega}\left[c_- + \frac{c_1}{\omega} + \cdots + \frac{c_n}{\omega^n} + \cdots\right]$$

$$+ e^{-j\omega\Delta}\left[d_- + \frac{d_1}{\omega} + \cdots + \frac{d_n}{\omega^n} + \cdots\right],$$

where for any n c_n and d_n are constants.

Proof. Necessity: If a scattering matrix is defined with a normalization of 1 at both ports, then $S(p) = S_{11}(p)$. Therefore Condition 1 follows from Theorem 4.3. The second condition is established by Eq. (4.10), since with $Z_\Delta = 1$, $\alpha(0) = \beta(0) = 1$ and $S_{11} = (\alpha - \beta)/(\alpha + \beta)$. From Lemma 4.4 we see (with $Z_\Delta = h_\Delta = 1$) that $S_{21}(p)$ is nonzero in the finite p plane. Therefore Condition 3 follows since $1 - S_{11}(p)S_{11}(-p) = S_{21}(p)S_{21}(-p)$. But $S_{22} = -S_{11}(-p)S_{21}(p)/S_{21}(-p)$ must be analytic in $\operatorname{Re} p > 0$ or, since S_{21} is nonzero, $S_{21}(-p)$ must have the same poles as $S_{11}(-p)$, and Condition 4 follows since $1 - S_{11}(p)S_{11}(-p) = S_{21}(p)S_{21}(-p)$. From Eqs. (4.8) and Corollary 4.4 one obtains $\alpha(p) = [1 + S_{11}(p)]/[A(p)e^{-\Delta p}]$ and $\beta(p) = [1 - S_{11}(p)]/[A(p)e^{-\Delta p}]$, and Condition 5 then follows from the required asymptotic behavior of these functions, as discussed in Section 4.2.1, particularly Lemma 4.3.

Now, in order to demonstrate the sufficiency we will show that a line exists with the given $S(p)$ when it is terminated in a unit resistor. For this purpose we will define $L(z)$ and $M(z)$ by using $K_1 = \infty$, $K_2 > 0$, and $\beta(p) = [1 - S(p)]/[A(p)e^{-\Delta p}]$; i.e., $L(z)$ will be given by Eq. (4.31) and $M(z)$ by Eq. (4.30). From the zeros of these two functions we will find uniquely the two sets $\{\lambda_i\}$ and $\{\beta_i\}$, and thus the set $\{\rho_i\}$. First we will demonstrate that λ_i, β_i, and ρ_i are all nonnegative. Now $\beta(p)$, with the Conditions 1–5, will be an entire function that is nonzero in $\operatorname{Re} p \geq 0$, and, moreover $\lim_{|p| \to \infty} |\beta(-p)/\beta(p)| = |d_-/c_-| < 1$ in $\operatorname{Re} p \geq 0$. Thus $\beta(-p)/\beta(p)$ is analytic and bounded by 1 in $\operatorname{Re} p \geq 0$, and $[\beta(p) + \beta(-p)]/[\beta(p) - \beta(-p)]$ is a positive-real Foster function. Therefore $pK_2 + \{[\beta(p) + \beta(-p)]/[\beta(p) - \beta(-p)]\}$ is a positive-real Foster function for $K_2 > 0$, and we conclude that the zeros of both $pK_2[\beta(p) - \beta(-p)] + \beta(p) + \beta(-p)$ and $p[\beta(p) - \beta(-p)]$ are all simple and restricted to lie on the $j\omega$ axis. Thus we conclude that the zeros of the functions $L(z)$ and $M(z)$ that were constructed from the given $S(p)$ are all simple, so that

$$\{\lambda_i\} = \{[L(z_i) = 0]^2\}$$

and

$$\{\beta_i\} = \{[M(z_i) = 0]^2\}$$

are sets of real, distinct, nonnegative numbers, and, in particular, $\lambda_0 = 0$. Now consider

$$\frac{L(z)}{M(z)} = \frac{k_1}{k_2} \frac{jz[\beta(jz) - \beta(-jz)]}{jzK_2[\beta(jz) - \beta(-jz)] + \beta(jz) + \beta(-jz)}$$

and note that $k_1/k_2 > 0$ because K_2 is positive and the asymptotic behaviors of $L(z)$ and $M(z)$ (which determine k_1 and k_2) are determined by the same $z[\beta(jz) - \beta(-jz)]$. Then with $\beta(jz) - \beta(-jz) = jA(z)$ and $\beta(jz) + \beta(-jz) = B(z)$ we have

$$\frac{d}{dz}\frac{L(z)}{M(z)} = \frac{k_1}{k_2}\frac{z[AB' - A'B] - AB}{[-zK_2A + B]^2}.$$

But since we have established that $[\beta(p) + \beta(-p)]/[\beta(p) - \beta(-p)]$ is a positive-real Foster function, we conclude that $(d/dz)[-B(z)/A(z)] > |(B/A)/z|$, except possibly at the zeros of A, and we obtain $d(L/M)/dz < 0$, or $L'(z)M(z) - M'(z)L(z) < 0$ for $z > 0$. In particular, this implies that the zeros of L and M will interlace, and because $L(z)$ is zero for $z = 0$ but $M(z)$ is not, we will obtain asymptotic expansions for their zeros (these will be shown to exist later, given Condition 5),

$$\sqrt{\lambda_i} = \frac{\pi n}{\Delta} + \frac{A_1}{\pi n} + O\left(\frac{1}{n^2}\right), \qquad \text{and} \qquad \sqrt{\beta_i} = \frac{\pi n}{\Delta} + \frac{B_1}{\pi n} + O\left(\frac{1}{n^2}\right),$$

such that $B_1 - A_1 > 0$. Finally, then, we may conclude that the ρ_i will all be positive integers, and thus we obtain from the given $S(p)$ two sets, $\{\lambda_i \geq 0\}$ and $\{\rho_i > 0\}$. Now, if the asymptotic behavior of these sets satisfy the conditions of the Gel'fand and Levitan theorem, we may conclude that they uniquely define a $Q(y)$ and a kernel $K(y, x)$ that are both infinitely differentiable functions. To show that this is the case, consider $M(z)$, and substitute the assumed asymptotic expansion for $\beta(j\omega) = [1 - S(j\omega)]/[A(j\omega)e^{-j\omega\Delta}]$ to obtain

$$M(z) = \left[k_1 z + \frac{k_2}{z} + \cdots\right] \sin z\Delta + \left[k_3 + \frac{k_4}{z^2} + \cdots\right] \cos z\Delta.$$

Thus the zeros of $M(z)$ are obtained from

$$\tan z\Delta = -\left[\frac{k_3 + (k_4/z^2) + \cdots}{k_1 z + (k_2/z) + \cdots}\right].$$

By approximating the tangent with a linear function, we obtain the zeros as solutions to

$$\Delta\left(z_n - \frac{\pi n}{\Delta}\right) = \frac{\alpha_2}{(\pi n/\Delta)^3} + \cdots,$$

or

$$z_n = \sqrt{\lambda_n} = \frac{\pi n}{\Delta} + \frac{b_1}{n} + \frac{b_2}{n^3} + \cdots.$$

Similarly, by using the assumed asymptotic expansion for $(1 - S)/Ae^{-\Delta p}$, we may demonstrate that the ρ_i also have the required asymptotic behavior.

At this point we have shown that an infinitely differentiable $Q(y)$ will exist, and, moreover, a kernel $K(y, x)$, so that we may represent the solution to the differential equation formed with this $Q(y)$ by

$$\Theta(\lambda, y) = \cos(y\sqrt{\lambda}) + \int_0^y K(y, t)\cos(t\sqrt{\lambda})\, dt.$$

In addition, using the fact that $\{\lambda_i\}$ and $\{\rho_i\}$ determine the set $\{\beta_i\}$ uniquely (as discussed previously), we know from the uniqueness of the inversion procedure used to find $Q(y)$ (say, Marčenko's algorithm) that the two spectra of the differential equation, $S_Q[h, H_1] = \{\lambda_i\}$ and $S_Q[h, H_2] = \{\beta_i\}$, will be precisely the two sets formed from the zeros of $L(z)$ and $M(z)$, which were uniquely determined by $\beta(p) = [1 - S(p)]/[A(p)e^{-\Delta p}]$.

Now, since $\lambda_0 = 0$ is one of the eigenvalues, then $\Theta(0, y) = 1 + \int_0^y K(y, t)\, dt$ is its corresponding eigenfunction, and it is well known [Inl, p. 235] that the eigenfunction corresponding to the smallest eigenvalue of a Sturm–Liouville operator is nonzero on $0 < y < \Delta$. But $\Theta(0, 0) = 1$, and, in addition, $\Theta(0, \Delta) \neq 0$, otherwise the boundary condition $\Theta'(0, \Delta) + H_1\Theta(0, \Delta) = 0$ would imply that Θ is identically zero for all y. We conclude that the $\Theta(0, y)$ obtained is greater than zero for $0 \le y \le \Delta$. Moreover, we note that

$$Q(y) = \frac{d^2\Theta(0, y)/dy^2}{\Theta(0, y)} = \frac{d^2[Z_0(y)]^{-1/2}/dy^2}{[Z_0(y)]^{-1/2}},$$

so that we may identify a $Z_0(y) = k/\Theta^2(0, y)$, where k is some constant. But from Marčenko's algorithm we also obtain, by using the given $L(z)$ and $M(z)$, a unique determination of

$$H_1 = -\left.\frac{(d/dy)\Theta(0, y)}{\Theta(0, y)}\right|_{y=\Delta} = \frac{\frac{1}{2}(dZ_0/dy)(\Delta)}{Z_0(\Delta)},$$

Since $\lambda_0 = 0$ corresponds to the boundary condition with $K = \infty$. Thus, since the differential equation for $Z_0(y)$ is second order, we conclude that the most general $Z_0(y)$ possible is of the form $k/\Theta^2(0, y)$. Moreover, since we identify B_1 and A_1 from the asymptotic expansions for λ_i and β_i, we determine a unique

$$Z_0(\Delta) = K_2(H_2 - H_1) = K_2(B_1 - A_1) > 0,$$

and then finally

$$Z_0(y) = \frac{K_2(B_1 - A_1)\Theta^2(0, \Delta)}{\Theta^2(0, y)}.$$

But since $\Theta(0, y)$ is nonzero and the existence of a continuous $K(y, x)$ implies that $\Theta(0, y)$ is bounded, we conclude that $Z_0(y)$ is both *positive* and *bounded.* Finally, since the kernel $K(y, x)$ is infinitely differentiable, then $\Theta(0, y)$ or $Z_0(y)$ is also infinitely differentiable.

We are left with one final task, namely, to show that a line formed with this $Z_0(y)$ will yield the given $S(p)$ when terminated in a unit resistor. First we note that the discussion given above shows that a $Z_0(y)$ is uniquely determined from the two sets $\{\lambda_i\}$ and $\{\beta_i\}$ obtained from $S(p)$. However, the poles of the input impedance of the line constructed with the $Z_0(y)$ must yield the same two sets $\{\lambda_i\}$ and $\{\beta_i\}$ when it is terminated in the given two loads; these poles are distinct and must be eigenvalues for the differential equation formed with the $Q(y)$, and two sets of eigenvalues uniquely determine $Q(y)$, or, as we showed, $Z_0(y)$. Thus we have shown that if $\hat{\beta}(p) = (1 - \hat{S})/\hat{A}e^{-\Delta p}$ is the function defined by the reflection factor $\hat{S}$ obtained from the constructed line, then $\hat{\beta}(p) - \hat{\beta}(-p)$ and $pK_2[\hat{\beta} - \hat{\beta}(-p)] + [\hat{\beta} + \hat{\beta}(-p)]$ have the same zeros as $\beta(p) - \beta(-p)$ and $pK_2[\beta - \beta(-p)] + [\beta + \beta(-p)]$, respectively, where $\beta(p)$ is obtained from the given $S(p)$ of the theorem. But since β or $\hat{\beta}$ are entire functions of exponential type Δ and are bounded on $p = j\omega$, the zeros of the functions $F_1 = \beta(p) - \beta(-p)$ and $F_2 = pK_2[\beta - \beta(-p)] + [\beta + \beta(-p)]$ together with the given asymptotic behavior of $\beta(p) = \{1 - S(p)\}/Ae^{-\Delta p}$ enable one to find F_1 and F_2 uniquely, and thus $\beta(p)$. Therefore $\hat{\beta}(p) = \beta(p)$. Finally, since $\alpha(p) = \{1 + S(p)\}/Ae^{-\Delta p}$ and $\beta(p)$ have the assumed asymptotic behavior stated in Condition 5a, we may show, given also that $S(p)$ satisfies Conditions 1–4, that α/β is positive-real, $\alpha(p)\beta(-p) + \alpha(-p)\beta(p) = 2$, and $\lim_{|p| \to \infty} (\alpha/\beta) = k > 0$ in $\operatorname{Re} p > 0$. Therefore $\alpha(p)$ is uniquely determined from $\beta(p)$, since

$$\left[\frac{\alpha}{\beta} + \frac{\alpha(-p)}{\beta(-p)}\right]_{p=j\omega} = \frac{2}{|\beta(j\omega)|^2} = 2 \operatorname{Re} \frac{\alpha}{\beta}\bigg|_{p=j\omega},$$

and the real part of a meromorphic positive-real function on the $j\omega$ axis uniquely determines the behavior of the function if $\lim_{|p| \to \infty} (\alpha/\beta) = k$ in $\operatorname{Re} p > 0$ (see Theorem 2.10). Therefore $S(p) = (\alpha - \beta)/(\alpha + \beta)$ is uniquely determined by $\beta(p)$ and we have shown that $\hat{S}(p) = S(p)$. QED

We noted earlier in our description of the Gel'fand–Levitan result that the necessary and sufficient conditions relating to the asymptotic expansions are only symmetric when $Q(y)$ or $Z_0(y)$ is infinitely differentiable. For this reason we stated Theorem 1 for this class of lines. However, one can obtain similar results using the results of Marčenko and Gel'fand and Levitan that were previously stated under weaker hypothesis; for example:

Corollary 4.5. Let $S(p)$ satisfy the conditions of Theorem 4.7 with the exception that for $|\omega| \to \infty$

$$\frac{1 - S(j\omega)}{A(j\omega)e^{-j\omega}} = e^{j\omega\Delta}\left[c_- + \frac{c_1}{\omega} + \frac{c_2}{\omega^2} + \frac{c_3}{\omega^3} + \frac{c_4}{\omega^4}\right]$$
$$+ e^{-j\omega\Delta}\left[d_- + \frac{d_1}{\omega} + \frac{d_2}{\omega^2} + \frac{d_3}{\omega^3} + \frac{d_4}{\omega^4}\right] + O\left(\frac{1}{\omega^5}\right);$$

then $S(p)$ is the input reflection factor of a line with a twice continuously differentiable, positive $Z_0(y)$.

As a final observation concerning the existence question, it is clear that aside from Conditions 1–4, which classify the function $S(p)$ for all p, the behavior of $S(p)$ in the finite portion of the p plane is not critical, since the position of a finite number of its poles and zeros will not control the asymptotic behavior of $\beta(p) = [1 - S(p)]/A(p)e^{-\Delta p}$.

The proof of Theorem 4.7 together with the conclusions of Theorem 4.5 also enables us to make definite statements concerning the unique relations between $S(p)$ and $Z_0(y)$. We single out one such statement as:

Corollary 4.6. Let $S(p)$ be the input reflection factor of a nonuniform lossless line of electrical length Δ terminated in a regular normalizing impedance $Z_\Delta(p)$. Then if the local characteristic impedance $Z_0(y)$ of the line is at least twice continuously differentiable, it is uniquely determined by $S(p)$, Δ, and $Z_\Delta(p)$.

Proof. From Theorem 4.5 we note that $S(p)$, Δ, and Z_Δ uniquely determine the input impedance of the line for arbitrary terminations, and, in particular, for inductive terminations. Thus we may assume that the poles of the input impedance with various inductive terminations are known, and we appeal to Lemma 4.5 to conclude that at least two spectra of the Sturm–Liouville system can be uniquely determined. The Marčenko result (Theorem 4.6) then establishes the fact that $Q(y)$, h, H_1, and H_2 are uniquely determined, so that we may use the arguments of the proof of Theorem 4.7 to conclude that $Z_0(y)$ is uniquely determined as

$$Z_0(y) = \frac{K_2(B_1 - A_1)\Theta^2(0, \Delta)}{\Theta^2(0, y)}.$$

We note that the assumption that $Z_0(y)$ is at least twice continuously differentiable was used to assure that $Q(y) = \{Z_0(y)\}^{1/2} d^2\{Z_0(y)\}^{-1/2}/dy^2$ was magnitude integrable, so that Marčenko's theorem was applicable. QED

Finally, by means of arguments similar to those already employed one may use Marčenko's uniqueness theorem concerning the spectra of a Sturm–Liouville operator to obtain:

Corollary 4.7. Let the poles and zeros of the input impedance $Z_{\text{in}}(p)$ of a line satisfying the conditions of Corollary 4.6 be known for *either* $Z_\Delta = 0$ or $Z_\Delta = \infty$, i.e., a termination of either a short circuit or an open circuit. Then $Z_0(y)$ is uniquely determined if in addition

$$\lim_{\substack{|p|\to\infty \\ (\text{Re } p \geq 0)}} Z_{\text{in}}(p)$$

is known.

This completes our discussion of the basic characterization of nonuniform lossless transmission lines. The following section collects some related results that are of general interest.

4.2.3. Related Topics and Illustrations

4.2.3.1. A Synthesis Procedure

Since Theorem 4.7 and its corollaries provide an existence theory for smooth lines, the next problem is one of finding the line from its terminal behavior. As noted above, knowledge of the input reflection factor corresponding to any "regular" passive termination enables us to find the $S(p)$ corresponding to a unit resistor load, and there is a direct procedure for accomplishing this (see Corollary 4.4). For our present purpose, then, we will assume that we have a unit-resistor termination. Utilizing the representations for $L(z)$ and $M(z)$ developed in the previous section, we may appeal to Marčenko's algorithm [Ma1] and obtain the following procedure for determining $Z_0(y)$ from $S(p)$, i.e., a synthesis procedure:

Given $S(p)$, we obtain $M(z)$ and $L(z)$ as in Eqs. (4.30) and (4.31), where K_2 is an arbitrary positive number, $\beta(p) = [1 - S(p)]/A(p)e^{-\Delta p}$, and $A(p)$ is its factorization as defined in Corollary 4.4. In addition, we find the constant $H_1 - H_2 = A_1 - B_1$ from the asymptotic behavior of the zeros of $L(z)$ and $M(z)$. Then, if $Z_0(y)$ is at least twice continuously differentiable, we may determine it as

$$Z_0(y) = K_2(H_2 - H_1)\left[\frac{1 + \int_0^\Delta K(y, t)\, dt}{1 + \int_0^y K(y, t)\, dt}\right],$$

where $K(y, x) = \lim_{n\to\infty} K_n(y, x)$, with

$$K_0(y, x) = F(y, x),$$

$$K_n(y, x) = F(y, x) - \int_y^{\Delta} K_{n-1}(t, y)K_{n-1}(t, x)\,dt,$$

and

$$F(y, x) = \left(\frac{1}{\Delta}\right)^2 \sum_{m=-\infty}^{\infty} \sum_{n=-\infty}^{\infty} G\left(\frac{n\pi}{\Delta}, \frac{m\pi}{\Delta}\right) \cos\left(\frac{n\pi}{\Delta} y\right) \cos\left(\frac{m\pi}{\Delta} x\right),$$

$$G(\lambda, u) = \frac{L(u)M(\lambda) - L(\lambda)M(u)}{(A_1 - B_1)(u^2 - \lambda^2)}$$

$$- \frac{u \sin u\Delta \cos \lambda\Delta - \lambda \sin \lambda\Delta \cos u\Delta}{u^2 - \lambda^2}.$$

One may argue with the utility of the procedure, but such an argument would have to be based on the assumption that there exists a closed-form expression for $Z_0(y)$ in terms of the terminal behavior of the line, and all of the literature on the subject seem to imply that some iterative or infinite series is the best that one can hope for. The primary virtue of the above procedure is its known convergence, assuming $S(p)$ is realizable, in the sense that the $K_n(y, x)$ converge uniformly, and the fact that all of the quantities are expressed directly in terms of $S(p)$.

4.2.3.2. Properties of the transfer Scattering Coefficient of Nonuniform Lines

In this section we obtain an existence theory for smooth lines in terms of the $S_{21} = S_{12}$ scattering coefficient of the line and then discuss the uniqueness and approximation theory associated with these functions.

This discussion starts with the representations obtained in Eqs. (4.8) for the S_{11} and S_{21} scattering coefficients of line normalized to 1 at both $y = 0$ and $y = \Delta$,

$$S_{11}(p) = (\alpha - \beta)/(\alpha + \beta), \qquad S_{21} = S_{12} = 2/(\alpha + \beta),$$

where α and β are entire functions of exponential type Δ. Now let us assume that we are given $S_{21}(p)$, or $\alpha + \beta$, but not α and β separately. Then from the lossless nature of the line we obtain

$$S_{11}(p)S_{11}(-p) = 1 - S_{21}(p)S_{21}(-p) = \frac{(\alpha + \beta)[\alpha(-p) + \beta(-p)] - 4}{(\alpha + \beta)[\alpha(-p) + \beta(-p)]}. \tag{4.32}$$

In order to identify the most general form of the corresponding $S_{11}(p)$, we would like to factor the numerator of Eq. (4.32). First we note that since $|S_{21}(j\omega)| \leq 1$,

$$F(j\omega) = [\alpha(j\omega) + \beta(j\omega)][\alpha(-j\omega) + \beta(-j\omega)] - 4 \geq 0.$$

In addition, it is known that $\alpha(j\omega)$ and $\beta(j\omega)$ must be bounded for all ω (being entire functions, they are bounded for all finite ω, and the asymptotic expressions of Section 4.2.1 with $Z_\Delta = 1$ then show that they remain bounded as $|\omega| \to \infty$). Finally, the asymptotic behaviors of both $\alpha(p)$ and $\beta(p)$ imply that $F(p)$ is an entire function of exponential type 2Δ. We may then appeal to the theory of entire functions [Le1, p. 437] to conclude that $F(p)$ may be factored into $F(p) = N(p)N(-p)$, where $N(p)$ is an entire function of exponential type Δ, and all of its zeros are in $\text{Re}\, p \leq 0$. Therefore we identify the most general $S_{11}(p)$ as

$$S_{11}(p) = \{N(p)/(\alpha + \beta)\}b(p),$$

where $b(p)$ is a regular all-pass. However, we know from Condition 4 of Theorem 4.7 that S_{11} and S_{21} must have the same poles, so that $b(p)$ must be formed with poles that are among the zeros of $N(p)$. Now if we note that the function $A(p)$ used in Theorem 4.7 is in fact given by

$$A(p) = e^{p\Delta} S_{21}(p),$$

then we may appeal to Theorem 4.7 and obtain:

Theorem 4.8. If $S_{21}(p)$ is to be the transfer scattering coefficient of a nonuniform lossless transmission line of length Δ, with a normalization of 1 at both $y = 0$ and $y = \Delta$, then the following conditions are both necessary and sufficient if the local characteristic impedance of the line, $Z_0(y)$, is to be infinitely differentiable:

1. $S_{21}(p) = 2/G(p)$ [where $G(p)$ is an entire function of exponential type Δ] must be analytic and bounded by 1 in $\text{Re}\, p \geq 0$, $S_{21}(0) = 1$, and $S_{21}(p^*) = S_{21}^*(p)$.

2. If $N(p)$ is that factorization of $G(p)G(-p) - 4$ that is nonzero in $\text{Re}\, p > 0$ and of exponential type Δ, then there must exist a regular all-pass, $b(p)$, whose poles are among the zeros of $N(p)$ such that (a)

$$\lim_{|p|\to\infty} \left| \frac{N(p)}{G(p)} b(p) \right| \leq 1 \quad \text{in} \quad \text{Re}\, p > 0$$

and (b)

$$S(p) = \frac{N(p)}{G(p)} b(p), \qquad A(p) = \frac{2e^{\Delta p}}{G(p)}$$

must satisfy Condition 5 of Theorem 4.7.

Proof. The preceding discussion established the direct connection between these statements and Theorem 4.7 with the exception of the condition 2(a), which is certainly necessary, since $S_{11}(p) = \{N(p)/G(p)\}b(p)$ must be bounded by 1 in Re $p \geq 0$. However, if this condition is satisfied, then one may establish directly using Condition 1 that $S(p) = \{N(p)/G(p)\}b(p)$ is uniformly bounded by 1 in Re $p \geq 0$, thereby satisfying Condition 1 of Theorem 4.7. QED

From the previous section we note that $S_{11}(p)$ cannot be uniquely determined from $S_{21}(p)$, although, as shown in Corollary 4.4, $S_{21}(p)$ can be uniquely determined from $S_{11}(p)$. Moreover, the role of the all-pass $b(p)$ in Theorem 4.8 is solely to satisfy certain asymptotic conditions, so that altering a finite number of its poles will not affect the realizability of the resulting $S_{11}(p)$. We may then conclude that $S_{21}(p)$ *does not uniquely determine the characteristic impedance of the line*; in fact, if $S_{21}(p)$ is the transfer scattering coefficient of some line, then there are an unlimited number of other lines with different $Z_0(y)$ each with the same $S_{21}(p)$. There is, however, a certain uniqueness that can be ascribed to $S_{21}(p)$, and it has definite bearing on measurement techniques.

Theorem 4.9. Assume that $|S_{21}(j\omega)|^2 = 1 - |S_{11}(j\omega)|^2$ is known for two different real, positive normalizations at $y = 0$, but with the same known "regular" passive normalization at $y = \Delta$; then the local characteristic impedance is uniquely determined.

Proof. By employing the same techniques as described in Section 4.2.1 it may be shown that

$$S_{21}(p) = \frac{2\sqrt{R}\, h_\Delta(p)}{\alpha(p) + R\beta(p)},$$

where R is the real, positive normalization at $y = 0$ and $h_\Delta(p)$ is the complex normalization factor corresponding to the "regular" passive normalization impedance $Z_\Delta(p)$. But in Corollary 4.4 it was shown that $S_{21}(p)$ may be uniquely determined from $h_\Delta(p)$ and $|S_{21}(j\omega)|$. Thus we obtain

$$\beta(p) = \frac{2h_\Delta[(\sqrt{R_1}/S_{21}^{(1)}) - (\sqrt{R_2}/S_{21}^{(2)})]}{R_1 - R_2}$$

and

$$\alpha(p) = \frac{2\sqrt{R_1}\, h_\Delta}{S_{21}^{(1)}} - R_1\beta(p),$$

where $S_{21}^{(1)}$ and $S_{21}^{(2)}$ correspond to R_1 and R_2, respectively, and, as noted above, they both are determined by their $j\omega$ axis magnitudes and the known

$h_\Delta(p)$. But then we may obtain the input reflection factor (normalized to, say, R_1) of the line terminated in $Z_\Delta(p)$ as

$$S_{11}(p) = \frac{\alpha(p) - R_1\beta(p)}{\alpha(p) + R_1\beta(p)},$$

and from Corollary 4.6 we conclude that the corresponding line defined by $Z_0(y)$ is uniquely determined. QED

As noted above, the asymptotic behavior of $S_{21}(p)$ controls its realizability, and this suggests a wide degree of latitude for the behavior of the function over any finite interval of the real frequency axis. In the following we describe two approximation techniques that depend heavily on the properties of $S_{21}(p)$ and that reflect the freedom associated with this function. The first result is in fact an interpolation procedure.

Lemma 4.6. If $S_{21}(p)$ is the transfer scattering coefficient of a smooth line of electrical length Δ, normalized to 1 at both $y = 0$ and $y = \Delta$, then for almost all ω

$$\frac{1}{S_{21}(j\omega)} = \sum_{n=-\infty}^{\infty} \left[\frac{1}{S_{21}(j\pi n/\Delta)} - (-1)^n(k_1 + k_2)\right] \frac{\sin(\Delta\omega - \pi n)}{\Delta\omega - \pi n}$$
$$+ k_1 e^{j\Delta\omega} + k_2 e^{-j\Delta\omega},$$

where k_1 and k_2 are determined from the asymptotic behavior of $1/S_{21}(p)$ as

$$1/S_{21}(p) = k_1 e^{p\Delta} + k_2 e^{-p\Delta} + e^{p\Delta}O(1/p) + e^{-p\Delta}O(1/p),$$

and the series converges in the mean-squared sense.

Proof. Since $1/S_{21}(p)$ is an entire function of exponential type Δ and it is known that, asymptotically, for a smooth line

$$\alpha(p) + \beta(p) = k_1 e^{p\Delta} + k_2 e^{-p\Delta} + e^{p\Delta}O(1/p) + e^{-p\Delta}O(1/p),$$

then

$$\{1/S_{21}(p)\} - [k_1 e^{p\Delta} + k_2 e^{-p\Delta}] = F(p)$$

is an entire function of exponential type Δ, and, moreover,

$$\int_{-\infty}^{\infty} |\{1/S_{21}(j\omega)\} - (k_1 e^{j\omega\Delta} + k_2 e^{-j\omega\Delta})|^2 \, d\omega < \infty$$

because $|F(j\omega)| = O(1/\omega)$ asymptotically. Appealing to the classic Paley–Wiener theorem leads us to the conclusion that $F(p)$ is the complex Fourier transform of a square integrable function, $f(t)$, that is zero a.e. for $|t| > \Delta$. Representing $f(t)$ by a Fourier series on this interval and taking its transform gives us the stated result, where we must interpret the convergence of the sum in the mean-squared sense, and thus obtain the representation for almost all values of ω. QED

Another approximation technique yields information concerning $S_{21}(p)$ from knowledge of its behavior on a finite interval only. In general, the behavior of an analytic function on any arc of finite length within its domain of analyticity uniquely defines the function throughout its domain of definition. However, there are no general techniques available for performing the analytic continuation from the given behavior on the arc. Fortunately, the properties of $S_{21}(p)$ are such that it is possible in this case.

Lemma 4.7. If $S_{21}(p)$ is the transfer scattering coefficient of a smooth line of some finite length (*which need not be known*), then for all finite p

$$1/S_{21}(p) = \sum_{n=0}^{\infty} k_n P_n(p),$$

where

$$k_n = \int_{\omega_0-a}^{\omega_0+a} \{1/S_{21}(j\omega)\} W(j\omega) P_n(j\omega)\, d\omega$$

and $P_n(p)$ are a set of Jacobi polynomials [*Sz*1] (which include the Legendre and Chebyshev polynomials) arranged to be orthonormal on the interval $\omega_0 - a < \omega < \omega_0 + a$ with respect to the weighting function $W(j\omega)$ determined by the polynomials.

Proof. Since $1/S_{21}(p)$ is an entire function, we may appeal to the theory of orthogonal polynomials to establish the result [e.g., Sz1, p. 243]. QED

4.2.3.3. Applicability of the Theory of Lossless Lines to the Study of RC or LG Lines

With appropriate transformations the theory developed above for lossless lines may be applied directly to the study of *RC* or *LG* lines. Consider, for example, an *RC* transmission line with defining equations

$$\partial V(z, p)/\partial z = -R(z)I(z, p) \qquad \partial I(z, p)/\partial z = -pC(z)V(z, p).$$

Let $I = \hat{I}\sqrt{P}$ and $S = \sqrt{p}$, where a single-valued definition of the square root is used. Then the above equations may be transformed to

$$\partial V/\partial z = -SR(z)\hat{I}, \qquad \partial \hat{I}/\partial z = -SC(z)V,$$

and, in particular,

$$\sqrt{p}(V/I) = V/\hat{I}.$$

Thus the problem of an *RC* line of electrical length

$$\Delta = \int_0^l (RC)^{1/2}\, d\eta$$

terminated in an impedance $Z_\Delta(p)$ may be transformed into the problem of a lossless line of length Δ with $\hat{L}(z) = R(z)$ and $\hat{C}(z) = C(z)$, and terminated in $\hat{Z}_\Delta(S) = SZ_\Delta(S^2)$, where the new transform variable is $S = \sqrt{p}$. The realizability theory and synthesis procedures of Section 4.2.4 may be applied, after the above transformation, if the terminating impedance $Z_\Delta = 1/\sqrt{p}$ (which is the characteristic impedance of an infinite, uniform *RC* line). In essence, our whole discussion is applicable if $\hat{Z}_\Delta(S) = SZ_\Delta(S^2)$ is a regular normalizing impedance in S.

4.2.3.4. An Inverse Scattering Problem in Quantum Mechanics

Many of the techniques and results given above have application to a wide range of physical problems. We will now consider one such application in the area of quantum mechanics. We will consider the one-dimensional time-independent Schrödinger equation describing the interaction of a particle of mass μ and energy E (E a real positive constant) with a material that is described by some potential function $V(x)$:

$$\frac{d^2\Psi(x)}{dx^2} + \frac{2\mu}{(h/2\pi)^2}[E - V(x)]\Psi(x) = 0,$$

where h is Planck's constant. The solution of this equation $\Psi(x)$ is called the wave function of the particle, and if it is normalized so that $\int_{-\infty}^{\infty} |\Psi(x)|\, dx = 1$, then its physical significance is that $\int_{x_1}^{x_2} |\Psi(x)|\, dx$ measures the probability that the particle will be located in the region of space between $x = x_1$ and $x = x_2$. Now let us assume that $V(x) = 0$ for $x < 0$ and for $x > \Delta$, and that on the interval $0 < x < \Delta$ it is some positive function of x. Then for a given E Schrödinger's equation has solutions of the form

$$\Psi(x) = A \exp -j\left\{\frac{2\mu}{(h/2\pi)^2} Ex\right\}^{1/2} + B \exp +j\left\{\frac{2\mu}{(h/2\pi)^2} Ex\right\}^{1/2}$$

in the regions $x < 0$ and $x > \Delta$. We may now impose boundary conditions at $x = \pm\infty$ that have definite physical interpretations. For example, if we set

$$\Psi(x) = \exp -j\left\{\frac{2\mu}{(h/2\pi)^2} Ex\right\}^{1/2} + B \exp +j\left\{\frac{2\mu}{(h/2\pi)^2} Ex\right\}^{1/2} \quad \text{for} \quad x < 0 \tag{4.33a}$$

and

$$\Psi(x) = C \exp -j\left\{\frac{2\mu}{(h/2\pi)^2} Ex\right\}^{1/2} \quad \text{for} \quad x > \Delta, \tag{4.33b}$$

then these solutions correspond to boundary conditions appropriate to a problem in which the particle is traveling toward the potential distribution $V(x)$ from the left, with first term of (4.33a) representing the incident particle, the second term of (4.33a) representing the reflected wave for $x < 0$, and (4.33b) representing the transmitted wave in the region $x > \Delta$, which is assumed to propagate with no further reflections in that region. If we now introduce the variables $p = \sigma + j\omega$,

$$Q(x) = \frac{2\mu}{(h/2\pi)^2} V(x) \quad \text{and} \quad \omega^2 = \frac{2\mu}{(h/2\pi)^2} E,$$

then Schrödinger's equation may be written as

$$\frac{d^2\Psi}{dx^2} - Q(x)\Psi - p^2\Psi = 0;$$

when $\sigma = 0$ we see that the above formulation is equivalent to the original system of equations. In addition, the boundary conditions discussed above may be restated in the form

$$\begin{aligned} &\frac{d\Psi}{dx}(\Delta) + p\Psi(\Delta) = 0, \\ &\frac{d\Psi}{dx}(0) + p\left(\frac{1-B}{1+B}\right)\Psi(0) = 0. \end{aligned} \tag{4.34}$$

Now consider the subsidiary variable Γ defined by the set of equations

$$\frac{d\Psi}{dx} = -p\Gamma, \qquad \frac{d\Gamma}{dx} = -\left[\frac{Q(x)}{p} + p\right]\Psi. \tag{4.35}$$

If we take as boundary conditions for this set of equations

$$\frac{\Psi}{-(d\Psi/dx)/p}(0) = \frac{1+B}{1-B} = Z(p) \tag{4.36a}$$

and

$$\frac{\Psi}{-(d\Psi/dx)/p}(\Delta) = 1, \tag{4.36b}$$

we may show directly that the corresponding solutions $\Psi(x)$ satisfy the basic Schrödinger equation and the boundary conditions of Eq. (4.34). The reason for the introduction of the subsidiary variable Γ is that the resultant set of equations in Ψ and Γ together with the boundary conditions of Eqs. (4.36a, b) are then equivalent to a transmission line with a voltage $V(x, p) = \Psi(x, p)$, a current $I(x, p) = \Gamma(x, p)$, and parameters $\hat{Z}_0 = p$ and $\hat{Y}_0 = \{Q(x)/p\} + p$. Moreover, we see from the boundary conditions that $Z = (1 + B)/(1 - B)$ is the input impedance of such a transmission line when it is terminated in a unit resistor at $x = \Delta$. We will now use the results of Sections 4.2.1. and 4.2.2 to show that if $B(p)$ is known for all $p = j\omega$, i.e., the reflection coefficient is known for all energies $E = (h/2\pi)^2(\omega^2/2\mu)$, then $Q(x)$, or $V(x)$, is uniquely defined and may in fact be found using Marčenko's algorithm of Section 4.2.2.

To this end consider the equivalent integral formulation of the analogous transmission line problem, which is obtained by substitution into Eq. (4.3) of Section 4.1,

$$r(x, p) = p \int_x^\Delta d\eta + 1 + \int_x^\Delta \left[\int_x^\tau d\eta\right][Q(\tau) + p^2]r(\tau, p)\, d\tau,$$

where we have used the identifications $\hat{Z}_0 = p$, $\hat{Y}_0 = (Q/p) + p$, and $Z_l = 1$ as discussed above. Using the arguments of the proof of Theorem 4.1, we may show directly that $r(x, p)$ is an entire function of p for each fixed value of x.† Thus if we define the scattering matrix of the transmission line normalized to 1 at both $x = 0$ and $x = \Delta$, we will obtain a matrix all of whose elements are meromorphic functions, as we see from Eqs. (4.4) and (4.5). Moreover, the matrix will be symmetric ($S_{12} = S_{21}$), as we may show using the technique described in the proof of Theorem 4.3. Further, we find directly from Eqs. (4.35) that for $0 < x < \Delta$

$$\frac{\partial}{\partial x}[\Psi(p)\Gamma(-p) + \Psi(-p)\Gamma(p)] = 0$$

and

$$\frac{\partial}{\partial x}[\Psi(p)\Gamma^*(p) + \Psi^*(p)\Gamma(p)] = -2 \operatorname{Re} p\left\{|\Gamma(p)|^2 + |\Psi(p)|^2\left[1 + \frac{Q(x)}{|p|^2}\right]\right\} \geq 0$$

in $\operatorname{Re} p > 0$, since, by assumption, $V(x)$, and thus $Q(x)$, is positive for $0 < x < \Delta$. We may now use the arguments of the proof of Theorem 4.3 to conclude that the scattering matrix is in fact bounded-real, and since it is meromorphic and unitary for $p = j\omega$, we have additionally that

$$[S(-p)][S(p)] = 1_n.$$

† We note that a direct application of Theorem 4.1 with $\hat{Z}_0 = p$ and $\hat{Y}_0 = (Q/P) + p$ would not show that $r(x, p)$ is an entire function; however, the presence of the factor p in the integral equation cancels the pole of $\hat{Y}_0$ at $p = 0$, so that the arguments of Theorem 4.1 may be applied.

Finally, we may show using the arguments of Section 4.2.1 that

$$S_{21}(p) = 2/(\alpha + \beta),$$

where α and β are entire functions of p determined by the solutions of the integral equations per Eq. (4.5). Therefore $S_{21}(p)$ is nonzero in $\operatorname{Re} p > 0$ and asymptotically $S_{21}(p) \sim e^{-p\Delta}$ as $|p| \to \infty$ in $\operatorname{Re} p > 0$. We may then use Corollary 4.4 to conclude that $S_{12}(p) = S_{21}(p)$ are determined uniquely from $|S_{11}(j\omega)|$, and so $S_{22}(p) = -S_{11}(-p)S_{12}(p)/S_{12}(-p)$ is thus also determined. But with the boundary condition (4.36b) we see that the variables **a** and **b** related by the scattering matrix defined by normalizing to 1 at both $x = 0$ and $x = \Delta$ will be such that $a_2 = 0$ and

$$\frac{b_1}{a_1} = S_{11}(p) = \frac{\Psi(0) - \Gamma(0)}{\Psi(0) + \Gamma(0)} = \frac{Z(p) - 1}{Z(p) + 1} = B(p).$$

Thus all the elements of the scattering matrix are uniquely determined by $B(p)$ and, in particular, as we saw in Section 4.2.2, when

$$\frac{\Psi}{-(d\Psi/dx)/p}(\Delta) = pK$$

we have

$$Z_{\text{in}}(j\omega) = \frac{\Psi(0) - \Gamma(0)}{\Psi(0) + \Gamma(0)} = \frac{j\omega K[\alpha(j\omega) + \alpha(-j\omega)] + [\alpha(j\omega) - \alpha(-j\omega)]}{j\omega K[\beta(j\omega) - \beta(-j\omega)] + [\beta(j\omega) + \beta(-j\omega)]},$$

where $\alpha(j\omega) = \{1 + S_{11}(j\omega)\}/S_{21}(j\omega)$ and $\beta(j\omega) = \{1 - S_{11}(j\omega)\}/S_{21}(j\omega)$. We may then find from the two sets of poles of $Z_{\text{in}}(j\omega)$, corresponding to selecting $K_1 = \infty$ and K_2 as some finite, real, positive constant, two spectra of the Sturm–Liouville operator defined by $Q(x)$. By employing Marčenko's algorithm (Section 4.2.2), we then find the unique $Q(x)$ and thus finally $V(x)$. We summarize our conclusions as follows: let the reflection factor $B(E)$ (a function of the energy of the incident particles) of a positive potential distribution $V(x)$ defined over $0 < x < \Delta$ be known for all positive values of E; then the potential function $V(x)$ is uniquely determined and may be found from $B(E)$ and Δ in an analytic fashion.

4.3. UNIFORM TRANSMISSION LINE NETWORKS

This discussion of nonuniform transmission lines presented in the previous section indicates the mathematical difficulties one might expect to encounter in the study of distributed networks. A good deal of the additional complexity is associated with the corresponding irrational behavior of the terminal description of such systems. Thus in lumped systems we could deal with rational, and thus finite-dimensional, functions, whereas in distributed systems one must consider irrational or transcendental functions. There are, however, certain distributed systems that can be studied using transformations that render the resultant system descriptions rational. In this section we will discuss one such class of systems that are also of considerable engineering interest, namely, those networks composed of the usual lumped elements of Chapter III and a finite number of uniform lossless transmission lines which by definition have local characteristic impedances that are not functions of position along the line, and are such that they all have the same electrical length, $\Delta = \int_0^l \{L(\eta)C(\eta)\}^{1/2}\, d\eta$.

If we define the scattering matrix of a uniform lossless transmission line of electrical length Δ whose characteristic impedance is a positive constant Z_0, then by selecting Z_0 as the normalization at both ports we obtain $S_{11} = S_{22} = 0$ and $S_{12} = S_{21} = e^{-p\Delta}$. We may now transform this representation using the variable $q = e^{-\Delta p}$ to obtain a rational scattering matrix in the q plane:

$$[S(q)] = \begin{bmatrix} 0 & q \\ q & 0 \end{bmatrix}.$$

Moreover, since $|q| < 1$ for $\operatorname{Re} p > 0$, we see that the transformation maps the half plane $\operatorname{Re} p > 0$ into the unit circle in the q plane, although not in a one-to-another manner. Specifically, the bounded-real property in the p plane is transformed into a bounded-real property in a unit circle, i.e., if $[S(p)]$ is bounded-real, then $[S(q)]$ is analytic and $1_n - [S(q)]^{*T}[S(q)]$ is nonnegative-definite in $|q| < 1$. We will utilize this transformation and its corresponding rational properties in the study of the uniform transmission line–lumped element networks. In particular, to focus attention on this transformation, we will adopt the approach described by Youla [Yo6] and consider the given network rearranged in a manner similar to that of Fig. 3.1, where we group all the transmission lines together and form a new network N that is composed solely of the lumped elements. Thus if the original network had n ports, we created a new network N having $n + 2m$ ports assuming that there are only m transmission lines in the original network. When the transmission lines are placed across the $2m$ ports the resultant system is then the original n-port. Now we know from Chapter III that since the network N is composed solely of lumped elements, its scattering matrix, assuming the normalizations used are constants, will be a rational bounded-real matrix. In particular, we will use 1 as the normalizing function at ports 1 through n (the original ports) and Z_{0j} at ports $n + 2j$ and $n + 2j - 1$, where Z_{0j} is the characteristic impedance of the transmission line connected across these two ports. The scattering matrix $[S_N]$ so defined then relates the parameters

$\mathbf{b}_N$ and $\mathbf{a}_N$, which are $(n + 2m)$-vectors, according to

$$\mathbf{b}_N = [S_N]\mathbf{a}_N.$$

But if $\mathbf{b}$ and $\mathbf{a}$ are the variables associated with the original n ports, we may write

$$\mathbf{a}_N = [\mathbf{a}/\hat{\mathbf{a}}], \qquad \mathbf{b}_N = [\mathbf{b}/\hat{\mathbf{b}}],$$

where $\hat{\mathbf{a}}$ and $\hat{\mathbf{b}}$ consist of the remaining $2m$ variables. If we partition the scattering matrix $[S_N]$, we may then write

$$\mathbf{b} = [S_{11}]\mathbf{a} + [S_{12}]\hat{\mathbf{a}} \qquad \text{and} \qquad \hat{\mathbf{b}} = [S_{21}]\mathbf{a} + [S_{22}]\hat{\mathbf{a}},$$

where

$$[S_N] = \begin{bmatrix} [S_{11}] & [S_{12}] \\ [S_{21}] & [S_{22}] \end{bmatrix}$$

and $[S_{jk}]$ are its partitioned submatrices. However, the pairs of variables in $\hat{\mathbf{a}}$ and $\hat{\mathbf{b}}$, corresponding to the pairs of terminals across which the individual transmission lines are placed, are related by the scattering matrices of the individual transmission lines. We noted that in the new variable q these scattering matrices are of the form

$$\begin{bmatrix} 0 & q \\ q & 0 \end{bmatrix},$$

so that we obtain

$$\hat{\mathbf{a}} = [Q]\hat{\mathbf{b}},$$

where

$$[Q] = \begin{bmatrix} 0 & q & 0 & 0 & & & \\ q & 0 & 0 & 0 & & & \\ 0 & 0 & 0 & q & & & \\ 0 & 0 & q & 0 & & & \\ & & & & \ddots & & \\ & & & & & 0 & q \\ & & & & & q & 0 \end{bmatrix},$$

since the normalization selected at these pairs of terminals was such that each transmission line was characterized by the same scattering matrix, given of course the fact that they all have the same electrical length $\Delta = \int_0^l (LC)^{1/2}\, d\eta$. Now we may find the scattering matrix representing the original n-port by first obtaining

$$\hat{\mathbf{b}} = [S_{21}]\mathbf{a} + [S_{22}][Q]\hat{\mathbf{b}},$$

or

$$\hat{\mathbf{b}} = (1_{2m} - [S_{22}][Q])^{-1}[S_{21}]\mathbf{a},$$

and then substituting to find

$$\mathbf{b} = \{[S_{11}] + [S_{12}][Q](1_{2m} - [S_{22}][Q])^{-1}[S_{21}]\}\mathbf{a} = [S_n]\mathbf{a}$$

and identifying

$$[S_n] = [S_{11}] + [S_{12}][Q](1_{2m} - [S_{22}][Q])^{-1}[S_{21}]$$

as the resulting scattering matrix. We may justify the various algebraic manipulations by noting that

$$\mathbf{y}^{*T}\mathbf{y} - \mathbf{y}^{*T}[Q(q)]^{*T}[Q(q)]\mathbf{y} > 0$$

in $|q| < 1$, and, moreover, since $[S_N]$ is the scattering matrix of a passive lumped network, its partition results in an $[S_{22}]$ that is a bounded-real matrix in p; using arguments similar to those of the proof of Theorem 2.16, we then show that

$$\det(1_{2m} - [S_{22}][Q]) \neq 0$$

when $\operatorname{Re} p > 0$ and $|q| < 1$. But since each element in $[S_{jk}]$ is analytic in $\operatorname{Re} p > 0$, we see that $[S_n]$ is a matrix of rational functions in the two variables p and q and, moreover, it is analytic in each separately when $\operatorname{Re} p > 0$ and $|q| < 1$. Now let us determine what additional limitations are imposed by passivity. First we observe that the network N constructed by removing the transmission lines is passive, so that for $\operatorname{Re} p > 0$

$$\mathbf{a}^{*T}\mathbf{a} - \mathbf{b}^{*T}\mathbf{b} + \hat{\mathbf{a}}^{*T}\hat{\mathbf{a}} - \hat{\mathbf{b}}^{*T}\hat{\mathbf{b}} \equiv \mathbf{a}_N^{*T}\mathbf{a}_N - \mathbf{b}_N^{*T}\mathbf{b}_N \geq 0,$$

or

$$\mathbf{a}^{*T}\mathbf{a} - \mathbf{b}^{*T}\mathbf{b} \geq -[\hat{\mathbf{a}}^{*T}\hat{\mathbf{a}} - \hat{\mathbf{b}}^{*T}\hat{\mathbf{b}}].$$

But if we consider q as an independent variable, we see directly from the expression $\hat{\mathbf{a}} = [Q(q)]\hat{\mathbf{b}}$, where $[Q(q)]$ is as indicated above, that

$$-[\hat{\mathbf{a}}^{*T}\hat{\mathbf{a}} - \hat{\mathbf{b}}^{*T}\hat{\mathbf{b}}] = (1 - |q|^2)\hat{\mathbf{b}}^{*T}\mathbf{b} \geq 0$$

for all $|q| < 1$. Thus with $\mathbf{b} = [S_n(p, q)]\mathbf{a}$ we see that

$$\mathbf{a}^{*T}\mathbf{a} - \mathbf{b}^{*T}\mathbf{b} = \mathbf{a}^{*T}(1_n - [S_n(p, q)]^{*T}[S_n(p, q)])\mathbf{a} \geq 0$$

whenever p and q satisfy $\operatorname{Re} p > 0$ and $|q| < 1$. Thus even though $q = e^{-\Delta p}$ or q is a function of p we see that the scattering matrix of a network of lumped elements and equal-length uniform transmission lines may be considered to be a function of two independent variables, and, in particular, the matrix $1_n - [S_n]^{*T}[S_n]$ is nonnegative-definite for any two unrelated values of p and q satisfying $\operatorname{Re} p > 0$ and $|q| < 1$; moreover, this matrix is rational in these two variables. We will refer to such a matrix as a two-variable.

rational, bounded-real matrix. In the special case when the lumped network N is lossless, i.e., the original network contained no resistors, then

$$\mathbf{a}_N^{*T}\mathbf{a}_N - \mathbf{b}_N^{*T}\mathbf{b}_N = 0$$

for $\operatorname{Re} p = 0$, so that we have

$$[S_n(p, q)]^{*T}[S_n(p, q)] = 1_n$$

for $\operatorname{Re} p = 0$ and $|q| = 1$. But if we write $q = e^{-\Delta p}$, then

$$[S_n(p)]^{*T}[S_n(p)] = 1_n$$

for $\operatorname{Re} p = 0$, and since $[S_n(p)]$ will have elements that are meromorphic functions of p, we conclude that

$$[S_n(-p)]^T[S_n(p)] = 1_n$$

for all p. However, when $p_0 = -p$, $q_0 = e^{-\Delta p_0} = e^{\Delta p} = 1/q$, so that the lossless condition corresponds to the two-variable statement

$$[S_n(-p, 1/q)][S_n(p, q)] = 1_n$$

for all p and q. Finally, we may show directly that if no gyrators are present in the network, so that $[S_N]^T = [S_N]$, then

$$[S_n]^T = [S_n].$$

We summarize all these statements by:

Theorem 4.10. If an n-port network contains a finite number of lumped passive elements and a finite number of uniform lossless transmission lines all of the same electrical length Δ, then the scattering matrix $[S(p, q)]$ of such a network exists and is a two-variable, rational, bounded-real matrix in p and $q = e^{-p\Delta}$. If the network contains no resistors, then

$$[S(-p, 1/q)][S(p, q)] = 1_n$$

for all p and q, while in the case where no gyrators are present $[S]^T = [S]$. Finally, in the case where neither inductors nor capacitors are present

$$[S(p, q)] = [S(q)].$$

We might note that if we considered a network composed of lumped passive elements and a finite number of uniform lossless transmission lines of differing lengths—say, $\Delta_1, \ldots, \Delta_n$—then by introducing the n variables $q_i = \exp(-p\Delta_i)$ and repeating the arguments leading to Theorem 4.10 we would obtain the conclusion that the scattering matrix of such a network is a rational bounded-real matrix in the $n+1$ variables $q_1, \ldots, q_n$, and p. It was recently shown by Koga [Ko1] that the converse of this statement is true—namely, if $[S(p, q_1, \ldots, q_n)]$ is an $(n+1)$-variable bounded-real matrix, then there exists a network composed of lumped elements and a finite number of uniform lossless transmission lines of lengths $\Delta_1, \ldots, \Delta_n$ whose scattering matrix is the given one. The question of finding a minimal network, in the sense that the smallest number of transmission lines is needed in the realization, is still under consideration, although some progress has been made (see, for example [Yo6]).

We will now consider the simpler situation in which the network contains no lumped inductors, capacitors, or gyrators. The first systematic study of such networks was made by Richards [Ri1], who introduced the rationalizing transformation $q = e^{-\Delta p}$ and then obtained various properties of the networks. Consider first a network in the form of a cascade of uniform transmission lines terminated in some positive-real impedance Z_L. Now, the impedance seen looking into the last transmission line toward the impedance Z_L may be directly determined as

$$Z_{\text{in}} = Z_{01}\left[\frac{1 + e^{-2\Delta p}S_L}{1 - e^{-2\Delta p}S_L}\right],$$

where Z_{01} is the characteristic impedance of the line and

$$S_L = (Z_L - Z_{01})/(Z_L + Z_{01}).$$

In a similar way we compute the impedance seen looking into the jth line of the cascade toward the terminating impedance Z_L as

$$Z_j = Z_{0j}\left[\frac{1 + e^{-\Delta 2p}S_{Lj-1}}{1 - e^{-2\Delta p}S_{Lj-1}}\right],$$

where $S_{Lj-1} = (Z_{j-1} - Z_{0j})/(Z_{j-1} + Z_{0j})$, with Z_{0j} the characteristic impedance of the jth line and Z_{j-1} the effective impedance terminating it. But since Z_{j-1} is the impedance seen looking into the $(j-1)$th line toward Z_L, we have obtained a recursive relation between Z_j and Z_{j-1} which can then be used to obtain the final impedance looking into the entire cascade of, say, n lines. We may also find a recursive relation between the reflection factors, i.e., the functions S_{Lk}, since $S_{Lj} = (Z_j - Z_{0j+1})/(Z_j + Z_{0j+1})$, and by direct substitution we have

$$S_{Lj} = \frac{Z_{0j}[(1 + e^{-2\Delta p}S_{Lj-1})/(1 - e^{-2\Delta p}S_{Lj-1})] - Z_{0j+1}}{Z_{0j}[(1 + e^{-2\Delta p}S_{Lj-1})/(1 - e^{-2\Delta p}S_{Lj-1})] + Z_{0j+1}}.$$

Now let us assume that we are given $S_{Ln} = S_{\text{in}}$ as the input reflection factor of a cascade of n lines, where Z_{0n+1}, the normalizing function, is given as 1, i.e., we are given

$$S_{\text{in}} = \frac{Z_{0n}[(1 + e^{-2\Delta p}S_{Ln-1})/(1 - e^{-2\Delta p}S_{Ln-1})] - 1}{Z_{0n}[(1 + e^{-2\Delta p}S_{Ln-1})/(1 - e^{-2\Delta p}S_{Ln-1})] + 1},$$

and we would like to determine if the given function S_{in} is the reflection factor of some cascade of lines terminated in a positive-real impedance. First we see by induction that each S_{Lj} is bounded-real, since it is of the form $\{Z(p) - k\}/\{Z(p) + k\}$ where the $Z(p)$ are positive-real functions of p, and k is a positive, real constant. Thus, in particular we have $|S_{Lj}| < 1$ in $\operatorname{Re} p > 0$, so that

$$\lim_{\substack{|p| \to \infty \\ (\operatorname{Re} p > 0)}} S_{\text{in}} = \frac{Z_{0n} - 1}{Z_{0n} + 1}$$

and we are then able to uniquely determine Z_{0n}. Moreover, with Z_{0n} determined we also have, as we find directly,

$$S_{Ln-1} = e^{2\Delta p} \frac{\{(1 + S_{\text{in}})/(1 - S_{\text{in}})\} - Z_{0n}}{\{(1 + S_{\text{in}})/(1 - S_{\text{in}})\} + Z_{0n}},$$

so that S_{Ln-1} may be found uniquely and the cycle repeated until we arrive at the function Z_L from

$$S_{L0} = (Z_L - Z_{01})/(Z_L + Z_{01}).$$

We see that given the lengths of the lines, Δ, Z_L, and S_{in}, the cascade is uniquely determined, and, moreover, we have found an algorithm for determining if a given function $f(p)$ can in fact be realized by such a cascade with a specific Δ and Z_L. We might compare this result with the comparable statement concerning nonuniform transmission lines, i.e., Theorem 4.7.

We will now return to the consideration of those networks composed of an arbitrary interconnection of uniform lossless transmission lines of equal length. In particular, we will consider the one-port problem and state the following result of Richards [Ri1]:

Theorem 4.11. Let $S(p)$ be the input reflection factor, normalized to 1, of an arbitrary (not necessarily a cascade) one-port network composed of transformers, gyrators, and uniform lossless lines of equal length Δ. Then with the transformation $q = e^{-\Delta p}$ (1) $S(q)$ is a real rational function of q analytic in $|q| < 1$, and (2) $S(1/q)S(q) = 1$. Now if in addition $S(q)$ is an even function of q, it is always possible to find a cascade of transmission lines which when terminated in either a short circuit or an open circuit has $S(q)$ as its input reflection factor.

Proof. The necessity of these statements were demonstrated in Theorem 4.10.

Now, as to the sufficiency of the conditions, we will show that a cascade of uniform lines will exist having $S(q)$ as its input reflection factor when the cascade is terminated in either a short or an open circuit. We first note in passing that the unit-circle analog of the rational Blaschke product representation of functions analytic in $\operatorname{Re} p > 0$ and with magnitude 1 when $p = j\omega$ is of the form

$$\pm q^{2m} \sum_{k=1}^{n} \left[\frac{q^2 - z_k}{1 - z_k{}^* q^2} \frac{q^2 - z_k{}^*}{1 - z_k q^2} \right], \tag{4.37}$$

i.e., any even rational function $f(q)$ analytic in $|q| < 1$ satisfying $|f(q)| = 1$ for $|q| = 1$ must be of the form of Eq. (4.37) with all $|z_k| < 1$. Thus the given function $S(q)$ will be of the form of Eq. (4.37). If we use the previously derived expression for the reflection factor S_{Lj} obtained at the input of a uniform line of length Δ and characteristic impedance Z_0 terminated in an impedance or reflection factor S_{Lj-1} we find, identifying $S(p)$ with S_{Lj},

$$S_{Ln-1}(q) = \frac{1}{q^2} \frac{(\{1 + S(q)\}/\{1 - S(q)\}) - Z_{0n}}{(\{1 + S(q)\}/\{1 - S(q)\}) + Z_{0n}},$$

where

$$\frac{Z_{0n} - 1}{Z_{0n} + 1} = S(q)\,|_{q=0} = \lim_{\substack{|p| \to \infty \\ (\operatorname{Re} p > 0)}} S(p),$$

and we note that Z_{0n} will be positive and real, since $|S(q)| < 1$ in $|q| < 1$ and $S(q)$ is a real function of q. We may then rewrite the expression for S_{Ln-1} as

$$S_{Ln-1} = \frac{1}{q^2} \frac{S(q) - S(0)}{1 - S(q)S(0)},$$

and we see that $S_{Ln-1}(q)S_{Ln-1}(1/q) = 1$, since $S(q)S(1/q) = 1$, and S_{Ln-1} is an even real rational function of q. In addition, we see that since $S(q)$ is an even function of q, $S(q) - S(0)$ has a zero of at least multiplicity two at $q = 0$, and, moreover, since $|S(q)| < 1$ in $|q| < 1$, we note that $1 - S(q)S(0)$ is nonzero in $|q| < 1$. Thus $S_{Ln-1}(q)$ is also an analytic function of q in $|q| < 1$. We have thus demonstrated that the S_{Ln-1} determined from $S(q)$ also satisfies the conditions stated in the theorem so that the process may be repeated again to obtain S_{Ln-2}, etc. But due to the cancellation of the factor q^2 in $1 - S(q)S(0)$ and the fact that $\lim_{q \to \infty} \{1 - S(q)S(0)\} = 0\,[S(1/q)S(q) = 1$ for all $q]$, we observe that the degree of both the numerator and denominator polynomials in S_{Ln-1} will be less than those of $S(q)$. Thus each time we repeat the process we will obtain an S_{Lj} that is of reduced degree, and finally we will arrive at a point where $S_{Lj} = \pm 1$, which is the reflection factor of an open circuit and a short circuit, respectively.

The special case where S_{Lj} has zeros at $q = 0$ is handled directly, since $S_{Lj}(0) = 0$, so that $Z_{0j} = Z_{0j+1}$ and we obtain $S_{Lj-1} = S_{Lj}/q^2$. QED

The requirement that $S(q)$ be an even function is a nontrivial restriction. For example, consider a three-port nonreciprocal circulator with scattering matrix

$$\begin{bmatrix} 0 & 0 & 1 \\ 1 & 0 & 0 \\ 0 & 1 & 0 \end{bmatrix}$$

and assume that ports two and three are terminated in a transmission line with scattering matrix

$$\begin{bmatrix} 0 & q \\ q & 0 \end{bmatrix}.$$

We find that

$$b_1/a_1 = q = S(q),$$

i.e., the input reflection factor of the resulting one-port is an odd function of q.

Now we state a theorem which describes the transfer scattering coefficient of a cascade of lines. The proof follows using similar arguments to the ones in Theorem 4.11 (see, for example, [Ca1]).

Theorem 4.12. The necessary and sufficient conditions such that $S(q)$ be the S_{12} coefficient of a cascade of uniform transmission lines of equal length are: (1) $S(q)$ be analytic and bounded by 1 in $|q| < 1$ and (2)

$$S(\lambda)S(-\lambda) = (1 - \lambda^2)^n / P_n(\lambda^2),$$

where P_n is an even polynomial of degree n in λ^2 and $\lambda = (1 - q^2)/(1 + q^2)$.

In summary, we have attempted to show by illustration that there exists a class of distributed networks whose terminal properties may be studied using essentially the theory developed in the study of rational passive networks. These quasi-lumped networks are not only of considerable practical importance, but in a sense they are also approximations to those distributed networks whose spatial variations are smooth, as, for example, the lossless nonuniform transmission line with a smooth characteristic impedance $Z_0(y)$.

4.4. ACTIVITY IN DISTRIBUTED NETWORKS

We saw in Chapter III that active lumped networks could be completely characterized if to the basic lumped passive elements one adds a single active element, the negative resistor. As we indicated in our introduction to distributed networks, the basic distributed building blocks or elements have not been discovered, if in fact they exist at all, so our study of active distributed networks must be similarly incomplete. We will, however, indicate how one might obtain distributed activity in a specific physical situation, so that our discussion will then be of an illustrative nature.

We have already seen in Section 4.1 that in certain cases the electromagnetic fields inside a waveguide containing some known material may be decomposed into modes, or eigenfunctions, and, moreover, the behavior of these modes is determined by differential equations of the form

$$\partial V/\partial z = -\hat{Z}_0(z, p)I, \qquad \partial I/\partial y = -\hat{Y}_0(z, p)V,$$

where $\hat{Z}_0$ and $\hat{Y}_0$ reflect the properties of the material—specifically, if the material is passive, then $\hat{Z}_0$ and $\hat{Y}_0$ must be positive-real functions of p for each fixed value of z. Conversely, if they are not positive-real, then on the basis of Theorem 4.2 we see that the material will be active. The best known physical situations in which such activity may be realized, at least in the microwave region, are those in which the fields interact with a stream or beam of charged particles. The corresponding devices that were developed by employing such active materials—in this case the beam of particles—are then utilized as amplifiers or oscillators and find wide practical application. More recently it has been found that similar phenomena are present in solid-state materials—specifically, semiconductors. We will discuss the latter source of activity after we have indicated some general conclusions [Wo7]. Our approach will duplicate some of our previous discussion, but this is necessitated by the fact that a simple modal analysis is not possible in these materials since the electromagnetic fields do not obey the form of Maxwell's equations given in Section 4.1.

Consider a material in which Maxwell's equations are of the form $\nabla \times \mathbf{E}(\mathbf{r}, t) = -\mu_0(\partial \mathbf{H}/\partial t)(\mathbf{r}, t)$ and $\nabla \times \mathbf{H}(\mathbf{r}, t) = \mathbf{J} + \varepsilon\, \partial \mathbf{E}(\mathbf{r}, t)/\partial t$, where $\mathbf{E}$ and $\mathbf{H}$ are the electromagnetic field vectors (measured in a stationary reference frame), which are functions of position ($\mathbf{r}$) and time (t). The explicit dependence of the current $\mathbf{J}$ on the fields will not be fixed at this point, but we have in mind that this current is the result of the interaction between the fields and the "plasma" present in the material. We shall be concerned only with the "linearized" form of these equations obtained by assuming that each variable may be written as $\mathbf{A} = \mathbf{A}_0 + \hat{\mathbf{A}}$, where $\mathbf{A}_0$ is not a function of time while $\hat{\mathbf{A}}$, the small-signal term, is time dependent. Thus, even though $\mathbf{J}$ is a nonlinear function of $\mathbf{E}$ and $\mathbf{H}$, we assume that the small-signal terms are sufficiently small so that $\hat{\mathbf{J}}$ is a linear function of $\hat{\mathbf{E}}$ and $\hat{\mathbf{H}}$, and $\nabla \times \hat{\mathbf{E}} = -\mu_0\, \partial \hat{\mathbf{H}}/\partial t$ and $\nabla \times \hat{\mathbf{H}} = \hat{\mathbf{J}} + \varepsilon\, \partial \hat{\mathbf{E}}/\partial t$.

To study the flow of electromagnetic energy in the material, we obtain, in the usual way, the small-signal energy integral of Maxwell's equation, which states that

$$-\int_{-\infty}^{t}\left[\int_V \nabla\cdot(\hat{\mathbf{E}}\times\hat{\mathbf{H}})\,dv\right]d\eta = \tfrac{1}{2}\int_V[\varepsilon\hat{\mathbf{E}}\cdot\hat{\mathbf{E}}+\mu_0\hat{\mathbf{H}}\cdot\hat{\mathbf{H}}]\,dv + \int_{-\infty}^{t}\left[\int_V \hat{\mathbf{J}}\cdot\hat{\mathbf{E}}\,dv\right]d\eta. \tag{4.38}$$

Now, if we assume that the field variables are real functions of time and position, then the first integral on the right of Eq. (4.38) is the stored energy in the small-signal electric and magnetic fields in the volume V, and the second integral represents the interaction energy between the fields and the plasma in which the current $\hat{\mathbf{J}}$ is flowing. Thus the integral

$$W(t) \doteq -\int_{-\infty}^{t}\left[\int_V \nabla\cdot(\hat{\mathbf{E}}\times\hat{\mathbf{H}})\,dv\right]d\eta$$

represents the net electromagnetic energy flow into the volume. If $W(t)\geq 0$, the electromagnetic fields have carried net energy into the volume, and this energy is either stored or transferred to the plasma. In this case we say that the material (including the plasma) is passive relative to the electromagnetic waves. However, if $W(t)<0$, then the electromagnetic fields have carried net energy out of the volume, and since $\frac{1}{2}\int_V[\varepsilon\,\hat{\mathbf{E}}\cdot\hat{\mathbf{E}}+\mu_0\hat{\mathbf{H}}\cdot\hat{\mathbf{H}}]\,dv\geq 0$, we must have $\int_{-\infty}^{t}[\int_V\hat{\mathbf{J}}\cdot\hat{\mathbf{E}}\,dv]\,d\eta<0$, or this energy was transferred from the plasma to the waves. In this case we say that the material (again including the plasma) is active. Moreover, for such a net energy transfer to occur there must be some physical interaction between the plasma and the waves, which we shall term an electromagnetic gain mechanism. Thus our criterion for the existence of an electromagnetic gain mechanism is that the material be capable of supporting some electromagnetic fields for which the corresponding $W(t)$ is less than zero for some t, i.e., that the material be active.

To facilitate the discussion, we shall consider particular volumes of material—specifically, those that are surrounded by perfectly conducting cylinders of arbitrary cross section, the only restriction being that if the $\mathbf{z}_0$ axis is directed along the cylinders, then cross sections obtained at any value of z must be identical. We assume that the small-signal electromagnetic waves may be represented as

$$\hat{\mathbf{E}}(\mathbf{r},t)=v(z,t)\mathbf{e}(x,y)+\mathbf{z}_0E_z,$$

$$\hat{\mathbf{H}}(\mathbf{r},t)=i(z,t)\mathbf{h}(x,y)+\mathbf{z}_0H_z,$$

where v and i are scalars and $\mathbf{e}$ and $\mathbf{h}$ are vectors that have no $\mathbf{z}_0$ component and are functions of neither z nor t. With this assumption we find that

$$\int_V \nabla \cdot [\hat{\mathbf{E}} \times \hat{\mathbf{H}}]\, dv = \int_V \nabla \cdot (vi\mathbf{e} \times \mathbf{h})\, dv + \int \nabla \cdot [iE_z \mathbf{z}_0 \times \mathbf{h} + vH_z \mathbf{e} \times \mathbf{z}_0]\, dv.$$

But since $\mathbf{z}_0 \times \mathbf{h}$ and $\mathbf{e} \times \mathbf{z}_0$ have no $\mathbf{z}_0$ components and since the boundary conditions imposed by the perfectly conducting cylinders are such that the tangential electric fields vanish at their surfaces, we may use the divergence theorem to conclude that

$$\int_V \nabla \cdot [iE_z \mathbf{z}_0 \times \mathbf{h} + vH_z \mathbf{e} \times \mathbf{z}_0]\, dv = \int [iE_z \mathbf{z}_0 \times \mathbf{h} + vH_z \mathbf{e} \times \mathbf{z}_0] \cdot \mathbf{n}\, ds = 0,$$

where $\mathbf{n}$ is the unit normal to the surfaces of the cylinders. Moreover, since $\mathbf{e} \times \mathbf{h}$ is a vector in the $\mathbf{z}_0$ direction, we have

$$\int_V \nabla \cdot (\hat{\mathbf{E}} \times \hat{\mathbf{H}})\, dv = \int_V \frac{\partial vi}{\partial z} \mathbf{z}_0 \cdot [\mathbf{e} \times \mathbf{h}]\, dv = \left[\int_0^l \frac{\partial vi}{\partial z}\, dz\right] \int_S \mathbf{z}_0 \cdot \mathbf{e} \times \mathbf{h}\, ds,$$

where we have assumed that the volume in question is bounded by $z = 0$ and $z = l$, and S is the cross-sectional area of the material. If we further assume that the various constants may be adjusted so that $\int_S \mathbf{z}_0 \cdot \mathbf{e} \times \mathbf{h}\, ds = 1$, we finally obtain

$$W(t) \equiv -\int_{-\infty}^{t} \int_V \nabla \cdot (\hat{\mathbf{E}} \times \hat{\mathbf{H}})\, dv\, d\eta = -\int_{-\infty}^{t} \int_0^l \frac{\partial vi}{\partial z}\, dz\, d\eta$$
$$= \int_{-\infty}^{t} v(0, \eta)i(0, \eta) + v(l, \eta)[-i(l, \eta)]\, d\eta. \tag{4.39}$$

In view of the expression for $W(t)$ given by Eq. (4.39) we are led to consider an equivalent network in which $v(0, t)$ and $i(0, t)$ are the voltage and current at one of the ports while $v(l, t)$ and $i(l, t)$ are the voltage and current at the other. Further, if we identify $v(0, t)$ and $v(l, t)$ as the independent excitations (which implies that in the original setting the transverse electric fields are specified at $z = 0$ and $z = l$), then the currents become the responses. If we assume that the various constants in Maxwell's equations are not functions of time, then, since the small-signal equations are linear, we obtain a network representation in terms of Laplace transforms that may be written as

$$\begin{bmatrix} I(0, s) \\ -I(l, s) \end{bmatrix} = \begin{bmatrix} Y_{11}(s) & Y_{12}(s) \\ Y_{21}(s) & Y_{22}(s) \end{bmatrix} \begin{bmatrix} V(0, s) \\ V(l, s) \end{bmatrix} = [Y] \begin{bmatrix} V(0, s) \\ V(l, s) \end{bmatrix},$$

where s is the complex variable used to define the Laplace transform, as, for example, $V(0, s) = \int_{-\infty}^{\infty} e^{-st} v(0, t)\, dt$. With the network thus defined we may invoke the extensive theory of passive networks, since $W(t)$ in Eq. (4.39) measures the energy flow into the network, and if $W(t) \geq 0$ for all t and all $v(0, t)$, $v(l, t)$, then the network is passive.

Thus, given that the various assumptions leading to the network formulation are satisfied, we see that the material will be active, and thus a gain mechanism exists, if and only if the $[Y]$ matrix fails to be positive-real. It is this criterion that we shall apply in the subsequent sections.

Further, we shall assume that the Laplace transform of $v(z, t)$ is of the form

$$\mathscr{L}[v(z, t)] = V(z, s) = a \exp\{k_1(s)z\} + b \exp\{k_2(s)z\},$$

where a and b are, in general, functions of s but otherwise independent variables. We assume that $k_1(s)$ and $k_2(s)$ are determined by the material. We note at this point that our approach is fundamentally different from the coupled-mode theory [Lo1], since even though the two components of $V(z, s)$, given by k_1 and k_2, are "modes" or eigenfunctions of the system, the amplitudes a and b are not coupled but are in fact independent variables. Our justification for considering such a model for $V(z, s)$ is that since the "small-signal" Maxwell's equations are linear, the sum of two solutions is also a solution. We shall refer to the present discussion as the "two-wave case," since $V(z, s)$ has these two independent components. Now, given the assumed form of $V(z, s)$, we see that $\partial v(z, t)/\partial z = -\mu_0\, \partial i(z, t)/\partial t$ implies that

$$\mathscr{L}[i(z, t)] = \frac{k_1(s)}{-s\mu_0}\, a \exp\{k_1(s)z\} + \frac{k_2(s)}{-s\mu_0}\, b \exp\{k_2(s)z\}.$$

We may then compute the elements of the $[Y]$ matrix, as, for example

$$Y_{11}(s) = \left.\frac{I(0, s)}{V(0, s)}\right|_{V(l, s)=0},$$

and thus obtain

$$Y_{11}(s) = \frac{1}{s\mu_0} \frac{k_1 \exp(k_2 l) - k_2 \exp(k_1 l)}{\exp(k_1 l) - \exp(k_2 l)},$$

$$Y_{12}(s) = \frac{1}{s\mu_0} \frac{k_2 - k_1}{\exp(k_1 l) - \exp(k_2 l)},$$

$$Y_{21}(s) = \{\exp(k_1 + k_2)l\} Y_{12}(s),$$

and

$$Y_{22}(s) = \frac{1}{s\mu_0} \frac{k_1 \exp(k_1 l) - k_2 \exp(k_2 l)}{\exp(k_1 l) - \exp(k_2 l)}.$$

The first two conditions for the positive-realness of $[Y]$ are thus satisfied if and only if $k_1(s^*) = k_2{}^*(s)$ and $k_2(s^*) = k_2{}^*(s)$, k_1 and k_2 are analytic in $\operatorname{Re} s > 0$, and $\exp(k_1 l) - \exp(k_2 l) \neq 0$ in $\operatorname{Re} s > 0$. To obtain a more concise statement of the third and last requirement, i.e., the nonnegative-definitness of $[Y^*]^T + [Y]$, we first note that various algebraic manipulations lead to

$$\begin{aligned}
M(s) &= \begin{bmatrix} V^*(0, s) \\ V^*(l, s) \end{bmatrix}^T \{[Y^*]^T + [Y]\} \begin{bmatrix} V(0, s) \\ V(l, s) \end{bmatrix} \\
&= V(0, s)I^*(0, s) + V^*(0, s)I(0, s) \\
&\quad + V(l, s)[-I^*(l, s)] + V^*(l, s)[-I(l, s)] \\
&= -\int_0^l \frac{\partial}{\partial z}[V(z, s)I^*(z, s) + V^*(z, s)I(z, s)]\,dz \\
&= aa^*[\{\exp(k_1 + k_1{}^*)l\} - 1]\left(\frac{k_1}{s\mu_0} + \frac{k_1{}^*}{s^*\mu_0}\right) \\
&\quad + bb^*[\{\exp(k_2 + k_2{}^*)l\} - 1]\left(\frac{k_2}{s\mu_0} + \frac{k_2{}^*}{s^*\mu_0}\right) \\
&\quad + a^*b[\{\exp(k_1{}^* + k_2)l\} - 1]\left(\frac{k_2}{s\mu_0} + \frac{k_1{}^*}{s^*\mu_0}\right) \\
&\quad + ab^*[\{\exp(k_1 + k_2{}^*)l\} - 1]\left(\frac{k_1}{s\mu_0} + \frac{k_2{}^*}{s^*\mu_0}\right).
\end{aligned} \tag{4.40}$$

Thus if $[Y]^{*T} + [Y]$ is to be nonnegative-definite, then $M(s) \geq 0$ in $\operatorname{Re} s > 0$ for all complex-valued pairs of $V(0, s)$ and $V(l, s)$. But since $V(0, s) = a + b$ and $V(l, s) = a \exp(k_1 l) + b \exp(k_2 l)$, we may always find a unique pair of a and b for each pair of $V(0, s)$ and $V(l, s)$ given that $\exp(k_1 l) - \exp(k_2 l) \neq 0$. Since the latter condition is necessary if $[Y]$ is to be positive-real, we see that $[Y^*]^T + [Y]$ will be nonnegative-definite if and only if the last expression in Eq. (4.40) is ≥ 0 in $\operatorname{Re} s > 0$ for all complex-valued pairs of a and b. Moreover, since this expression is a Hermitian form, this will be the case if and only if

$$\begin{aligned}
&[\{\exp(k_1 + k_1{}^*)l\} - 1]\left(\frac{k_1}{s\mu_0} + \frac{k_1^*}{s^*\mu_0}\right) \geq 0, \qquad \text{or} \\
&[\operatorname{Re} k_1(s)] \operatorname{Re}\left[\frac{k_1(s)}{s}\right] \geq 0,
\end{aligned} \tag{4.41}$$

$$\begin{aligned}
&[\{\exp(k_2 + k_2{}^*)l\} - 1]\left(\frac{k_2}{s\mu_0} + \frac{k_2{}^*}{s^*\mu_0}\right) \geq 0, \qquad \text{or} \\
&[\operatorname{Re} k_2(s)] \operatorname{Re}\left[\frac{k_2(s)}{s}\right] \geq 0,
\end{aligned} \tag{4.42}$$

and

$$[\{\exp(k_1 + k_1^*)l\} - 1][\{\exp(k_2 + k_2^*)l\} - 1]\left(\frac{k_1}{s\mu_0} + \frac{k_1^*}{s^*\mu_0}\right)\left(\frac{k_2}{s\mu_0} + \frac{k_2^*}{s^*\mu_0}\right)$$
$$- [\{\exp(k_1^* + k_2)l\} - 1][\{\exp(k_1 + k_2^*)l\} - 1]$$
$$\times \left(\frac{k_2}{s\mu_0} + \frac{k_1^*}{s^*\mu_0}\right)\left(\frac{k_1}{s\mu_0} + \frac{k_2^*}{s^*\mu_0}\right) \geq 0 \tag{4.43}$$

for all s in $\operatorname{Re} s > 0$. To summarize the two-wave case, we may state that for the material to be active it is both necessary and sufficient either that one of the inequalities (4.41)–(4.43) not be satisfied or that $k_1(s)$ or $k_2(s)$ fail to be analytic in $\operatorname{Re} s > 0$, or $\exp(k_1 l) - \exp(k_2 l) = 0$ at some point in $\operatorname{Re} s > 0$. Before we close this discussion of the two-wave case it is worth noting that the behavior of k_1 and k_2 when $s = j\omega$ can sometimes be used to show that the material is active. Thus, since the inequalities must hold for all $\operatorname{Re} s > 0$, then if k_1 and k_2 have limits as $\operatorname{Re} \to s0$ $(s \to j\omega)$, the inequalities must still hold. So with $k(\omega) = \alpha(\omega) + j\beta(\omega)$ we note that if there exists some ω for which either

$$\frac{\alpha_1(\omega)\beta_1(\omega)}{\omega} < 0 \quad \text{or} \quad \frac{\alpha_2(\omega)\beta_2(\omega)}{\omega} < 0$$

or

$$-\frac{1}{\omega^2}([\{\exp(k_1 + k_1^*)l\} - 1][\{\exp(k_2 + k_2^*)l\} - 1](k_1 - k_1^*)(k_2 - k_2^*)$$
$$- [\{\exp(k_1^* + k_2)l\} - 1][\{\exp(k_1 + k_2^*)l\} - 1]$$
$$\times (k_2 - k_1^*)(k_1 - k_2^*)) < 0,$$

then the material must be active.

The foregoing analysis may also be applied to the one-wave case, since by properly selecting $V(0, s)$ and $V(l, s)$, it is always possible to force either a or b to be zero and still keep the remaining variable, or wave, arbitrary. Therefore we see that when $V(z, s) = ce^{k(s)z}$, the material will be active if and only if there exists some point in $\operatorname{Re} s > 0$ at which either $[\operatorname{Re} k(s)] \operatorname{Re}[k(s)/s] < 0$ or $k(s)$ is not analytic.

To show the relevance of the preceding discussion to the study of electromagnetic gain mechanisms in solids we shall consider a "warm" plasma model of a semiconductor in which electrons (with a density n) and holes (with a density p) are present. From a Boltzmann theory development one finds that $\mathbf{J} = \mathbf{J}_n + \mathbf{J}_p$, where

$$\mathbf{J}_n + \tau_n \frac{\partial \mathbf{J}_n}{\partial t} = eu_n n\mathbf{E} - u_n \mathbf{J}_n \times \mu_0 \mathbf{H} + eD_n \nabla n,$$

$$\mathbf{J}_p + \tau_p \frac{\partial \mathbf{J}_p}{dt} = eu_p p\mathbf{E} + u_p \mathbf{J}_p \times \mu_0 \mathbf{H} - eD_p \nabla p,$$

and

$$\nabla \cdot \mathbf{J}_n - e(g - rnp) + e\frac{\partial n}{\partial t} = \nabla \cdot \mathbf{J}_p + e(g - rnp) - e\frac{\partial p}{\partial t} = 0,$$

where the u's, D's, and τ's are the mobilities, diffusion constants, and mean free times between collisions with the lattice for each type of particle and g and r are constants of the assumed recombination process between the holes and electrons. We may linearize these equations in the standard way so that the resulting $\hat{\mathbf{J}}$ is a linear function of $\hat{\mathbf{E}}$ and $\hat{\mathbf{H}}$. Now, assume that n_0 and p_0, the static values of n and p, are not functions of position, and, further, that a static electric field is applied along the $\mathbf{z}_0$ axis, causing static currents to flow of the form $J_{n0} = eu_n n_0 E_0$ and $J_{p0} = eu_p p_0 E_0$. If we neglect the static magnetic field caused by these currents, we may show that purely transverse electromagnetic waves are solutions of the small-signal equations. Specifically, we find that solutions exist of the form

$$\hat{\mathbf{E}} = v(z, t)\mathbf{e}, \qquad \hat{\mathbf{H}} = i(z, t)\mathbf{h},$$

where $\mathbf{z}_0 \times \mathbf{e} = \mathbf{h}$, $\partial v(z, t)/\partial z = -\mu_0 \, \partial i(z, t)/\partial t$, and $\nabla \times \mathbf{e} = 0$. We have assumed that the perfectly conducting cylinders surrounding the material admit field solutions satisfying $\nabla \times \mathbf{e} = 0$ with vanishing tangential components at the surfaces of the cylinder, as is the case, for example, whenever there is at least one cylinder within an outer one. Moreover, these waves are such that $V(z, s) = a \exp(k_1 z) + b \exp(k_2 z)$, with a and b arbitrary constants and k_1 and k_2 the two roots of the dispersion equation

$$k^2 + \mu_0\left[\frac{u_n J_{n0}}{1 + s\tau_n} - \frac{u_p J_{p0}}{1 + s\tau_p}\right]k - s\mu_0\left[s\varepsilon + \frac{eu_n n_0}{1 + s\tau_n} + \frac{eu_p p_0}{1 + s\tau_p}\right] = 0 \quad (4.44)$$

Neglecting the terms $s\tau_n$ and $s\tau_p$ in comparison to 1 in order to simplify the subsequent algebra, we find that $k_{1,2}(s^*) = k_{1,2}^*(s)$ and

$$k_{1,2} = -\tfrac{1}{2}\mu_0(u_n J_{n0} - u_p J_{p0}) \\ \pm \{\tfrac{1}{4}\mu_0{}^2(u_n J_{n0} - u_p J_{p0})^2 + s\mu_0(s\varepsilon + eu_n n_0 + eu_p p_0)\}^{1/2},$$

so that k_1 and k_2 are analytic functions of s in $\operatorname{Re} s > 0$, since with $eu_n n_0 + eu_p p_0 > 0$, the term under the square root is nonzero in $\operatorname{Re} s > 0$. In addition, $\exp(k_1 l) - \exp(k_2 l) \neq 0$ in $\operatorname{Re} s > 0$, since this would require that $k_1 - k_2 = j(2\pi n/l)$, or that the term under the square root be real and negative, which again cannot be the case in $\operatorname{Re} s > 0$. Thus the material supporting these two waves would be passive if the three inequalities of Eqs. (4.41)–(4.43) were satisfied. However, the first two inequalities are satisfied if and only if

$$\left[\frac{u_n J_{n0} - u_p J_{p0}}{eu_n n_0 + eu_p p_0}\right]^2 < \frac{1}{\mu_0 \varepsilon}, \tag{4.45}$$

i.e., if the static currents are small enough. Note that if, for example, $n_0 \gg p_0$, this would require that

$$(J_{n0}/en_0)^2 \equiv (v_n)^2 < 1/(\mu_0 \varepsilon),$$

or that the drift velocity of the electrons be less than the phase velocity of the electromagnetic waves in the material when no currents are present. Thus the material will be active if the currents are large enough to violate this inequality; if we restricted ourselves to the single-wave case, this would be the only way to achieve activity. Practically speaking, however, the stated requirement would necessitate currents so large that they would probably destroy the material through thermal effects. Moreover, there is considerable doubt as to the validity of the mathematical model assumed for these currents when they reach this level, so that this condition may be of only theoretical interest. The remaining inequality, Eq. (4.43), may or may not be satisfied, depending on l. For our present discussion, however, let us assume that l is small enough so that

$$\{\exp(k_1 + k_1^*)l\} - 1 \approx (k_1 + k_1^*)l$$

and

$$\{\exp(k_2 + k_2^*)l\} - 1 \approx (k_2 + k_2^*)l.$$

Now, if we let $s \to j\omega$, we obtain from Eq. (4.44) the requirement that

$$\frac{l^2}{-\omega^2}[(k_1{}^2 - k_1^{*2})(k_2{}^2 - k_2^{*2}) - (k_1^{*2} - k_2{}^2)(k_2^{*2} - k_1{}^2)]$$

$$= l^2 \frac{[|k_1|^4 + |k_2|^4 - 2 \operatorname{Re} k_1{}^2 k_2^{*2}]}{-\omega^2} \geq 0.$$

But

$$|k_1|^4 + |k_2|^4 - 2 \operatorname{Re} k_1{}^2 k_2^{*2} \geq |k_1|^4 + |k_2|^4 - 2|k_1|^2|k_2|^2$$
$$= (|k_1|^2 - |k_2|^2)^2 > 0$$

if $|k_1|^2 \neq |k_2|^2$. In fact, unless $k_1{}^2 = k_2{}^2$ we have

$$|k_1|^4 + |k_2|^4 - 2 \operatorname{Re} k_1{}^2 k_2^{*2} > 0$$

and Eq. (4.43) will not be satisfied. Moreover, we see from Eq. (4.44) that $k_1{}^2 \neq k_2{}^2$ as long as $u_n J_{n0} - u_p J_{p0} \neq 0$. Thus in the two-wave case, no matter how small the currents J_{n0} and J_{p0} become, a small length of material will be active, i.e., there exists some combination of wave amplitudes a and b such that the corresponding electromagnetic waves will gain net energy from the plasma.

The physical origin of the gain mechanism at work in the single-wave case is known and, as we see, requires a matching of velocities. To the author's knowledge, however, the gain mechanism that must be present in the two-wave case has not been studied and its physical origin is not evident to him.

The foregoing analysis also provides an example that demonstrates that activity and instability (both temporal and spatial) are distinct effects. To illustrate this, we shall employ a well-known stability criterion that has most recently been discussed by Briggs [Br2]. Since in his formalism one deals with waves having a spatial variation of the form $e^{-j\hat{k}z}$, we shall substitute $k = -j\hat{k}$ into Eq. (4.44) to obtain the following dispersion equation:

$$\hat{k}^2 + j(\mu_0 u_n J_{n0} - \mu_0 u_p J_{p0})\hat{k} + j\omega\mu_0[j\omega\varepsilon + eu_n n_0 + eu_p p_0] = 0,$$

which can be solved for $j\omega$, yielding

$$(j\omega)^2\mu_0 \varepsilon + j\omega\mu_0(eu_n n_0 + eu_p p_0) + \hat{k}^2 + j(\mu_0 u_n J_{n0} - \mu_0 u_p J_{p0})\hat{k} = 0. \quad (4.46)$$

If we view ω as a complex variable, then the simplest conditions governing stability may be stated as follows: If the two ω's obtained from the dispersion equation have positive imaginary parts when $\hat{k}$ is real, then the waves are both temporally (absolutely) and spatially stable. Now, with $s = j\omega$ we note that ω will have a negative imaginary part only if s has a positive real part. Applying the generalized Routh criterion to Eq. (4.46), with $s = j\omega$, we may show that when

$$K = \left[\frac{u_n J_{n0} - u_p J_{p0}}{eu_n n_0 + eu_p p_0}\right]^2 < \frac{1}{\mu_0 \varepsilon}$$

both roots have negative real parts, and consequently the ω's have positive imaginary parts, implying stability. But if $K > 1/(\mu_0 \varepsilon)$, then there is at least one ω with a negative imaginary part, and the stability criterion indicates that there will be at least one unstable wave. Now, from our previous discussion of activity in this material we see that in the one-wave case activity is present only when an instability exists (although it is not clear that this will always be the case for other situations). However, if the currents J_{n0} and J_{p0} are small enough that $0 < K < 1/(\mu_0 \varepsilon)$, the waves are both temporally and spatially stable but with two waves present the material is active. Thus instability criteria are not in general correct indicators of the existence of gain mechanisms.

Once we have shown that a specific volume of material is active, as, for example, either of the two previous situations, then the problem of utilizing this activity to develop amplifiers, oscillators, etc. may be approached in a manner analogous to that described in Section 3.4 for the lumped negative resistor, i.e., we search for a passive system into which we place the material in such a way that the desirable gain characteristics are achieved.

4.5. SUMMARY

Our discussion of distributed networks reflects the fact that in comparison to the case for lumped or rational systems the available theory is sketchy and in many instances unsatisfactory. This state of affairs is a consequence of the additional parameter, namely physical distance, which is introduced in these studies. At best, the network functions are simple extensions of rational functions—namely, meromorphic functions. There are many investigations which indicate that as long as the physical dimensions of the system remain finite and the material parameters remain entire functions the network functions will be meromorphic. However, even in such cases the simple rational theory of passive systems is inadequate, and we must employ many of the generalized results which we developed in Chapters I and II.

The discussion in this chapter is dominated by the simple nonuniform transmission line, and, in particular, the lossless line. The terminal properties of these lines were ascertained, and then by employing complex normalization techniques together with some relatively new work of mathematicians studying the Sturm–Liouville problem, we were able to obtain a complete theory for the realizability and synthesis of the lossless nonuniform line. In some ways this theory is not completely satisfactory, e.g., the inversion or synthesis problem has been transformed to the task of obtaining a solution of an integral equation in an iterative manner. This type of result typifies the dilemma one faces when attempting to find closed form solutions of linear differential equations with variable coefficients. The lack of closed form solutions to these problems will, in the author's opinion, require future investigators to redefine their goals and place their primary attention on efficient approximate synthesis techniques. In this context the analysis presented in this chapter serves mainly to focus attention on those variables which uniquely characterize the transmission line and to indicate an inversion procedure or algorithm with known convergence properties. Fortunately, there are many physical problems, such as quantum scattering and the propagation of electromagnetic waves in nonhomogeneous dielectrics, which can be studied by means of the equivalent transmission line, and thus we can justify the expenditure of additional effort in the study of these problems.

We also discussed the problem of characterizing those distributed systems that are constructed from uniform transmission lines. If these networks contain a finite number of such building blocks, then, by means of a conformal transformation, the problem may be recast as a study of certain rational functions. Fortunately, many of the structural properties of such systems are directly reflected into simple properties of these rational functions. One avenue of approach to the continuous distributed network is by means of these piecewise-constant distributed networks. The obvious virtue of this approach is the tractability of the analysis and the finite-step algorithms that can be developed for the synthesis problem.

The characterization of gain or activity in distributed systems was discussed briefly. In particular, those models which evolve naturally from physical systems that are capable of activity were described. The aim of this preliminary analysis is to formulate the structure of certain active distributed networks whose subsequent study is pertinent to physically realizable situations.

The need for continuing research in distributed networks is great. Not only is our current knowledge severely limited, but the attention of both the engineering and scientific communities are being focused toward this area. The preliminary studies summarized in this chapter generated many other questions, some of which are summarized in Appendix III.

V

Topics in Optimization Theory and their Applications in Network Theory

INTRODUCTION

Although the primary role of analysis in the study of passive systems is to deduce properties of such systems and obtain representations for them, e.g., their characterization via positive-real or bounded-real functions, the techniques that were utilized in the analysis phase have generally been the ones used in the study of the inverse, or synthesis, problem, i.e., the problem in which one is given the terminal behavior of the system and then attempts to find its internal structure. Unfortunately, many of the synthesis procedures that have been developed are extremely difficult to implement, e.g., the procedure described in Chapter IV for the synthesis of nonuniform lossless transmission lines. Moreover, if one attempts to impose additional constraints on the synthesis problem, such as restricting either the form of the synthesized network or the allowable numerical values of the elements to be used in the network, then no general synthesis procedures are available, nor does it seem feasible to develop them. In addition, the synthesis procedures presuppose that the desired terminal performance is specified in some analytic form; this form is generally obtained by some preliminary approximation procedure. The topic of this chapter, optimization theory, is one that offers great promise as an alternative approach to the classic approximation-synthesis problem. Its basic premise is that one should phrase the problem such that we are searching for that optimal system whose terminal behavior best satisfies some performance criteria, i.e., find the system that minimizes, in some appropriate sense, the difference between its actual terminal performance and the desired one. The application of optimization theory to such minimization problems generally yields sets of conditions which must be satisfied by the system if it is optimal. If one finds all possible systems which satisfy these conditions (in general, these conditions are not sufficient to guarantee optimality, so that many solutions may exist), then one may restrict his attention to these candidates for optimality and thus refine the search for the optimal system. Optimization theory also indicates how one might develop algorithms that will enable one to proceed in a numerical manner through successive iterations toward the optimal system. By their very nature these techniques are only feasible if one has access to a high-speed computer that can accomplish a large number of such iterations in a short time. The purpose of this chapter is to describe in a brief fashion some of the background material from the calculus of variations that forms the mathematical basis of optimization theory and then indicate how one may develop some algorithms that will allow direct numerical computation techniques to search for the optimal system. Finally, we will illustrate these approaches through some specific examples.

Before we embark on this program it might be instructive to indicate the formulation of one problem that falls within the framework of our present discussion. Thus let us assume that we have a given voltage source with a specified open circuit voltage $v(t)$ and an output impedance whose impulse response $Z_g(t)$ is also specified. We will also assume that we would like to connect some other impedance across the source so that the energy delivered to its is maximized. If we assume that this unknown impedance has an impulse response $Z_L(t)$, then when it is connected across the voltage source a current $i(t)$ will flow that is given as the solution of the equation

$$v(t) - [Z_L(t) + Z_g(t)] * i(t) = 0, \tag{5.1}$$

with $*$ denoting convolution. The energy delivered to $Z_L(t)$ is expressed as

$$E = \int_{-\infty}^{\infty} \hat{v}(t) i(t)\, dt = \int_{-\infty}^{\infty} [i(t) * Z_L(t)] i(t)\, dt, \tag{5.2}$$

where $\hat{v}(t) = i(t) * Z_L(t)$ is the voltage that will be produced across $Z_L(t)$. We may now phrase our problem as follows: find that $Z_L(t)$ such that the current $i(t)$ obtained from Eq. (5.1) with $v(t)$ and $Z_g(t)$ prescribed maximizes E of Eq. (5.2). We might also want to impose some limitations on the allowable load impedance, as, for example, that it be passive. In this case the maximization of E must be obtained subject to the additional constraint that the Laplace transform of $Z_L(t)$ be a positive real function. In any event, problems of this type are within the scope of the following discussion.

5.1. SOME TOPICS IN OPTIMIZATION THEORY

Let $\mathbf{x}$ denote some n-vector all of whose elements are functions of t, and assume that $\mathbf{x}$ is an element of some Banach space E with some prescribed norm $\|\mathbf{x}\|$.† Assume that we are also given some functional $f(\cdot)$ that assigns a real number to each $\mathbf{x} \in E$, i.e., $f(\mathbf{x}) = a$, where a is real. Consider then the problem of finding a minimum (maximum) of the functional $f(\mathbf{x})$ subject to the constraint that $\varphi(\mathbf{x}) = \varnothing$, where $\varphi(\cdot)$ is some operator on E and $\varnothing$ is the null element of E. To be specific, we will say that $f(\mathbf{x})$ has a minimum at $\mathbf{x}_0$ if $f(\mathbf{x}) \geq f(\mathbf{x}_0)$ for every $\mathbf{x}$ such that $0 < \|\mathbf{x} - \mathbf{x}_0\| < \varepsilon$, i.e., if we can find some nonzero neighborhood of the point $\mathbf{x}_0$ defined by $\|\mathbf{x} - \mathbf{x}_0\| < \varepsilon$ throughout which $f(\mathbf{x}) \geq f(\mathbf{x}_0)$. If we have the constraint $\varphi(\mathbf{x}) = \varnothing$ imposed, then a minimum exists at $\mathbf{x}_0$ subject to this constraint if the inequality holds for all $\mathbf{x}$, in some neighborhood of $\mathbf{x}_0$, that additionally satisfy $\varphi(\mathbf{x}) = \varnothing$. An effective tool in the study of such minima (maxima) is the strong or Frechet derivative of an operator (see, for example, [Li1]), which may be defined in the following way: if $\mathbf{h}$ and $\mathbf{x}$ are elements of E and if in addition there exists a *linear* operator $L(\cdot)$, in general depending on $\mathbf{x}_0$, such that

$$F(\mathbf{x}_0 + \mathbf{h}) - F(\mathbf{x}_0) = L(\mathbf{h}) + \alpha(\mathbf{x}_0, \mathbf{h}),$$

where $\|\alpha(\mathbf{x}_0, \mathbf{h})\|/\|\mathbf{h}\| \to 0$ as $\|h\| \to 0$, then $L(\mathbf{h})$ is called the differential of the operator $F(\cdot)$ at the point $\mathbf{x}_0$ for the increment $\mathbf{h}$, and the linear operator $L(\cdot)$ is its Frechet derivative. We will adopt the notation $dF(\mathbf{x}_0, \mathbf{h}) \equiv F'(\mathbf{h})$ for the differential and $F'(\cdot)$ for the Frechet derivative. Moreover, if it exists, the differential may be calculated as

$$dF(\mathbf{x}_0, \mathbf{h}) = (d/d\eta)F(\mathbf{x}_0 + \eta\mathbf{h})\big|_{\eta=0},$$

where η is an arbitrary one-dimensional variable. We may obtain a more detailed computational formula for this differential by noting that $F'(\cdot)$ is by definition linear, and thus

$$dF(\mathbf{x}_0, \mathbf{h}) \equiv F'(\mathbf{h}) = \sum_{i=1}^{n} F'(\mathbf{h}_i),$$

where $\mathbf{h} = \sum_{i=1}^{n} \mathbf{h}_i$ with $\mathbf{h}_i$ an n-vector all of whose elements are zero except h_i, which is the ith element of $\mathbf{h}$. But we now observe that we may compute each of the differentials in the sum as

$$F'(\mathbf{h}_i) = (d/d\eta)F(\mathbf{x}_0 + \eta\mathbf{h}_i)\big|_{\eta=0} = (d/d\eta)F(x_1, \ldots, x_i + \eta h_i, \ldots, x_n)\big|_{\eta=0}.$$

Thus formally we obtain

$$dF(\mathbf{x}_0, \mathbf{h}) = \sum_{i=1}^{n} F_{x_i}(h_i),$$

† A Banach space is a normed linear space that is complete in the sense of convergence in its norm. For example, the collection of all $\mathbf{x}$ such that $x_j \in L_2$, with the norm $\|\mathbf{x}\| = (\sum_{j=1}^{n} \int x_j^2\, dt)^{1/2}$, is such a space.

where the "partial differentials" $F_{x_i}(h_i)$ are defined by

$$F_{x_i}(h_i) \equiv (d/d\eta)F(x_1, \ldots, x_i + \eta h_i, \ldots, x_n)\big|_{\eta=0} \equiv F'(\mathbf{h}_i)$$

and the linear operators $F_{x_i}(\cdot)$ are referred to as partial derivatives of $F(\cdot)$.

The following theorem of Liusternik [Li1, p. 209] enables us to apply these differentials to the study of the constrained minimization problem:

Theorem 5.1. If a function $f(\mathbf{x})$ (where $\mathbf{x}$ is an n-dimensional element of some Banach space E) has a maximum or a minimum at $\mathbf{x}_0$ subject to the constraint $\varphi(\mathbf{x}) = \varnothing$, and if $f(\cdot)$ and $\varphi(\cdot)$ are differentiable, then there exists a linear functional $l(\cdot)$ such that the augmented functional $g(\mathbf{x}) = f(\mathbf{x}) - l[\varphi(\mathbf{x})]$ satisfies

$$dg(\mathbf{x}_0, \mathbf{h}) = 0$$

for all $\mathbf{h} \in E$.

By a direct extension of this theorem we may consider the situation in which there are m constraints of the form $\varphi^i(\mathbf{x}) = \varnothing$, $i = 1, \ldots, m$, and obtain the requirement that m linear functionals $l^i(\cdot)$ must exist such that $dg(\mathbf{x}_0, \mathbf{h}) = 0$, where now $g(\mathbf{x}) = f(\mathbf{x}) - \sum_{i=1}^{m} l^i[\varphi^i(\mathbf{x})]$. Moreover, by employing the partial differentials discussed previously we see that

$$dg(\mathbf{x}_0, \mathbf{h}) \equiv \sum_{j=1}^{n} \left\{ f_{x_j}(h_j) - \sum_{i=1}^{m} l^i[\varphi^i_{x_j}(h_j)] \right\} = 0.$$

Here we have employed the functional analog of the chain rule of differentiation and the fact that the Frechet derivative of a linear operator is the same operator [with $l(\cdot)$ linear we see that

$$dl(x, h) = (d/d\eta)[l(\mathbf{x} + \eta\mathbf{h})]\big|_{\eta=0} = (d/d\eta)[l(\mathbf{x}) + \eta l(\mathbf{h})]\big|_{\eta=0} = l(\mathbf{h}),$$

so that $l'(\cdot) = l(\cdot)$]. Finally, since $dg(\mathbf{x}_0, \mathbf{h})$ must vanish for all $\mathbf{h} \in E$, then each term in the sum over n must vanish identically, and we have thus established:

Corollary 5.1. If in Theorem 5.1 there are m constraints of the form $\varphi^i(\mathbf{x}) = \varnothing$, $i = 1, \ldots, m$, then it is necessary that m linear functionals $l^i(\cdot)$ exist such that at $\mathbf{x} = \mathbf{x}_0$

$$f_{x_j}(h_j) - \sum_{i=1}^{m} l^i[\varphi^i_{x_j}(h_j)] = 0, \qquad j = 1, \ldots, n,$$

for all h_j contained in the space of x_j.

If we return for a moment to the example described in the introduction to this section, then by identifying x_1 with $i(t)$ and x_2 with $Z_L(t)$ we see that the problem posed was the maximization of

$$f(x_1, x_2) = \int [i(t) * Z_L(t)]i(t)\,dt$$

subject to the constraint

$$\varphi(x_1, x_2) = v(t) - (Z_L + Z_g) * i = 0.$$

Applying Theorem 5.1 and its corollary, we obtain the following necessary conditions: if $E = \int (i * Z_L)i\,dt$ has a maximum at some point $i(t)$, $Z_L(t)$, then there must exist a linear functional $l(\cdot)$ such that

$$\int h_1(t)[i(t) * Z_L(t)] + i(t)[h_1(t) * Z_L(t)]\,dt - l[(Z_L + Z_g) * h_1] = 0 \quad (5.3)$$

and

$$\int [i(t) * h_2(t)]i(t)\,dt - l[h_2 * i] = 0 \quad (5.4)$$

for all h_1 contained in the space of $i(t)$ and all h_2 contained in the space of $Z_L(t)$. If we combine these two equations with the constraint equation

$$v(t) - (Z_L + Z_g) * i = 0,$$

we see that we have three equations with which we must find the linear functional $l(\cdot)$ and the two variables $i(t)$ and $Z_L(t)$. In general, such sets of equal tions do not possess unique solutions, reflecting the fact that the definition of maximum or minimum we have adopted is a local one, i.e., $f(\mathbf{x}) \geq f(\mathbf{x}_0)$ in some neighborhood of $\mathbf{x}_0$, and it is very possible that many such local maxima or minima exist. Moreover, the conditions of Theorem 5.1 are not sufficient to guarantee that a maximum or minimum exists. Thus if we pursue the approach indicated in the solution of such necessary conditions, we will in effect be determining sets of functions [in the example above these are the pairs $i(t)$, $Z_L(t)$] that are candidates for optimality, and if we are successful in finding all the candidates then we must look over this set to determine the desired optimal. Thus this approach is only realistic if the necessary conditions are restrictive enough so that they possess only a relatively few number of solutions. In general, however, one never knows *a priori* if this will be the case, and so there is considerable interest in developing alternative ways of searching for the optimal solution. One such way is the iterative approach, in which one starts from some solution of $\varphi(\mathbf{x}_0) = \varnothing$ having some specific value of $f(\mathbf{x}_0)$ and attempts to change $\mathbf{x}$ in a way that will increase the value of $f(\mathbf{x})$, assuming we are looking for a maximum. If a systematic procedure is developed which will enable one to determine how to vary $\mathbf{x}$ so that $f(\mathbf{x})$ is always increased after each successive iteration, then one could proceed computationally to search for the maximum at least with the assurance that even though one does not attain the true maxima, the performance after n iterations, as measured by $f(\mathbf{x})$, is better than at the starting point $f(\mathbf{x}_0)$.

One method of generating iterative methods to search for optimal systems is based on the Taylor series representation of a function in the neighborhood of a point, and is generally referred to as a gradient procedure. In basic concept the procedure may be described as follows. Assume that $P(\mathbf{x})$ is a functional on $\mathbf{x}$ so that $P(\mathbf{x}_0 + \alpha\mathbf{h}) = f(\alpha)$ for a fixed $\mathbf{x}_0$ and $\mathbf{h}$, where $f(\cdot)$ is a function of the one-dimensional variable α. Assuming the function $f(\cdot)$ is differentiable at $\alpha = 0$, we then employ the Taylor series

$$f(\alpha) = f(0) + \alpha f'(0) + \frac{\alpha^2 f''(0)}{2!} + \cdots$$

to obtain the approximation representation

$$f(\alpha) - f(0) \approx \alpha f'(0),$$

which is only valid for α sufficiently close to zero. Now noting the fact that

$$f'(0) = (d/d\alpha)P(\mathbf{x}_0 + \alpha\mathbf{h})\big|_{\alpha=0} = P'(\mathbf{h}),$$

where $P'(\mathbf{h})$ is the Frechet differential evaluated for the given $\mathbf{x}_0$, we see that for α sufficiently small

$$P(\mathbf{x}_0 + \alpha\mathbf{h}) - P(\mathbf{x}_0) \approx \alpha P'(\mathbf{h})$$

independent of $\mathbf{h}$. Let us assume further that we could find at least one $\mathbf{h}$ such that $P'(\mathbf{h}) > 0$ [since $P(\cdot)$ was a functional, $P'(\cdot)$ is also a functional, so that for any $\mathbf{h}$, $P'(\mathbf{h})$ is some real number]. Then we see that by selecting $\alpha > 0$

$$P(\mathbf{x}_0 + \alpha\mathbf{h}) - P(\mathbf{x}_0) \approx \alpha P'(\mathbf{h}) > 0,$$

and we would have succeeded in finding some new $\mathbf{x}$, specifically $\mathbf{x} = \mathbf{x}_0 + \alpha\mathbf{h}$, such that $P(\mathbf{x})$ is greater than $P(\mathbf{x}_0)$. We note that the justification for this statement involves keeping α sufficiently close to zero so that the new $\mathbf{x}$ will be close to $\mathbf{x}_0$. Thus we will only be able to search for improvements in $P(\mathbf{x}_0)$ in some neighborhood of $\mathbf{x}_0$, and if we want to search some larger region, we will have to reapply the above procedure in an iterative fashion. In any event, we may apply the technique to the class of problems under consideration in a straightforward manner. If we assume that the problem is to maximize with respect to $\mathbf{u}$ the functional $f(\mathbf{x}, \mathbf{u})$, where $\mathbf{x}$ is uniquely determined from $\mathbf{u}$ via an operator equation $\varphi(\mathbf{x}, \mathbf{u}) = \varnothing$, then we will determine explicit expressions for the required increments. Thus with $\mathbf{x}_0$ one solution of $\varphi(\mathbf{x}, \mathbf{u}) = \varnothing$

corresponding to $\mathbf{u}_0$, and $\mathbf{x}_0 + \Delta\mathbf{x}$ the solution corresponding to the $\mathbf{u}_0 + \alpha\mathbf{h}$, we must obtain an approximate expression for

$$\Delta f = f(\mathbf{x}_0 + \Delta\mathbf{x}, \mathbf{u}_0 + \alpha\mathbf{h}) - f(\mathbf{x}_0, \mathbf{u}_0).$$

For this purpose we will employ the generalized Taylor series representation of an operator (see [Li1]), in particular, the "linear" terms. Thus in such series it may be shown that if $\|h_1\|$ and $\|h_2\|$ are sufficiently close to zero,

$$k(z_1 + h_1, x_2 + h_2) - k(x_1, x_2) \approx k_{x_1}(h_1) + k_{x_2}(h_2),$$

where $k(x_1, x_2)$ is the given operator and $k_{x_1}(\cdot)$ and $k_{x_2}(\cdot)$ are its corresponding partial derivatives. Employing such an approximation, we see that

$$\Delta f \approx f_{\mathbf{x}}(\Delta\mathbf{x}) + f_{\mathbf{u}}(\alpha\mathbf{h})$$

if α and $\|\Delta\mathbf{x}\|$ are sufficiently small. However, since $\Delta\mathbf{x}$ is determined from $\alpha\mathbf{h}$ by the operator $\varphi(x_0 + \Delta\mathbf{x}, \mathbf{u} + \alpha\mathbf{h}) = \varnothing$, we also have

$$\varphi_{\mathbf{x}}(\Delta\mathbf{x}) + \varphi_{\mathbf{u}}(\alpha\mathbf{h}) = \varnothing,$$

and if the linear operator $\varphi_{\mathbf{x}}(\cdot)$ is invertible, we obtain

$$\Delta\mathbf{x} = -\varphi_{\mathbf{x}}^{-1}[\varphi_{\mathbf{u}}(\alpha\mathbf{h})].$$

Note that if $\varphi_{\mathbf{x}}^{-1}$ exists it also will be linear, so that $\Delta\mathbf{x} = -\alpha\varphi_{\mathbf{x}}^{-1}[\varphi_{\mathbf{u}}(\mathbf{h})]$. Combining these expressions, we find

$$\Delta f = -f_{\mathbf{x}}\{\varphi_{\mathbf{x}}^{-1}[\varphi_{\mathbf{u}}(\alpha\mathbf{h})]\} + f_{\mathbf{u}}(\alpha\mathbf{h}) = \alpha\hat{f}(\mathbf{h}),$$

where $\hat{f}(\cdot)$ is the linear functional $-f_{\mathbf{x}}\varphi_{\mathbf{x}}^{-1}\varphi_{\mathbf{u}}(\cdot) + f_{\mathbf{x}}(\cdot)$. We may go one step further by appealing to Theorem 1.1 to argue that the linear functional $\hat{f}(\cdot)$ may always be represented in terms of some fixed kernel $[K]$, and we then have

$$\Delta f = \alpha\int [K(\eta)]\mathbf{h}(\eta)\,d\eta = \alpha\int \sum_{i=1}^{n} K_i(\eta)h_i(\eta)\,d\eta.$$

In general, this representation will involve distributions as the K_i and will be restricted to those h_i that are infinitely differentiable and zero outside some finite interval. However, in those cases where the K_i are square integrable functions the representation is valid for all h_i that are square integrable, and we see that selecting

$$h_i(t) = K_i(t)$$

and $\alpha > 0$ will result in

$$\Delta f = \alpha\sum_{i=1}^{n}\int K_i^2(t)\,dt > 0.$$

Therefore for α sufficiently small we will have achieved an increased value of $f(\mathbf{x}, \mathbf{u})$ if we set $\mathbf{u} = \mathbf{u}_0 + \alpha\mathbf{h}$.

Although the preceding discussion is conceptually satisfying, the practicalities involved with obtaining the inverse operator $\varphi_{\mathbf{x}}^{-1}(\cdot)$ many times rule out this direct implementation of the gradient procedure, and we would like to obtain alternate procedures which do not require such an inversion. We will describe one such alternative, and at the same time will enlarge the scope of the discussion by allowing for more than one operator constraint. In particular, we will assume that $\mathbf{x}$ is an m-dimensional variable, that $\mathbf{u}$ is of dimension q, and that there are m operator constraints of the form $\varphi^i(\mathbf{x}, \mathbf{u}) = \varnothing$, $i = 1, \ldots, m$, which determine $\mathbf{x}$ once $\mathbf{u}$ is prescribed. From Theorem 5.1 and its corollary we know that necessary conditions for the existence of a maximum or minimum of the functional $f(\mathbf{x}, \mathbf{u})$ are that there exist m linear functionals $l^i(\cdot)$ such that

$$f_{x_j}(h_j) - \sum_{i=1}^{m} l^i[\varphi^i_{x_j}(h_j)] = 0, \qquad j = 1, \ldots, m,$$

and

$$f_{u_k}(h_k) - \sum_{i=1}^{m} l^i[\varphi^i_{u_k}(h_k)] = 0, \qquad k = 1, \ldots, q,$$

for all h_j and h_k in the space of x_j and u_k, respectively (that these statements follow may be seen by applying Theorem 5.1 to $\hat{\mathbf{x}}$, where $\hat{x}_k = x$, $k = 1, \ldots, m$ and $\hat{x}_k = u_k$, $k = m + 1, \ldots, m + q$). Again let us assume that for some $\mathbf{u} = \mathbf{u}_0$ we have determined the $\mathbf{x} = \mathbf{x}_0$ which satisfies the m constraint equations and that we then let $\mathbf{u} = \mathbf{u}_0 + \alpha\mathbf{h}$, producing a new $\mathbf{x} = \mathbf{x}_0 + \Delta\mathbf{x}$. Let us further assume that we have found m linear functionals $l^i(\cdot)$ such that at $\mathbf{x}_0$, $\mathbf{u}_0$

$$f_{x_j}(\Delta x_j) - \sum_{i=1}^{m} l^i[\varphi^i_{x_j}(\Delta x_j)] = 0, \qquad j = 1, \ldots, m,$$

for all Δx_j. Now, by employing the Taylor series approximation, we have

$$\Delta f = f(\mathbf{x}_0 + \Delta\mathbf{x}, \mathbf{u}_0 + \alpha\mathbf{h}) - f(\mathbf{x}_0, \mathbf{u}_0) \approx \sum_{j=1}^{m} f_{x_j}(\Delta x_j) + \sum_{k=1}^{q} f_{u_k}(\alpha h_k),$$

or, by adding and subtracting $\sum_{k=1}^{q} \sum_{i=1}^{m} l^i[\varphi^i_{u_k}(\alpha h_k)]$ we see that

$$\Delta f = \sum_{k=1}^{q} \left\{ f_{u_k}(\alpha h_k) - \sum_{i=1}^{m} l^i[\varphi^i_{u_k}(\alpha h_k)] \right\}$$

$$+ \sum_{j=1}^{m} f_{x_j}(\Delta x_j) + \sum_{k=1}^{q} \sum_{i=1}^{m} l^i[\varphi^i_{u_k}(\alpha h_k)].$$

However, we have assumed that the linear functionals $l^i(\cdot)$ have been determined so that

$$f_{x_j}(\Delta x_j) - \sum_{i=1}^{m} l^i[\varphi^i_{x_j}(\Delta x_j)] = 0,$$

and thus

$$\sum_{j=1}^{m} f_{x_j}(\Delta x_j) + \sum_{k=1}^{q} \sum_{i=1}^{m} l^i[\varphi^i_{u_k}(\alpha h_k)] = \sum_{i=1}^{m} l^i\left[\sum_{j=1}^{m} \varphi^i_{x_j}(\Delta x_j) + \sum_{k=1}^{q} \varphi^i_{u_k}(\alpha h_k)\right].$$

Further, since the m operator equations have been used to determine $\mathbf{x}_0 + \Delta\mathbf{x}$ from $\mathbf{u}_0 + \alpha\mathbf{h}$, i.e., $\varphi^i(\mathbf{x}_0 + \Delta\mathbf{x}, \mathbf{u}_0 + \alpha\mathbf{h}) = \varnothing$, we see that

$$\varphi_{\mathbf{x}}{}^i(\Delta\mathbf{x}) + \varphi_{\mathbf{u}}{}^i(\alpha\mathbf{h}) = \sum_{j=1}^{m} \varphi^i_{x_j}(\Delta x_j) + \sum_{k=1}^{q} \varphi^i_{u_k}(\alpha h_k) = \varnothing.$$

Combining these statements, we see that

$$\Delta f \approx \alpha \sum_{k=1}^{q} \left\{ f_{u_k}(h_k) - \sum_{i=1}^{m} l^i[\varphi^i_{u_k}(h_k)] \right\},$$

which may be written as

$$\Delta f = \alpha \int \sum_{k=1}^{q} m_k(\eta) h_k(\eta)\, d\eta,$$

where the $m_k(\eta)$ are the kernels of the representation of the linear functional $f_{u_k}(\cdot) - \sum_{i=1}^{m} l^i[\varphi^i_{u_k}(\cdot)]$. Thus, as before, with $\alpha > 0$ and $h_k(t) = m_k(t)$ we have $\Delta f > 0$. The difference between the present approach and the direct one is the fact that previously it was necessary to obtain the inverse operator $\varphi_{\mathbf{x}}^{-1}(\cdot)$, whereas in the present case one must find the m linear functionals $l^i(\cdot)$.

In the general case the determination of the linear functionals that are required in the above gradient algorithm may pose a difficult problem even though it may be simpler than inverting the operator $\varphi_{\mathbf{x}}(\cdot)$. It is thus worth describing at least one situation in which the procedure can be described explicitly. For this purpose we assume that the constraint equations may be written as

$$\varphi^i(\mathbf{x}, \mathbf{u}) = (dx_i/dt) - f^i(\mathbf{x}, \mathbf{u}, t) = \varnothing$$

subject to $x_j(0)$ being prescribed. Here we assume that $f^i(\cdot)$ is for each fixed t a function of $\mathbf{x}(t)$ and $\mathbf{y}(t)$. These equations may then be written as the set of ordinary differential equations

$$d\mathbf{x}/dt = f(\mathbf{x}, \mathbf{u}, t)$$

with $\mathbf{x}(0)$ prescribed. Let us further assume that the functional to be maximized is of the form

$$f(\mathbf{x}, \mathbf{u}) = g[\mathbf{x}(t_f)],$$

i.e., it is some function of the values of x_j at some fixed $t = t_f$. In this case we see that

$$f_{x_j}(h_j) = \frac{d}{d\eta}\{g[x_1(t_f), \ldots, x_j(t_f) + \eta h_j(t_f), \ldots, x_m(t_f)]\}\big|_{\eta=0}$$
$$= \frac{\partial g}{\partial x_j(t_f)} h_j(t_f);$$

$$f_{u_k}(h_k) = 0;$$

$$\varphi^i_{x_j}(h_j) = \frac{d}{d\eta}\begin{Bmatrix} \dfrac{dx_i}{dt}, & i \neq j \\ \dfrac{d[x_j + \eta h_j]}{dt}, & i = j \end{Bmatrix} - f^i(x_1, \ldots, x_j + \eta h_j, \ldots, x_m)\big|_{\eta=0}$$
$$= \frac{dh_j}{dt}\delta_{ij} - \frac{\partial f^i}{\partial x_j} h_j, \qquad \delta_{ij} = 1, \quad i = j, \quad \text{and zero otherwise};$$

$$\varphi^i_{u_k}(h_k) = -\frac{\partial f^i}{\partial u_k} h_k;$$

thus the necessary conditions for optimality become

$$\frac{\partial g}{\partial x_j(t_f)} h_j(t_f) - \sum_{i=1}^{m} l^i\left[-\frac{\partial f^i}{\partial x_j} h_j + \frac{dh_j}{dt}\delta_{ij}\right] = 0, \qquad j = 1, \ldots, m,$$

for all h_j such that $h_j(0) = 0$, and

$$\sum_{i=1}^{m} l^i\left[\frac{\partial f^i}{\partial u_k} h_k\right] = 0, \qquad k = 1, \ldots, q,$$

for all h_k. Note that the restriction $h_j(0) = 0$ must be made if in the calculation of $\varphi^i_{x_j}(h_j)$ the $x_j + \eta h_j$, are to satisfy the given boundary conditions, i.e., $x_j(0)$ being specified. Now let us assume that the m linear functions $l^i(\cdot)$ have representations as

$$l^i(h) = \int_0^{t_f} \lambda_i(\eta) h(\eta)\, d\eta,$$

where the range of integration is over the range of interest of the functions. Granting this, one then obtains the requirement

$$\frac{\partial g}{\partial x_j(t_f)} h_j(t_f) - \int_0^{t_f} \lambda_j(\eta)\frac{dh_j}{d\eta}\, d\eta + \sum_{i=1}^{m}\int_0^{t_f} \lambda_i(\eta)\frac{\partial f^i}{\partial x_j} h_j\, d\eta = 0, \qquad j = 1, \ldots, m,$$

and

$$\sum_{i=1}^{m}\int_0^{t_f} \lambda_i(\eta)\frac{\partial f^i}{\partial u_k} h_k\, d\eta = 0, \qquad k = 1, \ldots, q.$$

Integrating by parts, we see that

$$\int_0^{t_f} \lambda_j(\eta) \frac{dh_j}{d\eta}\, d\eta = \lambda_j(t_f)h_j(t_f) - \int_0^{t_f} h_j(\eta) \frac{d\lambda_i}{d\eta}\, d\eta,$$

since $h_j(0) = 0$. The first m of the necessary conditions then may be written as

$$\left[\frac{\partial g}{\partial x_j(t_f)} - \lambda_j(t_f)\right] h_j(t_f) + \int_0^{t_f} \left\{\frac{\partial \lambda_j}{\partial \eta} + \sum_{i=1}^{m} \lambda_i(\eta) \frac{\partial f^i}{\partial x_j}\right\} h_j(\eta)\, d\eta = 0.$$

Thus in the gradient procedure described previously we may meet the requirement that the linear functionals cause

$$f_{x_j}(\Delta x_j) - \sum_{i=1}^{m} l^i[\varphi^i_{x_j}(\Delta x_j)] = 0$$

by selecting the λ_i as solutions of the linear differential equations

$$\frac{d\lambda_j}{dt} = -\sum_{i=1}^{m} \lambda_i(t) \frac{\partial f^i}{\partial x_j}, \qquad j = 1, \ldots, m,$$

subject to the boundary conditions $\lambda_j(t_f) = \partial g / \partial x_j(t_f)$. We have thus established the following:

Theorem 5.2. Given: some function $g[\mathbf{x}(t_f)]$ and some q-dimensional $\mathbf{u}_0(t)$. Let $\mathbf{x}_0(t)$ and $\boldsymbol{\lambda}_0(t)$ satisfy the differential equations

$$\frac{dx_{0i}}{dt} = f^i(\mathbf{x}_0, \mathbf{u}_0, t), \qquad i = 1, \ldots, m, \qquad \text{with} \quad x_{0i}(0) \quad \text{prescribed},$$

$$\frac{d\lambda_j}{dt} = -\sum_{i=1}^{m} \lambda_i(t) \frac{\partial f^i}{\partial x_j}\bigg|_{\mathbf{x}=\mathbf{x}_0,\, \mathbf{u}=\mathbf{u}_0}, \qquad \lambda_j(t_f) = \frac{\partial g}{\partial x_j(t_f)}.$$

Then with α sufficiently small

$$g[\mathbf{x}_0(t_f) + \Delta\mathbf{x}(t_f)] - g[\mathbf{x}_0(t_f)] \approx \alpha \sum_{k=1}^{q} \int_0^{t_f} \left\{\sum_{i=1}^{m} \lambda_i \frac{\partial f^i}{\partial u_k}\bigg|_{\mathbf{x}=\mathbf{x}_0,\, \mathbf{u}=\mathbf{u}_0}\right\} h_k\, d\eta,$$

where $\mathbf{x}_0(t_f) + \Delta\mathbf{x}(t_f)$ is the solution corresponding to $\mathbf{u} = \mathbf{u}_0 + \alpha\mathbf{h}$. Moreover, if $\alpha > 0$, then

$$g[\mathbf{x}_0(t_f) + \Delta\mathbf{x}_0(t_f)] \geq g[\mathbf{x}_0(t_f)]$$

when

$$h_k(t) = \sum_{i=1}^{m} \lambda_i \frac{\partial f^i}{\partial u_k}\bigg|_{\mathbf{x}=\mathbf{x}_0,\, \mathbf{u}=\mathbf{u}_0}.$$

A few comments are in order concerning the algorithm which we may develop on the basis of Theorem 5.2, although these comments apply to most iterative techniques that have been developed in optimization theory. First, throughout we have tacitly assumed that the various derivatives exist, and, specifically, that they exist at the maxima or minima of $f(\mathbf{x})$. This need not be the case, as, for example, when the constraints imposed on the problem are of the form $\|\mathbf{x}\| \le C$, as might occur if we imposed *a priori* limitation on either the independent variables $\mathbf{u}$, the dependent variables $\mathbf{x}$, or both. The computational procedure need not be impaired by such limitations, since in many cases we may adjust α so as to keep the variables within the allowable range. However, in those cases where such constraints will not allow the algorithm to proceed one may attempt to solve a simpler but related problem in which we drop the constraints on the variables but consider as the performance function

$$P(\mathbf{x}) = f(\mathbf{x}) - |K|\, g(\mathbf{x}),$$

where $g(\mathbf{x}) > 0$ if $\mathbf{x}$ exceeds the constraints and $g(\mathbf{x}) = 0$ if not. If we obtain a maximum of P for some fixed $|K|$, then we see that this has been achieved in spite of the fact that we have penalized ourselves, through the term $|K|\, g(\mathbf{x})$. By arbitrarily picking $|K|$ large enough, the penalty for using $\mathbf{x}$'s that fall outside the desired range will be so great that we would expect that the $\mathbf{x}$ that maximizes $P(\mathbf{x})$ will fall within the allowable range. This "penalty function" approach will often allow a direct application of the gradient techniques, and is therefore of considerable practical interest.

Another of the limitations inherent in these gradient algorithms may be seen from the expression for the change in the functional $f(\mathbf{x}, \mathbf{u})$:

$$\Delta f \approx \sum_{k=1}^{q} \left\{ f_{u_k}(\alpha h_k) - \sum_{i=1}^{m} l^i[\varphi^i_{u_k}(\alpha h_k)] \right\}.$$

Since this expression is in fact one of the necessary conditions that must be satisfied at maximum or minimum, i.e.,

$$f_{u_k}(\alpha h_k) - \sum_{i=1}^{m} l^i[\varphi^i_{u_k}(\alpha h_k)] = 0$$

for all h_k, then $\Delta f = 0$ at a maximum or minimum. Thus as we approach such a maximum, the gradient procedure inherently "slows down," and in fact never will reach the true maximum. One may consider going to higher terms in the Taylor series approximation near the maximum as one way of avoiding this problem. These and other alternate iterative procedures are described in [Ba1].

Before terminating this section we will state another set of necessary conditions for optimality which can be established for those systems whose constraint operators are ordinary differential equations. From the discussion preceding Theorem 5.2 we see that necessary conditions for the existence of a maximum or minimum are that

$$\int_0^{t_f} \sum_{i=1}^{m} \lambda_i(\eta) \frac{\partial f^i}{\partial u_k} h_k(\eta)\, d\eta = 0$$

for all $h_k(t)$. Assuming that the integrand contains continuous functions of η, this can only be true if

$$\sum_{i=1}^{m} \lambda_i(t)\, \partial f^i / \partial u_k = 0$$

for all t in the interval $0 - t_f$. One might be led to conjecture that the function

$$H(t) = \sum_{i=1}^{m} \lambda_i(t) f^i(\mathbf{x}, \mathbf{u}, t)$$

also must possess a maximum or minimum with respect to $\mathbf{u}$, where the $\lambda_i(t)$ are fixed at the values determined by $\mathbf{x}_0$ and $\mathbf{u}_0$. This may be shown to be the case, and we state this maximum principle of Pontryagin as follows: let $\lambda_i(t)$ be obtained from $\mathbf{x}_0$ and $\mathbf{u}_0$; then if the functional $g[\mathbf{x}(t_f)]$ is maximized (minimized) by $\mathbf{u}_0$, $\mathbf{x}_0$, it is necessary that $H(t)$ considered as a function of $\mathbf{u}$ also be maximized (minimized) for all t in $(0, t_f)$ by $\mathbf{u} = \mathbf{u}_0$. This result is of considerable interest, since it can be established under much weaker conditions than Theorem 5.2; specifically, one need not assume the existence of the various derivatives at the point where the maximum exists. For a proof and an elaboration of this result we refer the reader to [Po1].

5.2. AN EXAMPLE

We will illustrate the use of the gradient algorithm described above by considering the problem of designing a nonuniform lossless transmission line filter. For our current purposes the description of such transmission lines as given in Chapter IV is not convenient, and we will first obtain an alternative description in terms of the ratio of the voltage and current at various points along the line. Thus we define $Z(z, \omega) = -V(z, \omega)/I(z, \omega)$, and then from the differential equations of the line,

$$\partial V(z, \omega)/\partial z = -j\omega L(z) I(z, \omega)$$

and

$$\partial I(z, \omega)/\partial z = -j\omega C(z) V(z, \omega),$$

we obtain

$$\frac{\partial V}{\partial z} = -\left[\frac{\partial Z}{\partial z} I + Z \frac{\partial I}{\partial z}\right] = -\left[\frac{\partial Z}{\partial z} I - j\omega CVZ\right] = -j\omega LI,$$

or

$$-\frac{\partial Z}{\partial z} + j\omega C \frac{V}{I} Z = -j\omega L,$$

and, finally,

$$dZ(z, \omega)/dz = -j\omega C(z)Z^2(z, \omega) + j\omega L(z).$$

If we further define $Z(z, \omega) \equiv R(z, \omega) + jX(z, \omega)$, where R and X are real functions of z and ω, then we obtain

$$dR(z, \omega)/dz = 2\omega C(z)R(z, \omega)X(z, \omega)$$

and

$$dX(z, \omega)/dz = \omega L(z) - \omega C(z)[R^2(z, \omega) - X^2(z, \omega)].$$

If we impose boundary conditions at $z = 0$ of the form $V(0, \omega)/I(0, \omega) = -R_L$, i.e., we place a load resistor R_L at $z = 0$, and excite the resultant system with a source, $E_s(\omega)$, having an output impedance R_g at $z = l$, then we may define an appropriate scattering matrix for the transmission line normalized to R_L at port two ($z = 0$) and R_g at port one ($z = l$) and obtain

$$S_{11}(\omega) = \frac{-[V(l, \omega)/I(l, \omega)] - R_g}{-[V(l, \omega)/I(l, \omega)] + R_g} = \frac{Z(l, \omega) - R_g}{Z(l, \omega) + R_g}.$$

Moreover, since $L(z)$ and $C(z)$ are real and nonnegative, we know from Chapter IV that the transmission line is lossless, so that

$$|S_{11}(j\omega)|^2 + |S_{12}(j\omega)|^2 = 1.$$

Let us now consider the resultant voltage transfer function between the source E_s and the voltage that appears across the load R_L, i.e., $V(0, \omega)$. As we saw in Section 2.4.1, the magnitude of this transfer function is given as

$$\left|\frac{V(0, \omega)}{E_s(\omega)}\right|^2 = \frac{1}{4}\frac{R_L}{R_g}|S_{12}|^2 = \frac{1}{4}\frac{R_L}{R_g}\{1 - |S_{11}(\omega)|^2\}.$$

Thus we see that this transfer function may be found in terms of $|S_{11}(\omega)|^2$, or, since $S_{11}(\omega) = \{Z(l, \omega) - R_g\}/\{Z(l, \omega) + R_g\}$, it may be computed from the values of $Z(l, \omega) = R(l, \omega) + jX(l, \omega)$ which are obtained as solutions of the differential equations in R and X subject to the boundary conditions $R(0, \omega) = R_L$, $X(0, \omega) = 0$. Let us now assume that we wish to obtain some desired variation of $|V(0)/E_s|$ over the range of frequencies from ω_1 to ω_2, and designate this as $T_d(\omega)$. Moreover, assume that we are willing to measure our success in achieving this desired performance in terms of the integral

$$P = \int_{\omega_1}^{\omega_2} [|V(0, \omega)/E_s(\omega)|^2 - T_d(\omega)]^2 \, d\omega,$$

which we will approximate by the finite sum

$$P_N = \frac{1}{N} \sum_{i=1}^{N} [|V(0, \omega_i)/E_s(\omega_i)|^2 - T_d(\omega_i)]^2,$$

with $\omega_i = \omega_1 + (i/N)$. We have now arrived at a problem formulation for the design of the transmission line filter in which we search for those values of $L(z)$ and $C(z)$ defining the transmission line such that P_N [obtained by solving the differential equations in R and X at each ω_i subject to $R(0, \omega_i) = R_L$, $X(0, \omega_i) = 0$] is minimized. Due to the presence of the parameter ω_i we have $2N$ constraint equations in the form of the $2N$ differential equations for $R(z, \omega_i)$ and $X(z, \omega_i)$. Thus, in applying the gradient algorithm, we will require $2N$ linear functionals, or, as we saw in Theorem 5.2, $2N$ variables λ_j. For convenience in notation we will associate the variables λ_R with the differential equation $dR/dz = 2\omega CRX$ and the variables λ_X with the equation $dx/dz = \omega L - \omega C[R^2 - X^2]$. By making the proper identifications in Theorem 5.2 we then see that these variables are defined as solutions of the differential equations

$$\frac{d\lambda_R(z, \omega_i)}{dz} = -\{2\lambda_R(z, \omega_i)[\omega_i X(z, \omega_i)C(z)] - 2\lambda_X(z, \omega_i)[C(z)\omega_i R(z, \omega_i)]\},$$
$$i = 1, \ldots, N,$$

$$\frac{d\lambda_X(z, \omega_i)}{dz} = -\{2\lambda_R(z, \omega_i)[\omega_i R(z, \omega_i)C(z)] + 2\lambda_X(z, \omega_i)[\omega_i C(z)X(z, \omega_i)]\},$$
$$i = 1, \ldots, N$$

subject to the boundary conditions

$$\lambda_R(l, \omega_i) = \partial P_N/\partial R(l, \omega_i), \qquad i = 1, \ldots, N,$$

and

$$\lambda_X(l, \omega_i) = \partial P_N/\partial X(l, \omega_i), \qquad i = 1, \ldots, N.$$

The algorithm then proceeds by varying $L(z)$ and $C(z)$ by increments given as

$$\Delta C(z) = -\alpha \sum_{i=1}^{N} \{\lambda_R(z, \omega_i) 2\omega_i R(z, \omega_i)X(z, \omega_i) - \lambda_X(z, \omega_i)\omega_i[R^2(z, \omega_i) - X^2(z, \omega_i)]\}$$

and

$$\Delta L(z) = -\alpha \sum_{i=1}^{N} \omega_i \lambda_X(z, \omega_i),$$

which we obtained again by making the proper identifications in Theorem 5.2.

Consider now the specific problem of designing a filter in which $R_g = R_L = 50$, and for which

$$\begin{aligned} T_d(\omega) &= 0, & 2 \text{ GHz} < \omega < 8 \text{ GHz}, \\ &= 1, & 8 \text{ GHz} < \omega < 14 \text{ GHz}, \\ &= 0, & 14 \text{ GHz} < \omega < 20 \text{ GHz}. \end{aligned}$$

Assume that $L(z)$ is fixed for all z at a value $L = u_0 = 4\pi \times 10^{-7}$ and that $C(z)$ is restricted to lie in the range

$$\frac{1}{36\pi} \times 10^{-9} < C < \frac{5}{36\pi} \times 10^{-9},$$

which physically corresponds to an allowable dielectric constant variation of five to one. If we normalize our parameters so that $\omega = 4\pi \times 10^9 \omega'$, $I(z, \omega) = 10^{-2} I'$, and $z = 10^{-2} z'$, then the new parameters become $L' = 4\pi \times 10^5 L = 16\pi^2 \times 10^{-2}$, $1/9 < C' < 5/9$, $l' = 10^{-2} l$, $Z_L' = Z_g' = 0.5$, and

$$\begin{aligned} T_d(\omega') &= 0, & 1 < \omega' < 4, \\ &= 1, & 4 < \omega' < 7, \\ &= 0, & 7 < \omega' < 10. \end{aligned}$$

The computer program used ten integer values for ω_i' between 1 and 10, and the differential equations were integrated by a Newton–Raphson technique that employed a fixed integration interval of $\Delta z' = 0.002$. Two filters were designed as shown in Fig. 5.1, with $l' = 0.635$ and 1.27. The initial choice for $C'(z')$ in the first problem was the constant $\frac{1}{3}$, whereas in the second case the initial $C'(z)$ for $z' < 0.635$ was the one that resulted from the converged solution of the first problem, and for $z' > 0.635$ it was selected as the constant $\frac{1}{9}$. The computing times on the IBM 7094-II computer were 4 and 7 min, respectively, although these numbers should only be considered relatively, since no particular effort was made to "streamline" the program. Other details may be found in [Mo1]. We note that the solutions obtained in this example are such that the "control" variable $C(z)$ is equal to its upper bound over portions of the interval. We may correlate this fact with the Pontryagin maximum principle, which in this example requires that

$$\begin{aligned} H(z) = \sum_{i=1}^{N} \{ & \lambda_X(\omega_i)\omega_i L + C(z) \\ & \times [\lambda_R(\omega_i) 2\omega_i R(\omega_i) X(\omega_i) - \lambda_x(\omega_i)\omega_i (R^2(\omega_i) - X^2(\omega_i))]\} \end{aligned}$$

be minimized for all z with respect to $C(z)$. If the bracket

$$Q(z) = [\lambda_R(\omega_i) 2\omega_i R(\omega_i) X(\omega_i) - \lambda_X(\omega_i)\omega_i (R^2(\omega_i) - X^2(\omega_i))] \equiv \partial Q / \partial C$$

is nonzero, then this requires that

$$\begin{aligned} C(z) &= C_{\min} & \text{if} \quad Q(z) > 0, \\ &= C_{\max} & \text{if} \quad Q(z) < 0, \end{aligned}$$

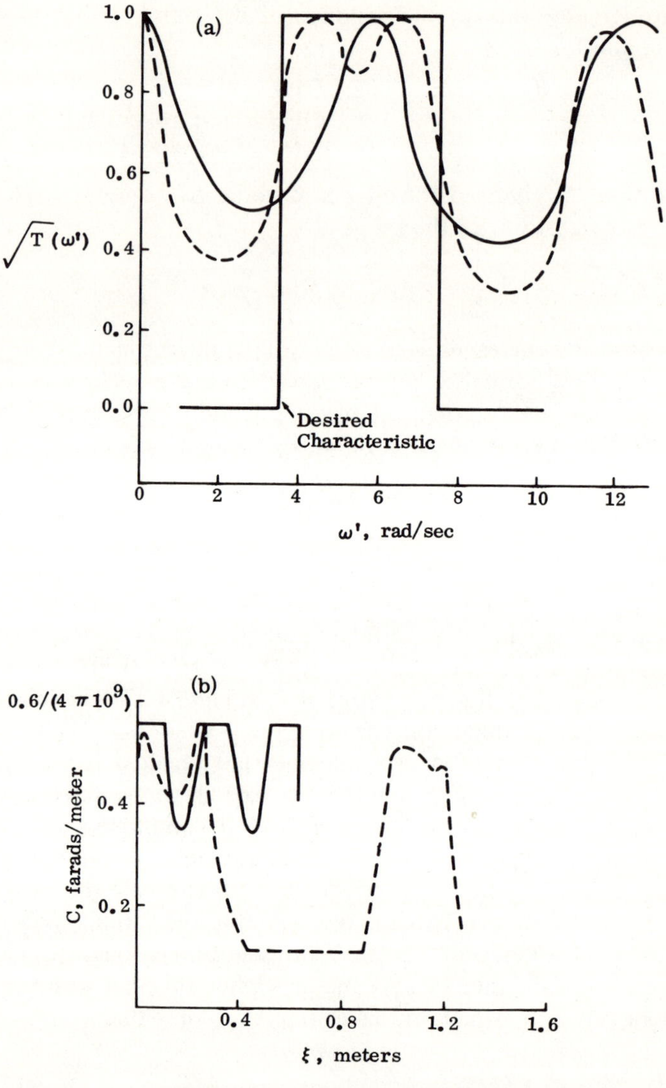

FIG. 5.1 Nonuniform transmission line filter. (a) Voltage gain versus frequency. (b) Capacitance versus line length. Solid lines: $l' = 0.635$. Dashed lines: $l' = 1.27$.

i.e., the "control" $C(z)$ can only take on its extreme values C_{min} and C_{max}. Such solutions have been called bang-bang solutions. On the other hand, situations may be encountered in which $Q(z)$ may vanish over some nonzero interval of z, in which case the Pontryagin maximum principle is satisfied even though $C(z)$ takes on some intermediate values in the interval. In the illustrative problem we have just presented it appears that the solution is only partially bang-bang, although the intermediate values occurring over the remaining intervals may only be due to the fact that the solution we presented is the result of only a finite number of iterations and is thus not a true minimum.

5.3. SUMMARY

The basic philosophy of optimization theory was described in this chapter. In network terms one abandons the classic approximation-realization cycle that dominated the classical filter design procedure and instead attempts to arrive at a design or synthesis in a direct manner. At the moment, one must first select a network topology and then search for those values of the various elements which will force the terminal properties of the network to come as close as possible to some desired performance. In general, the solution of such problems cannot be obtained in closed form, and so the aim of the designer is to employ those numerical search procedures which produce networks with near-optimum performance using a minimum number of computations.

We described the so-called "gradient" algorithm, which is one of the more basic that have been evolved in optimization theory, and then showed how it could be applied to a problem of general interest.

One of the basic difficulties in the optimization approach is the inability of the designer to be assured that the final design (final in the sense that it is the one that is the result of a finite number of applications of a computational algorithm) is the best possible. The first aspect of this problem is that the computational technique, by its very nature, searches only for local minimum rather than the absolute or global minimum. The second aspect is the general lack of knowledge concerning the rate of convergence to a minimum that can be expected with any one algorithm. However, the flexibility of these techniques, e.g., the ability to impose the constraint that the element values lie between certain limits, coupled with the ability, now provided by high-speed computers, to implement them, amply justifies their future study for network problems.

ACKNOWLEDGMENTS

The author would like to acknowledge the fact that Dr. R. Kopp was one of the first to realize that these computational techniques which he and others had developed were applicable in network theory, and that portions of this chapter are based on work which he directed. The functional analytic approach which was used to describe the background material from the calculus of variations was developed jointly by the author and Mr. P. Kenneth.

APPENDIX **I**

A Brief Survey of Distribution Theory

A function $f(t)$ is in reality a mapping between sets of numbers, i.e., for every value of a number t, $f(\cdot)$ is the rule for assigning another number $g = f(t)$. The difficulties associated with any attempt to extend this definition to generalized functions is perhaps best illustrated by the Dirac delta, $\delta(t)$. If $\delta(t)$ is considered to be a function in the classic sense, then it is zero for every value of t except $t = 0$, and it is thus equivalent to zero in the classic sense almost everywhere (a.e.). However, if we extend the definition of function by means of an integral representation, then we can, for example, define $\delta(t)$ as $\int \delta(t)\varphi(t)\, dt = \varphi(0)$ for any continuous function $\varphi(t)$. This definition certainly distinguishes $\delta(t)$ from zero. Although this approach seems very reasonable, there are still some doubts that could be raised concerning the ability of such a definition to uniquely specify a generalized function. The resolution of this doubt can only be found in a systematic development of distribution theory. In particular, it is found that such questions may be resolved by considering classes of functions, called testing functions, whose properties are such as to display the differences between different generalized functions in a concrete manner. To be specific we will consider one such set of testing functions. Thus, consider the set of all infinitely differentiable (C^∞) functions that are zero outside some interval, where the interval may depend upon the specific function. We say that these functions have compact support and we designate the collection of them as $C_0{}^\infty$. A standard example of such a function is

$$\begin{aligned} \varphi(x) &= a \exp\left(\frac{-1}{1 - x^2}\right), \qquad |x| \le 1, \\ &= 0, \qquad |x| > 1, \end{aligned}$$

and we note that $\lim_{|x|\to 1} d^n\varphi/dx^n = 0$ if we approach from either direction $|x| < 1$ or $|x| > 1$. We define a topology on C_0^∞ (called the topology of $\mathscr{D}$) such that a sequence of functions contained in C_0^∞ converges to zero if each member of the sequence is zero outside some fixed interval, and the functions converge to zero uniformly on that interval together with their derivatives of any order. We will then adopt as a working definition of a distribution the following: a generalized function $f(t)$ is a distribution, or member of the set $\mathscr{D}'$, if there exists a linear mapping dependent upon $f(t)$ that assigns to each $\varphi(t) \in C_0^\infty$ a single number; in addition, this mapping is such that if a sequence of φ_n converges to zero in the topology of $\mathscr{D}$, then the corresponding sequence of numbers determined by the mapping converge to zero, i.e., the mapping is continuous in the above sense on the set C_0^∞. Although we are taking liberties with established notation, we will rephrase the definition as follows: $f \in \mathscr{D}'$ if

$$\int f(t)[a\varphi_1(t) + b\varphi_2(t)]\, dt = a\int f(t)\varphi_1(t)\, dt + b\int f(t)\varphi_2(t)\, dt$$
$$= a\lambda_1 + b\lambda_2$$

for all $\varphi \in C_0^\infty$, where a, b, λ_1, and λ_2 are constants, and in addition

$$\lim_{n\to\infty} \int f(t)\varphi_n(t)\, dt = 0,$$

where φ_n is a sequence of functions in C_0^∞ that converge to zero in the topology of $\mathscr{D}$. We will also say for brevity that a distribution f is contained in $\mathscr{D}'$ if it is a linear continuous mapping defined on the set of testing functions denoted as $\mathscr{D}$, to incorporate specifically the topology we have introduced. We justify the use of the integral symbol in lieu of the inner product symbol $\langle f, \varphi \rangle$ on the basis of its intuitive value and in the understanding that it is not to be interpreted in either the Riemann or Lebesgue sense, but merely as a way of indicating the linear operation by which the distribution assigns a single number to each testing function $\varphi(t)$. Now, returning to the question of uniqueness, we see that two distributions are equal if $\int (f_1 - f_2)\varphi\, dt = 0$ for all $\varphi \in C_0^\infty$, and, in particular, two distributions are equal on some open interval K if the above integral vanishes for all testing functions that are zero except on K. Since distribution theory is to be a generalization of ordinary function theory, we should expect that this definition applies to ordinary functions, and in fact it does, since one may prove that the above definition of equality implies that the two functions will be equal pointwise a.e. when they are both members of L_p (functions whose pth power is integrable from $-\infty$ to $+\infty$).

If we consider the operations that are possible with distributions, we see first that addition is always defined via $\int (f_1 + f_2)\varphi\, dt = \int f_1\varphi\, dt + \int f_2\varphi\, dt$.

However, multiplication is not always defined between arbitrary distributions, except in the case where one of them is infinitely differentiable. In that case $f(t)\beta(t)$ is defined (assuming $\beta \in C^\infty$) as:

$$\int (f\beta)\varphi\, dt = \int f(\beta\varphi)\, dt,$$

since $\beta\varphi \in C_0^\infty$ when $\varphi \in C_0^\infty$ and $\beta \in C^\infty$. The convergence of a sequence of distributions is defined and referred to as weak convergence. Thus a sequence of distributions $f_n \in \mathscr{D}'$ converges to zero weakly, i.e., $f_n \xrightarrow{w} 0$, if

$$\lim_{n\to\infty} \int f_n(t)\varphi(t)\, dt = 0$$

for all $\varphi \in C_0^\infty$. Differentiation may always be defined using the analog of the classic integration by parts. Thus we have, $g = d^n f/dt^n$, where

$$\int g\varphi\, dt = \int \frac{d^n f}{dt^n}\varphi\, dt = (-1)^n \int f \frac{d^n\varphi}{dt^n}\, dt$$

when $f \in \mathscr{D}'$, since the second integral is well defined when $\varphi \in C_0^\infty$ since $d^n\varphi/dt^n$ exists and is also in C_0^∞. In this generalized sense all distributions are infinitely differentiable, and we will have to indicate specifically when we are assuming a classical definition of differentiation, e.g., by $f \in C^\infty$ we mean that f possesses derivatives of all orders in a classical sense.

A distribution is said to belong to the set $\mathscr{E}'$ if it is a member of $\mathscr{D}'$, and, moreover, it is zero outside some finite interval. One of the more fundamental results of distribution theory is the fact that any $f \in \mathscr{E}'$ may be represented as the mth generalized derivative of a continuous function, i.e., $f \in \mathscr{E}'$ if and only if there exists a continuous function $\hat{f}(t)$ such that $f = d^m\hat{f}/dt^m$ for some finite m. We also note that $\int f\varphi\, dt$ is well defined when $f \in \mathscr{E}'$ and $\varphi \in C^\infty$, i.e., φ need not be zero outside some finite interval. One reason for the study of the set $\mathscr{E}'$ is that a weighted multiplication, specifically convolution, is always defined between two distributions whenever one of them is in $\mathscr{E}'$. The classical definition of convolution is

$$f * g = \int f(\tau)g(t-\tau)\, d\tau = \int f(t-\tau)g(\tau)\, d\tau$$

and we may reasonably extend it as follows: convolution between two distributions f and g is defined if we can give meaning to

$$\int (f*g)\varphi dt = \int g(x)\, dx \int f(t)\varphi(t+x)\, dt = \int f(x)\, dx \int g(t)\varphi(t+x)\, dt$$

for all $\varphi \in C_0^\infty$. Thus, if for example, $g \in \mathscr{E}'$, then the function

$$\beta(x) = \int g(t)\varphi(t+x)\, dt$$

not only exists for all $\varphi \in C_0^{\infty}$, but it is an infinitely differentiable (in the classic sense) function of x. Moreover, since $g \in \mathscr{E}'$, we note that $\beta(x)$ will be zero outside some finite interval, and we conclude that $\beta \in C_0^{\infty}$, so that $\int f(x)\beta(x)\,dx$ is well defined for $f \in \mathscr{D}'$. To complete the above demonstration, we also show that with $f \in \mathscr{D}'$

$$\int f(t)\varphi(t+x)\,dt = \psi(x)$$

exists for all $\varphi \in C_0^{\infty}$, is infinitely differentiable (classically), but it is not zero outside some finite interval. However, since g was assumed to be in $\mathscr{E}'$, $\int g\psi\,dx$ is still defined, since $\psi \in C^{\infty}$. We note that convolution has, among others, the following useful properties:

1. $(f_1 + f_2) * g = f_1 * g + f_2 * g$.
2. $f * \delta(t-a) = f(t-a)$, since

$$\int (f * \delta(t-a))\varphi\,dt = \int f(x)\,dx \int \delta(t-a)\varphi(t+x)\,dt$$

$$= \int f(x)\varphi(x+a)\,dx = \int f(x-a)\varphi(x)\,dx.$$

3. $g = f * \varphi \in C^{\infty}$ when $\varphi \in C_0^{\infty}$.
4. $(d^n/dt^n)(f * g) = (d^n f/dt^n) * g = f * (d^n g/dt^n)$ when $f * g$ is defined, since

$$\int \frac{d^n(f * g)}{dt^n}\varphi\,dt = (-1)^n \int (f * g)\frac{d^n\varphi}{dt^n}\,dt$$

$$= \int g(x)\,dx \int f(t)(-1)^n \frac{d^n\varphi(t+x)}{dt^n}\,dt$$

$$= \int g(x)\,dx \int \frac{d^n f}{dt^n}\varphi(t+x)\,dt.$$

A property of convolution which we will make extensive use of is its continuity. Thus if $f \in \mathscr{D}'$, we find that for $\varphi_n \in C_0^{\infty}$

$$\lim_{n\to\infty} \int (f * \varphi_n)\beta\,dt = \lim_{n\to\infty} \int f(x)\,dx \int \varphi_n(t)\beta(t+x)\,dt$$

$$= \lim_{n\to\infty} \int f(x)\psi_n(x)\,dx,$$

where $\psi_n(x) \in C_0^{\infty}$. Now if φ_n converges to zero in the topology of $\mathscr{D}$, ψ_n will also converge to zero in $\mathscr{D}$, and we see that the operator $f * (\cdot)$ is continuous in the topology of $\mathscr{D}$. One also shows in essentially the same way that if g_n is a sequence of distributions in $\mathscr{E}'$, that all zero outside some fixed interval, and are such that

$$\lim_{n\to\infty} \int g_n \varphi \, dt = 0$$

for all $\varphi \in C_0^{\infty}$, i.e., $g_n \xrightarrow{w} 0$; then

$$\lim_{n\to\infty} \int (f * g_n)\varphi \, dt = 0$$

for any fixed $f \in \mathscr{D}'$, i.e., convolution is continuous in the weak topology of $\mathscr{E}'$.

We will also have occasion to consider some specialized classes of distributions, those indicated by the symbol $\mathscr{D}'_{L_p}$. For our purposes we will define these sets as follows: $f \in \mathscr{D}'_{L_p}$ if and only if f can be written as a finite sum of generalized derivatives of functions contained in L_p. It is a simple matter to show that if $f \in \mathscr{D}'_{L_p}$, then $f * \varphi = g \in L_p$ for all $\varphi \in C_0^{\infty}$; thus

$$f * \varphi = \left(\sum \frac{d^k f_k}{dt^k}\right) * \varphi = \sum \left(f_k * \frac{d^k \varphi}{dt^k}\right) = \sum g_k,$$

where $g_k \in L_p$ since $f_k \in L_p$ and $d^k\varphi/dt^k \in L_1$ and it is well known that if $u \in L_p$, $v \in L_g$, $u * v \in L_r$, where $1/r = (1/p) + (1/g) - 1$. It is a much more difficult matter to establish the converse, namely, that if $f * \varphi \in L_p$ for all $\varphi \in C_0^{\infty}$, then $f \in \mathscr{D}'_{Lp}$, and we will refer the reader to [Sc3]. One may also show in a direct manner that if $f \in \mathscr{D}'_{L_p}$ and $g \in \mathscr{D}'_{L_g}$, then $f * g = u$ is well defined if $(1/p) + (1/g) \geq 1$, and, moreover, $u \in \mathscr{D}'_{L_r}$, where $1/r = (1/p) + (1/g) - 1$. In this case we admit $r = \infty$ or L_∞ as the set of all bounded functions and $\mathscr{D}'_{L_\infty}$ as the set of all distributions that can be written as a finite sum of generalized derivatives of bounded functions.

We will make extensive use of Fourier and Laplace transform techniques, and as a preliminary to their definition we introduce the set of testing functions denoted by $\mathscr{S}$ as the collection of all infinitely differentiable functions $\varphi(t)$ such that

$$\lim_{|t|\to\infty} |t|^m \left|\frac{d^n\varphi}{dt^n}\right| = 0$$

for all positive integers m and n. A topology is also introduced on $\mathscr{S}$ by defining

$$p_{m,l}(\varphi) = \sup_{-\infty < t < \infty} \left| t^m \frac{d^l\varphi}{dt^l}\right|,$$

and a sequence φ_n is then said to converge in the topology of $\mathscr{S}$ if $\lim_{n\to\infty} p_{m,l}(\varphi_n) = 0$ for all m and l. The set of all distributions defined as linear continuous mappings on the set $\mathscr{S}$ is designated as $\mathscr{S}'$. In particular, it may be shown that $f \in \mathscr{S}'$ if and only if there exist constants m and n and a bounded continuous function $\hat{f}$ such that

$$f = (d^n/dt^n)[(1 + t^2)^m \hat{f}],$$

where the derivative operation is to be interpreted distributionally. Now we may define the generalized Fourier transform by analogy with the classic definition. Thus if $f \in \mathscr{S}'$, then its Fourier transform $\mathscr{F}(f) = F(\omega)$ is another distribution contained in $\mathscr{S}'$ defined as

$$\int F(\omega)\varphi(\omega)\, d\omega = \int f(\omega)\Phi(\omega)\, d\omega = \int f(\omega)\, d\omega \int \varphi(t)e^{-j\omega t}\, dt,$$

where $\Phi(\omega) = \int_{-\infty}^{\infty} \varphi(t)e^{-j\omega t}\, dt$ and $\varphi(\omega) \in \mathscr{S}$. The definition is only meaningful for $\varphi \in \mathscr{S}$, since $\Phi(\omega)$, as we may show directly, will then also be contained in $\mathscr{S}$, as required if $\int f\Phi\, d\omega$ is to be defined. One may show that the generalized Fourier transform has many of the properties of its classical counterpart. Thus

$$\mathscr{F}[(d^n/dt^n)f] = (j\omega)^n \mathscr{F}(f) = (j\omega)^n F(\omega),$$

since

$$\int \frac{d^n f}{d\omega^n}\Phi(\omega)\, d\omega = (-1)^n \int f(\omega)\frac{d^n\Phi(\omega)}{d\omega^n}\, d\omega = \int f(\omega)\mathscr{F}[(jt)^n\varphi(t)]\, d\omega$$

$$\equiv \int F(\omega)(j\omega)^n\varphi(\omega)\, d\omega.$$

Similarly, one obtains $\mathscr{F}[t^n f] = (j)^n\, d^n F/d\omega^n$, and in both this and the previous formula the generalized derivative is to be understood. We also have

$$\mathscr{F}(u * v) = \mathscr{F}(u)\mathscr{F}(v)$$

if either v or $u \in \mathscr{E}'$. For example, if $u \in \mathscr{E}'$, then $u = d^m\hat{u}/dt^m$, or $\mathscr{F}(u) = (j\omega)^m\mathscr{F}(\hat{u})$, and one may always select $\hat{u}$ to be zero outside some finite interval K given that $u \in \mathscr{E}'$, so that

$$\frac{d^n\mathscr{F}(\hat{u})}{d\omega^n} = \int_K (-jt)^n e^{-j\omega t}\hat{u}\, dt$$

exists for all n since, as we noted above, $\hat{u}$ is also bounded. Another form of the same result is

$$\mathscr{F}(uv) = \tfrac{1}{2}\pi\mathscr{F}(u) * \mathscr{F}(v),$$

and we note in both cases that one must first ascertain that either the product uv [or $\mathscr{F}(u)\mathscr{F}(v)$] or the convolution exists in order for such relationships to hold. For example, if $\mathscr{F}(u)$ and $\mathscr{F}(v) \in \mathscr{D}'_{L_2}$, then $\mathscr{F}(u) * \mathscr{F}(v)$ is, as we noted above, well defined, since $\frac{1}{2} + \frac{1}{2} - 1 = 0$. In fact, u and v can be written as a finite sum of factors of the form $t^n f_n$, with $f_n \in L_2$, since u may be written as a finite sum of derivatives of L_2 functions. An interesting application of this result is in the study of those distributions that are zero for $t < 0$. First one shows that

$$\mathscr{F}(\tfrac{1}{2} \operatorname{sgn} t) = pv(1/j\omega)$$

where

$$\begin{aligned} \operatorname{sgn} t &= 1, \qquad t > 0, \\ &= -1, \qquad t < 0, \end{aligned}$$

and $pv(1/\omega)$ is defined by

$$\begin{aligned} \int pv \frac{1}{\omega} \varphi(\omega)\, d\omega &\equiv pv \int \frac{\varphi(\omega)}{\omega}\, d\omega \\ &= \lim_{\varepsilon \to 0} \left[\int_{|\varepsilon|}^{\infty} \frac{\varphi(\omega)}{\varepsilon}\, d\omega + \int_{-\infty}^{-|\varepsilon|} \frac{\varphi(\omega)}{\omega}\, d\omega \right] \end{aligned}$$

for all $\varphi \in C_0^{\infty}$. Then one obtains

$$\mathscr{F}(u_0(t)) = \mathscr{F}[\tfrac{1}{2}(1 + \operatorname{sgn} t)] = \pi\delta(\omega) + pv \frac{1}{j\omega},$$

where $u_0(t)$ is the Heaviside step function and $\mathscr{F}[1] = 2\pi\delta(\omega)$, as we may show directly. Now, if $f(t) = 0$ for $t < 0$ and $\mathscr{F}(f) \in \mathscr{D}'_{L_2}$, we have $fu_0(t) = f$ (this is well defined, since, as noted above, f will be an ordinary function) and

$$\begin{aligned} F(\omega) = \mathscr{F}(f) = \mathscr{F}(fu_0) &= \frac{1}{2\pi} \mathscr{F}(f) * \left[\pi\delta + pv \frac{1}{j\omega} \right] \\ &= \frac{F(\omega)}{2} + \frac{1}{2\pi} F * pv \frac{1}{j\omega}, \end{aligned}$$

or

$$F(\omega) = \frac{1}{\pi} F(\omega) * pv \frac{1}{j\omega}.$$

Thus we have shown that the fact that $f(t) = 0$ for $t < 0$ imposes a convolution constraint on its Fourier transform [at least in the case where $F(\omega) \in \mathscr{D}'_{L_2}$] It should be noted that the convolution is well defined, since $u_0 = (1 + t^2)[u_0/(1 + t^2)]$, and thus $\mathscr{F}(u_0) = \hat{U} - (d^2\hat{U}/d\omega^2) \in \mathscr{D}'_{L_2}$, since $\hat{U} = \mathscr{F}(u_0/[1 + t^2]) \in L_2$. As one final application of these results, we observe that if f and $g \in L_2$, then

$$\mathscr{F}(fg) = \tfrac{1}{2}\pi F * G = \tfrac{1}{2}\pi \int F(x)G(\omega - x)\, dx,$$

since F and G, the Fourier transforms of L_2 functions, will also be L_2 functions, as is known classically (see, for example [Ti1]), and the convolution may be interpreted in the Lebesgue sense. Now, by setting $\omega = 0$, we obtain

$$\mathscr{F}(fg)\big|_{\omega=0} = \int fg\, dt = \tfrac{1}{2}\pi \int F(x)G(-x)\, dx.$$

In the special case where g is a real function of time $G(-x) = G^*(x)$, so that we may write

$$\int fg\, dt = \tfrac{1}{2}\pi \int F(x)G^*(x)\, dx,$$

and, finally, in the case where $f = g$ we obtain Plancherel's equality,

$$\int f^2\, dt = \tfrac{1}{2}\pi \int |F(x)|^2\, dx.$$

The complex Fourier or Laplace transform, $\mathscr{L}(\cdot)$, can be defined most directly for these distributions $f(t)$ that are zero for t less than zero, and such that $fe^{-at} \in \mathscr{S}'$ for $a \geq a_0$. Then $\mathscr{L}(f) = \mathscr{F}(e^{-at}f)$, or, with $p = \sigma + j\omega$,

$$F(p) = \mathscr{L}(f) = \int fe^{-pt}\, dt,$$

where the integral may be defined directly for all p such that $\operatorname{Re} p > a_0$ using the following artifice: consider the function

$$g_0(t) = \frac{m(t+\varepsilon)}{m(t+\varepsilon) + m(-t)},$$

where

$$\left.\begin{aligned} m(t) &= 0, && t < 0, \\ &= e^{-1/t}, && t \geq 0, \end{aligned}\right\} \in C^\infty,$$

and note that $g_0(t) \in C^\infty$, and, moreover,

$$\begin{aligned} g_0(t) &= 1, && t \geq 0, \\ &= 0, && t \leq -\varepsilon. \end{aligned}$$

Now we see that $g_0(t)e^{-pt} \in \mathscr{S}$ for $\operatorname{Re} p > 0$, and, moreover, since $f = 0$ for $t < 0$,

$$\int fe^{-pt}\, dt \equiv \int fg_0 e^{-pt}\, dt = \int f(\exp - a_0 t) g_0 \exp[-(p - a_0)t]\, dt$$

which is defined for $\operatorname{Re} p > a_0$, since, by assumption, $f\exp(-a_0 t) \in \mathscr{S}'$. One then shows that $dF(p)/dp = -\int tf(t)e^{-pt}\, dt$ also exists for $\operatorname{Re} p > a_0$, or that $F(p)$ is an analytic function of p in the half plane $\operatorname{Re} p > a_0$.

The connection between the Laplace and Fourier transforms has been explored in both the classic and generalized settings. A typical result of these studies may be obtained as follows if we assume $\mathscr{F}(f) \in \mathscr{D}'_{L_2}$, and, moreover, that $f = 0$ for $t < 0$: thus $F(p) = \mathscr{L}(f) = \mathscr{F}(fe^{-\sigma t}) = \mathscr{F}[f(e^{-\sigma t}u_0(t))]$ exists for $\operatorname{Re} p > 0$, and we find explicitly that

$$F(p) = \frac{1}{2\pi} F * \mathscr{F}(e^{-\sigma t}u_0(t)) = \frac{1}{2\pi} \int \frac{F(\zeta)}{p - j\zeta}\, d\zeta,$$

since $\mathscr{F}(e^{-\sigma t}u_0) = 1/(\sigma + j\omega)$, when $\sigma = \operatorname{Re} p > 0$. We may write out the convolution in an explicit manner, since for $\operatorname{Re} p > 0$, $1/(p - j\zeta) \in C^\infty$ (with respect to ζ) and it, together with all of its derivatives, are L_2 functions, so that with $F \in \mathscr{D}'_{L_2}$ the integral is well defined, and in fact is the analog of the classic Cauchy integral.

It is hoped that the foregoing brief and sketchy outline of those aspects of distribution theory which will be of interest to us may serve the purpose for which it was intended: to show that many of the tools and techniques which have been used in a formal way may actually be placed on a sound basis. In this way one is able to make definitive statements as opposed to what are essentially conjectures. Although many of the results described either above or in other portions of this book may not come as a great surprise, rigorous proofs of their statements were either obtained in a classic setting or in fact they were conjectured to hold in the larger setting. What should be noted is that the generality of their present setting is only possible in the context of distribution theory. Moreover, it is the author's opinion that many of the arguments used in these distributional proofs are actually more natural to the problems at hand, and in fact these proofs are much more illuminating than their classic counterparts because one need not become encumbered with many of the detailed arguments necessary in the classic setting.

For a much more thorough discussion of these and related topics in distribution theory the reader is referred to [Be1], [Ze1], and [Br1], and, for even greater depth, to the now classic works in the subject of distributions, [Sc3] and [Ge1].

APPENDIX **II**

The Inversion of a Sturm–Liouville Operator

In our discussion of nonuniform lossless transmission lines in Chapter IV we used the work of Marčenko concerning the inversion of a Sturm–Liouville operator in an essential way. This appendix presents a nonrigorous discussion of the essential features of this inversion process. With the exception of a brief discussion in Levin's book [Le1], there are no other expositions of Marčenko's work in English, at least to the author's knowledge, and for this reason the following discussion, although brief and presented without formal proofs, seems appropriate.

The starting point in our discussion is a Sturm-Liouville system consisting of a differential equation

$$\frac{d^2\Theta(\lambda, x)}{dx^2} - q(x)\Theta(\lambda, x) + \lambda\Theta(\lambda, x) = 0$$

subject to the boundary conditions $\Theta(\lambda, 0) = 1$ and $(d\Theta/dx)(\lambda, 0) = h$. One of the more significant results concerning this system is the fact that there exists a kernel $K(x, y)$ which is independent of λ such that each solution of the system may be represented as

$$\Theta(\lambda, x) = \cos(x\sqrt{\lambda}) + \int_0^x K(x, y)\cos(y\sqrt{\lambda})\,dy$$

for all λ, and, moreover, this kernel satisfies $K(x, y) = 0$, $y > x$. For a proof of this statement we refer the reader to [Le1]. If we now impose an additional boundary condition of the form

$$\frac{d\Theta}{dx}(\lambda, a) + H_n\Theta(\lambda, a) = 0,$$

then we may show that there will be only a specific set of discrete values of λ

for which the corresponding functions $\Theta(\lambda, x)$ satisfy such a condition. We designate the set $\{\lambda_i\}$ of such eigenvalues λ by $S_q[h, H_n]$ and refer to it as the spectra of the system defined by $q(x)$, h, and H_n. Now consider the function

$$L(z) = \frac{d\Theta}{dx}(z^2, a) + H_n \Theta(z^2, a)$$

and note, from the integral representation of Θ in terms of $K(x, y)$, that $L(z)$ will be an entire function of z. In addition, the zeros of $L(z)$ are precisely those values for which the second boundary condition imposed on the Sturm–Liouville system is satisfied, i.e., if $L(z_i) = 0$, then $z_i^2 = \lambda_i$. One then shows that the asymptotic behavior of $\Theta(z^2, a)$ for $|z| \to \infty$ is such that $L(z)$ is of exponential type in the entire z plane. But it is known (see again, for example, [Le1]) that entire functions of exponential type are uniquely determined up to a scale factor by their zeros. Thus one shows that

$$L(z) = a(\lambda_0 - z^2) \prod_{n=1}^{\infty} \frac{\lambda_n - z^2}{(\pi/a)^2 n^2},$$

where $\{\lambda_i\} = S_q[h, H_n]$.

Now let us assume that two boundary conditions of the form $(d\Theta/dx) + H_n \Theta = 0$ are imposed at $z = a$ with two different constants H_1 and H_2. For each H_n we obtain a different set of eigenvalues—say, $\{\lambda_i\} = S_q[h, H_1]$ and $\{\beta_i\} = S_q[h, H_2]$, where we note that the boundary conditions at $x = 0$ $[\Theta(0) = 1, (d\Theta/dx)(0) = h]$ are the same for both cases. In particular, we have two uniquely determined entire functions corresponding to these two spectra —say, $L(z)$ corresponding to $S_q[h, H_1]$ and $M(z)$ corresponding to $S_q[h, H_2]$. Now, if we return to the basic differential equation of the Sturm–Liouville system, we obtain directly that

$$\Theta(\beta^2, x)\frac{d^2\Theta(\lambda^2, x)}{dx^2} - q(x)\Theta(\beta^2, x)\Theta(\lambda^2, x) + \lambda^2\Theta(\beta^2, x)\Theta(\lambda^2, x)$$

$$- \left[\Theta(\lambda^2, x)\frac{d^2\Theta(\beta^2, x)}{dx^2} - q(x)\Theta(\lambda^2, x)\Theta(\beta^2, x) + \beta^2\Theta(\lambda^2, x)\Theta(\beta^2, x)\right] = 0$$

or, by solving for $\Theta(\lambda^2, x)\Theta(\beta^2, x)$ and then integrating we obtain

$$\begin{aligned}(\beta^2 - \lambda^2)\int_0^a \Theta(\lambda^2, x)\Theta(\beta^2, x)\,dx &= \int_0^a \left[\Theta(\beta^2, x)\frac{d^2\Theta(\lambda^2, x)}{dx^2} - \Theta(\lambda^2, x)\frac{d^2\Theta(\beta^2, x)}{dx^2}\right] dx \\ &= \frac{d\Theta}{dx}(\lambda^2, a)\Theta(\beta^2, a) - \frac{d\Theta}{dx}(\beta^2, a)\Theta(\lambda^2, a),\end{aligned}$$

where we have imposed the condition that $\Theta(\beta^2, 0) = \Theta(\lambda^2, 0) = 1$ and

$(d\Theta/dx)(\beta^2, 0) = (d\Theta/dx)(\lambda^2, 0) = h$. However, from their definitions we see that

$$L(\beta)M(\lambda) - L(\lambda)M(\beta) = (H_1 - H_2)\left[\frac{d\Theta}{dx}(\lambda^2, a)\Theta(\beta^2, a) - \frac{d\Theta}{dx}(\beta^2, a)\Theta(\lambda^2, a)\right],$$

and we may conclude that

$$\int_0^a \Theta(\lambda^2, x)\Theta(\beta^2, x)\,dx = \frac{L(\beta)M(\lambda) - L(\lambda)M(\beta)}{(H_1 - H_2)(\beta^2 - \lambda^2)}$$

for all values of λ and β, and, in particular, for the $\lambda = \lambda_i$ and $\beta = \beta_i$ corresponding to the eigenvalues of the two boundary conditions imposed at $x = a$. However, we may also use the integral representation of Θ in terms of the kernel $K(x, y)$ to obtain

$$\int_0^a \Theta(\lambda^2, x)\Theta(\beta^2, x)\,dx = \int_0^a \cos\lambda x \cos\beta x\,dx + \int_0^a\int_0^a K(x, t)\cos\beta t \cos\lambda x\,dt\,dx$$
$$+ \int_0^a\int_0^a K(x, t)\cos\lambda t\cos\beta x\,dt\,dx$$
$$+ \int_0^a\left\{\left[\int_0^a K(x, t)\cos\lambda t\,dt\right]\left[\int_0^a K(x, t)\cos\beta t\,dt\right]\right\}dx$$

or, equivalently,

$$\frac{L(\beta)M(\lambda) - L(\lambda)M(\beta)}{(H_1 - H_2)(\beta^2 - \lambda^2)} - \int_0^a \cos\lambda x\cos\beta x\,dx$$
$$\equiv \int_0^a \Theta(\lambda^2, x)\Theta(\beta^2, x)\,dx - \int_0^a \cos\lambda x\cos\beta x\,dx$$
$$= \int_0^a\int_0^a \left[K(x, y) + K(y, x) + \int_0^a K(z, x)K(z, y)\,dz\right]\cos\lambda x\cos\beta y\,dx\,dy.$$

If we define

$$F(x, y) = K(x, y) + K(y, x) + \int_0^a K(z, x)K(z, y)\,dz,$$

then we may write

$$\int_0^a\int_0^a F(x, y)\cos\lambda x\cos\beta y\,dx\,dy = \frac{L(\beta)M(\lambda) - L(\lambda)M(\beta)}{(H_1 - H_2)(\beta^2 - \lambda^2)}$$
$$- \frac{\beta(\sin\beta a)(\cos\lambda a) - \lambda(\sin\lambda a)(\cos\beta a)}{\beta^2 - \lambda^2}$$
$$= G(\lambda, \beta),$$

where we have simplified the expression by evaluating the integral $\int_0^a \cos \lambda x \cos \beta x \, dx$. But we recognize that the last expression, $G(\lambda, \beta)$, is the two-dimensional Fourier transform of the function $F(x, y)$ defined on the rectangle $0 \le x \le a, 0 \le y \le a$. We may then find $F(x, y)$ from $G(\lambda, \beta)$ using either the inverse Fourier transform or the Fourier series representation. If we elect the latter, we see that on the rectangle

$$F(x, y) = \frac{1}{a^2} \sum_m \sum_n G\left(\frac{n\pi}{a}, \frac{m\pi}{a}\right) \cos\left(\frac{n\pi}{a} x\right) \cos\left(\frac{m\pi}{a} y\right)$$

so that $F(x, y)$ is uniquely determined from $L(z)$ and $M(z)$. Finally, we note that since $K(x, y) = 0$ for $y > x$, we have

$$F(x, y) = K(x, y) + K(y, x) + \int_0^a K(z, x)K(z, y)\, dz \equiv K(x, y) + \int_x^a K(z, x)K(z, y)\, dz$$

for $x > y$. If this nonlinear integral equation has a unique solution $K(x, y)$, then we see that $\Theta(\lambda, x)$ may be found uniquely from its integral representation in terms of $K(x, y)$, and, moreover,

$$q(x) = \frac{(d^2\Theta/dx)(0, x)}{\Theta(0, x)}, \qquad h = \frac{d\Theta}{dx}(\lambda, 0),$$

$$H_1 = \frac{-(d\Theta/dx)(\lambda_i, a)}{\Theta(\lambda_i, a)}, \quad \text{and} \quad H_2 = \frac{-(d\Theta/dx)(\beta_i, a)}{\Theta(\beta_i, a)}$$

are all uniqeuly determined. Marčenko was able to prove that this integral equation does possess a unique solution which in fact can be obtained in an iterative manner using successive approximations, and he thus established [Ma1]:

Theorem. Denote by $S_q[h, H_n]$ the set of all λ such that there exists a non-zero solution of

$$\frac{d^2u}{dx^2} - q(x)u + \lambda u = 0$$

that satisfies the boundary conditions

$$\frac{du}{dx}(0) - hu(0) = 0, \qquad \frac{du}{dy}(a) + H_n u(a) = 0$$

(it is assumed that $\int_0^a |q|\, dx < \infty$). Then the two sets $S_q[h, H_1]$ and $S_q[h, H_2]$

corresponding to the same h but with $H_1 \neq H_2$ uniquely determine h, H_1, H_2, and $q(x)$.

It might be noted that in our previous discussion we have assumed that $H_1 - H_2$ was a known quantity. We will demonstrate that in fact it can be uniquely determined from $S_q[h, H_1]$ and $S_q[h, H_2]$. First one may show directly that if Θ is a solution of the integral equation

$$\Theta(\lambda, x) = \cos(x\sqrt{\lambda}) + \frac{h}{\sqrt{\lambda}} \sin(x\sqrt{\lambda}) + \frac{1}{\sqrt{\lambda}} \int_0^x \sin[(x - t)\sqrt{\lambda}]q(t)\Theta(\lambda, t)\, dt,$$

then $\Theta(\lambda, 0) = 1$, $(d\Theta/dx)(\lambda, 0) = h$, and Θ is a solution of the differential equation $(d^2\Theta/dx^2) - q\Theta + \lambda\Theta = 0$. But the integral equation may be iterated successively, and in the process one observes that if $q(t)$ is a continuous function of t, then

$$\Theta(\lambda, x) = \cos\left(\sqrt{\lambda}x\right)\left[1 + O\left(\frac{1}{\lambda}\right)\right] + \sin\left(\sqrt{\lambda}x\right)\frac{N(x)}{\sqrt{\lambda}} + O\left(\frac{1}{\lambda}\right)\Bigg],$$

where $N(x) = h + \frac{1}{2}\int_0^x q(t)\, dt$. If we substitute this into the boundary condition $(d\Theta/dx) + H_n\Theta = 0$, we obtain, after simplification,

$$\tan(a\sqrt{\lambda}) = \frac{h + H_n + \frac{1}{2}\int_0^a (qt)\, dt + O(1/\sqrt{\lambda})}{\sqrt{\lambda} + O(1/\sqrt{\lambda})}.$$

If we expand the tangent about the points $\sqrt{\lambda} = \pi n/a$ and retain only the linear term, we see that

$$a\left(\sqrt{\lambda} - \frac{\pi n}{a}\right) = \frac{h + H_n + \frac{1}{2}\int_0^a q(t)\, dt}{\pi n/a} + O\left(\frac{1}{n^2}\right).$$

so that

$$\sqrt{\lambda} = \frac{\pi n}{a} + \frac{h + H_n + \frac{1}{2}\int_0^a q(t)\, dt}{\pi n} + O\left(\frac{1}{n^2}\right) = \frac{\pi n}{a} + \frac{A_1}{\pi n} + O\left(\frac{1}{n^2}\right),$$

which expresses the asymptotic behavior of the location of the eigenvalues λ. In particular, we see that if two such sets of eigenvalues are obtained for the same h and $q(x)$ but with different values of H_n then

$$A_1 - B_1 = H_1 - H_2,$$

where A_1 corresponds to the set generated by H_1 and B_2 to the set generated by H_2.

The Gel'fand–Levitan theorem noted in Chapter IV provides a means for determining if two given sets of real numbers are in fact two spectra of a Sturm–Liouville system. Their proof does not make direct use of the Marčenko integral equation to determine $K(x, y)$, but proceeds by finding an alternate linear integral equation which they then prove is uniquely soluble given certain conditions. However, their sufficient conditions are directly applicable to the Marčenko procedure, since Marčenko established his theorem by assuming that the given two sets of numbers were two spectra of some Sturm–Liouville system, and his concern was to find the variables h, H_1, H_2, and $q(x)$ which defined the system.

APPENDIX III

Research Problems

There are many open problems in the theory of linear networks. Some have a long history of evading solution, while others are relatively new and have been prompted by the emergence of new physical effects or devices. We will describe some of these problems and in certain cases indicate an approach which hopefully might lead to their solution. To the author's knowledge, none of these problems has been completely solved, and by no means are they to be considered as exercises. In fact, our purpose in describing these problems is to indicate the broad range of ideas and challenges that face the network theorist, and also to encourage others to contribute to the area.

GENERAL PROBLEMS

Time-Varying Networks

1. Develop procedures for determining if a time-varying network is passive, i.e., develop conditions for determining whether the kernel $K(x, y) = \delta(x - y) - \int h(t, x)h(t, y)\, dt$ is positive in the sense that $\iint K(x, y)\varphi(x)\varphi(y)\, dx\, dy \geq 0$. The outline presented in Section 2.5 suggests a distributional approach that would focus attention on the eigenvalues of $K(x, y)$. There are, however, some indications that the two-dimensional Laplace transform might enable one to rephrase the problem into a more tractable form.

2. Consider those time-varying networks whose terminal behavior is time-invariant, with the major emphasis being on the realization of active networks. For example, consider a time-varying transformer defined by $v_2 = \alpha e^{t/RK} v_1$ and $i_2 = (-1/\alpha)e^{-t/RK}$, and assume that the second port is terminated in a capacitor with capacitance $C(t) = (K/\alpha^2)e^{-2t/RK}$. Then one may show directly that the resulting one-port is time-invariant and its admittance is $(-1/R) + pK$, i.e., the network is equivalent to a negative resistor

shunted by a fixed capacitor. A basic question in this area is the following: given a network composed of time-varying lumped elements, can we realize an input impedance of $-R$? If the answer is no, then what limitations exist on these input impedances, assuming they are time-invariant, and, in particular, what properties do they possess if the variations of the elements are bounded functions of time?

Complex Normalization Techniques

3. Develop alternate conditions for Theorem 2.17, i.e., conditions which will enable one to determine if a given matrix is the scattering matrix of a passive system normalized to a given $[Z]$ matrix. Since the conditions of the present Theorem 2.17 form the basis of our approach to such problems as broadband matching, alternate and perhaps more tractable conditions would alter our viewpoint on this and many other problems.

4. If the matrix $[Z]$ is not diagonal, but still positive-real and rational in p, then one can still normalize a scattering matrix to $[Z]$ using Rohrer's technique [Ro1]. One problem is to obtain an analogous Theorem 2.17. Another problem is to extend these techniques to those $[Z]$ that are not positive-real. Since one of the basic problems here is to obtain a factorization in the form $[Z(p)] + [Z(-p)]^T = [H(-p)]^T[H(p)]$, one might have to accept irrational $[H(p)]$. A real-frequency ($p = j\omega$) procedure has been developed [Wo9] which may indicate a possible direction to the more difficult, but essential, complex-frequency problem.

Darlington's Problem

5. The example given in Section 3.1.2 demonstrates that in general a positive-real function cannot be realized by a lossless two-port terminated in a resistor, i.e., we cannot always isolate the dissipation in a single element. It is of physical interest to know what the largest class of positive-real functions is for which Darlington's result still holds. For example, if the impedance is meromorphic, does the result hold? The analogous question in active networks is also of interest, i.e., what is the largest class of functions that can be realized using a lossless three-port, one positive resistor, and one negative resistor?

Activity, Causality, and Stability

6. If we terminate a uniform lossless transmission line of electrical length Δ in a resistor that is equal to the *negative* of the characteristic impedance of the line, then the only bounded solutions that exist are those for which the

voltage transfer factor is $e^{+\Delta p}$, i.e., the impulse response of the resulting two-port is $\delta(t + \Delta)$. Since Δ is positive, we conclude that the system is *not* causal. If we restrict our attention solely to lumped networks, then even though they are active we can always find solutions that correspond to causal systems, i.e., a rational function of p can always be identified as the Laplace transform of a function of time that vanishes for $t < 0$. The simple example given previously demonstrates that the interconnection of causal elements can result in a noncausal system if we demand that bounded solutions exist. Thus in the analysis of distributed networks we must verify the causality of the system, i.e., the compatibility of the individual elements in our network. Once the question of causality has been resolved we may turn our attention to the stability question. But here again when we consider distributed networks or irrational functions we must exercise caution. For example, the function $e^{1/p}/p$, which is analytic in $\operatorname{Re} p > 0$ and bounded in any half-plane $\operatorname{Re} p \geq \sigma_0 > 0$, is the Laplace transform of the Bessel function $I_0(2\sqrt{t})u(t)$, which is zero for $t < 0$ but behaves like $e^{+\sqrt{t}}$ as $t \to +\infty$. Such examples indicate the need for a reevaluation of our stability criteria when distributed systems are considered. A basic problem on which all of these examples bear is the need for an axiomatic theory of active networks. Perhaps one should first consider the problem of obtaining meaningful definitions of stability and then discuss axioms for stable active systems.

Interpolation

7. Youla and Saito have shown [Yo8] that a positive-real function exists such that $Z(p_i) = Z_i$ and $Z(p_i^*) = Z_i^*$, $i = 1, \ldots, n$, where $\operatorname{Re} p_i > 0$, if and only if the $2n \times 2n$ matrix A, with $A_{rk} = (Z_r^* + Z_k)/(p_r^* + p_k)$ and p_i^* considered as a separate point, is nonnegative-definite. The algorithm presented in Section 3.2 provides an indirect test for determining if such an interpolating function exists. But this algorithm is of independent interest because it also allows one to consider the problem of interpolating to

$$Z(p_i) = Z_i, \qquad \frac{dZ}{dp}(p_i) = Z_i^{(1)}, \ldots, \frac{d^nZ}{dp^n}(p_i) = Z_i^{(n)}$$

(assuming that all the derivatives up to $Z_i^{(n)}$ are prescribed). The first problem is to extend the Youla–Saito result to obtain necessary and sufficient conditions for the case where derivatives are specified. The second and more general problem is to extend the results to include the case where we wish to interpolate to prescribed $n \times n$ matrices with positive-real matrices at points in $\operatorname{Re} p > 0$. A tool for demonstrating the sufficiency of such conditions might be found in the n-port cascade theory of Youla (see [Yo3] and [Ne2]).

LUMPED NETWORKS

8. Perhaps the most outstanding problem in the area of lumped networks is the synthesis of such networks without the use of transformers. Based on the representation of the scattering matrix of such systems as given in Theorem 3.3, we see that the problem can be reduced to the question of determining a lossless network with a scattering matrix

$$\hat{S} = \begin{bmatrix} S_{11} & S_{12} \\ S_{21} & S_{22} \end{bmatrix}$$

in which no transformers are present. The first problem is then one of finding necessary and sufficient conditions such that a given $\hat{S}$ is the scattering matrix of *wires*. One may then phrase the n-port resistor network problem, using Theorem 3.3, as follows: under what conditions can a given $n \times n$ matrix be the S_{11} partition of the scattering matrix of wires? This approach breaks the problem into seemingly simpler problems, i.e., a network of wires has a commonality that is analogous to a resistor network composed only of 1–Ω resistors. Moreover, the scattering matrix allows us to focus directly on the topological interconnecting problem, i.e., the use of wires versus transformers.

9. For both pedagogical interest and because it might indicate alternate representations it would be highly desirable to have a direct algebraic proof of Theorem 3.3. Moreover, since the proof would probably indicate a method of identifying the partitioned scattering matrix $[\hat{S}]$ of the lossless network of wires, transformers, and gyrators, the proof would also provide a direct synthesis procedure by reducing the problem to the synethesis of the real, constant, lossless scattering matrix $[\hat{S}]$.

10. In many microelectronic applications it seems possible to obtain elements whose terminal performance is close to that of a gyrator. Since two cascaded gyrators can be made equivalent to a transformer, there is considerable interest in the synthesis of lumped networks using gyrators in place of transformers. Moreover, since inductors can be realized as the input impedance of a gyrator that is terminated in a capacitor, one is led to the study of lumped networks composed of resistors, capacitors, and gyrators, and the pertinent question is the minimum number of gyrators required.

11. The problem of broadband matching an arbitrary Z_L and Z_g is still unsolved, and the dual interpolation problem which results from the use of complex normalization techniques (see Section 3.3.2) provides one avenue of approach. However, problem 3 listed above also bears on this question, since it might yield alternate formulations of the problem.

12. Techniques have been developed for the study of active one-port devices and their incorporation into amplifiers (see [Yo9] and [Ku1]), but the problems involved in the study of active n-ports have not been resolved.

The basic approach is to consider a passive $(n + 2)$-port to act as a coupling or matching network between the source, load, and active n-port. If one can extend the complex normalization technique to active n-ports (see problem 4), then this would provide a means for attacking the problem. A real-frequency normalization is available [Wo9], but when one considers the question of stability it becomes apparent that a complex-frequency technique is necessary. One may also consider simpler network configurations. For example, if we consider active two-ports, then we might assume that the coupling network consists of two uncoupled passive two-ports that are cascaded on either side of the active network. This configuration is of practical interest in the microwave region, but if a reasonably simple procedure can be found to design stable amplifiers with this configuration it would be attractive in all regions of the spectrum.

DISTRIBUTED NETWORKS

Synthesis of Smooth Lines with $dV/dz = -Z(z, p)I$ and $dI/dz = -Y(z, p)V$

13. The ultimate approach to the design of distributed networks is through the design of the materials in which the electromagentic fields are propagating. Since the properties of the material can be reflected most directly, in the case of transverse wave propagation, into properties of $Z(z, p)$ and $Y(z, p)$, we may formulate physically significant network problems in terms of these variables. We have shown that when Z and Y possess continuous partial derivatives with respect to z and p in the entire finite p plane the terminal properties of these networks will be meromorphic functions of p. One would like to know under what conditions isolated essential singularities or branch points are introduced into these terminal variables and by what classes of Z and Y. One would also like to be able to establish equivalence classes of lines in terms of certain functions of Z and Y. For example, when $Z = pL(z)$ and $Y = pC(z)$ it was shown that all lines having the same electrical length, $\int_0^l (LC)^{1/2}\, dz$, and the same local characteristic impedance function, $\{(L/C)(y)\}^{1/2}$, will have the same terminal properties, and, moreover, the local characteristic impedance variation can be uniquely determined from the terminal performance. Thus in the case of more general Z and Y we must first look for equivalence classes and then phrase the inversion or synthesis problem in terms of those functions of Z and Y that are appropriate.

14. When Z and Y are positive-real functions of p for each fixed z we have shown that the resulting network formed by a finite length of line will be passive. However, it is not known whether this condition is necessary, i.e., there may exist certain Z and Y which are not positive-real for each z along a given length of line, but the network is still passive when viewed at its

terminals. Thus one would like to have necessary and sufficient conditions for the passivity of such networks. Such conditions would then provide a criterion in the search for activity in such materials.

15. Develop "reasonable" synthesis procedures for the lossless (L, C) line. In particular, one might consider using the Marčenko algorithm as the basis for an iterative procedure if one could be sure that after a finite number of iterations the resulting approximation to the actual line is realizable, i.e., has a positive $Z_0(y)$. An alternate approach would be the use of stepped lines [lines with piecewise constant $Z_0(y)$] to approximate the smooth line. Here we must find a way of approximating the terminal performance (as we have shown, one need only deal with the input reflection factor) of the smooth line by the stepped line. In particular, we would want an approximation that converges to the smooth-line performance as the number of steps is allowed to approach infinity. As a preliminary result in the study of such convergence, it would be highly desirable to have a simple means of determining both the maximum and minimum values of $Z_0(y)$ from the input reflection factor of a stepped line (at present, one must actually carry out the synthesis procedure to obtain these bounds).

16. Is the following conjecture true? $F(p)$ is the S_{21} element of a smooth lossless line if $F(p)$ is analytic and bounded by 1 in $\operatorname{Re} p > 0$, $F(0) = 1$, and $1/F(p)$ is an entire function of exponential growth. In view of Theorem 4.8 we see that in order to prove this conjecture we must show that we can always find the regular all-pass specified there.

17. Studies of electromagnetic waves in plasmas (particularly solid-state plasmas) suggest that a reasonable network model for such materials is a transmission line in which

$$\partial V/\partial z = -Z(z, p)I$$

and

$$(\partial I/\partial z) + J(z, p)I = -Y(z, p)V.$$

One can show that such a network is nonreciprocal if $J \neq 0$, and, moreover, that the network may be active even though Z and Y are positive-real functions of p. However, the terminal properties of such networks have not been fully explored, and, in particular, one would like to have necessary and sufficient conditions for their passivity (activity).

OPTIMIZATION TECHNIQUES

18. Since most numerical optimization procedures seek local minima, it would be highly desirable to establish that certain classes of networks together

with their performance functions have only one such local minima. However, this problem is in general intractable, at least in the current state of optimization theory, and one must be willing to study sufficiently restricted classes of networks in order to obtain any such result.

Many of the directions which recent investigations are taking in this area are summarized in the special issue of the Proceedings of the IEEE on Computer-Aided Design, Vol. 55, No. 11, November 1967.

Glossary of Symbols

$\mathbf{a}$	an $n \times 1$ matrix or column vector
$[A]$	an $r \times q$ matrix with elements A_{jk}
$[A]^T$	the transpose of the matrix $[A]$
1_n	the unit $n \times n$ matrix with $A_{jk} = 0, j \neq k$, and $A_{jj} = 1$
$[S_z]$	see p. 65
$r[S]$	see p. 91
$\delta[S]$	see p. 91
$j = \sqrt{-1}$	
f^*	the conjugate of f
$\tilde{f}(t) = f^*(-t)$	
$\|f\|$	the magnitude of f
$\mathrm{Re}\, f$	the real part of f
$\mathrm{Im}\, f$	the imaginary part of f
$\delta(t)$	the delta function defined by $\int f\delta\, dt = f(0)$
$\mathscr{L}(f)$	the Laplace transform of f
$p = \sigma + j\omega$	the complex variable of the Laplace transform
$\mathscr{F}(f)$	the Fourier transform of f
$j\omega$	the variable of the Fourier transform
$u * v$	the convolution between $u(t)$ and $v(t)$
$f_n \xrightarrow{w} f$	the weak convergence of f_n to f
$\nabla \times$	the curl operator
$\nabla \cdot$	the divergence operator
∇^2	the Laplacian operator
$O(1/p^n)$	$f(p) = O(1/p^n)$ as $\|p\| \to \infty$ if $\|f(p)\| < K/\|p^n\|$, for some fixed K when $\|p\|$ is sufficiently close to $+\infty$
C_0^∞	p. 2
$\mathscr{D}$	p. 3
$\mathscr{D}'$	p. 2
$\mathscr{D}'_{L_p}$	p. 210
$\mathscr{D}'_{t_0}$	p. 10
$\mathscr{S}$	p. 2
$\mathscr{S}'$	p. 2
$\mathscr{E}'$	p. 2

$pv \int f(x)\, dx = \lim_{\varepsilon \to 0} [\int_{|\varepsilon|}^{\infty} f(x)\, dx + \int_{-\infty}^{-|\varepsilon|} f(x)\, dx]$; the principal value of the integral

Bibliography

[Ba1] Balakrishnan, A. V., and L. W. Neustadt, eds., "Computing Methods in Optimization Problems." Academic Press, New York, 1964.

[Be1] Beltrami, E. J., and M. Wohlers, "Distributions and the Boundary Values of Analytic Functions." Academic Press, New York, 1966.

[Bo1] Boas, R. P., "Entire Functions." Academic Press, New York, 1954.

[Br1] Bremermann, H. J., "Distributions, Complex Variables, and Fourier Transforms." Addison-Wesley, Reading, Massachusetts, 1965.

[Br2] Briggs, R., "Electron-Stream Interactions with Plasmas." M.I.T. Press, Cambridge, Massachusetts, 1964.

[Br3] Brune, O., Synthesis of a finite 2-terminal network whose driving point impedance is a prescribed function of Frequency, *J. Math. and Phys.* **10**, 191 (1930).

[Ca1] Carlin, H. J., Cascade transmission line synthesis, MRI Rept. No. 889–61, Polytechnic Institute of Brooklyn, 1961.

[Ca2] Carlin, H. J., and A. B. Giordano, "Network Theory—An Introduction to Reciprocal and Nonreciprocal Circuits." Prentice-Hall Englewood Cliffs, New Jersey, 1964.

[Ca3] Carlin, H. J., and D. C. Youla, Network synthesis with negative resistors, in *Proc. Polytechnic Inst. of Brooklyn Symp. on Active Networks and Feedback Systems*. Polytechnic Press, Brooklyn, 1960.

[Ca4] Cauer, W., Die verwirklichung von wechselstrom widerstandon vorgeschriebeuer frequenz abhangigkeit, *Arch. Elektrotech.* **17**, No. 4, 355–388 (1926).

[Ca5] Cauer, W., The Poisson Integral for functions with Positive real parts, *Bull. Amer. Math. Soc.* **38**, 713–714 (1932).

[Co1] Courant, R., and D. Hilbert, "Methods of Mathematical Physics," Vol. 1. Wiley (Interscience), New York, 1953.

[Da1] Darlington, S., Synthesis of reactance 4 poles, *J. Math. and Phys.* **18**, 257–353 (1939).

[Fo1] Foster, R. M., A reactance theorem, *Bell System Tech. J.* **3**, 259–267 (1924).

[Ge1] Gel'fand, I. M., *et al.*, "Generalized Functions," Vols. 1–5. Academic Press, New York, 1964–1968.

[Ge2] Gel'fand, I. M., and B. M. Levitan, On the determination of a differential equation from its spectral function, *Izv. Akad. Nauk. SSSR Ser. Mat.* **15**, 309–350 (1951); English transl., *Amer. Math. Soc. Transl.* (2) **1**, 253–304 (1955).

[Ge3] Gel'fand, I. M., and N. Vilenkin, "Generalized Functions—Applications of Harmonic Analysis," Vol. 4. Academic Press, New York, 1964.

[Hi1] Hille, E., "Analytic Function Theory," Vols. I and II. Ginn, New York, 1959.

[Ho1] Ho, C., and N. Balabanian, *IEEE Trans. Circuit Theory* **CT–14** (June, 1967).

[In1] Ince, E. L., "Ordinary Differential Equations." Dover, New York, 1956.

[Ko1] Koga, T., Synthesis of finite passive *n*-ports with prescribed positive real matrices of several variables, *IEEE Trans. Circuit Theory* **CT–15** (March, 1968).

[Ko2] Kolmogorov, A. N., and S. V. Fomin, "Elements of the Theory of Functions and Functional Analysis," Vol. I. Graylock Press, Rochester, New York, 1957.

[Ko3] König, H., and A. Zemanian, Necessary and sufficient conditions for a matrix distribution to have a positive-real Laplace transform, *SIAM J.* **13**, No. 4, 1036–1040 (1965).

[Ku1] Kuh, E. S., and R. A. Rohrer, "Theory of Linear Active Networks." Holden-Day, San Francisco, 1967.

[Le1] Levin, B. Ja., "Distribution of Zeros of Entire Functions." *Amer. Math. Soc. Transl.* **5**. Amer. Math. Soc., Providence, Rhode Island, 1964.

[Li1] Liusternik, L. A., and V. J. Sobolev, "Elements of Functional Analysis." Ungar, New York, 1961.

[Lo1] Louisell, W., "Coupled Mode and Parametric Electronics." Wiley, New York, 1960.

[Ma1] Marčenko, V. A., Some questions of the theory of one-dimensional linear Differential operators of the second order, I, II, *Trudy Moskov. Mat. Obšč.* **1**, 327–420 (1952); **2**, 3–83 (1953) [in Russian].

[Mo1] Moyer, H. G., M. R. Wohlers, and R. E. Kopp, Computational aspects of the design of optimal distributed components, in *Proc.* 1966 *Electronic Components Conference, Washington, D. C.*

[Na1] Natanson, I. P., "Theory of Functions of a Real Variable," Vol. I. New York, 1961.

[Ne1] Nevanlinna, R., "Eindeutige Analytische Funktionen." Springer-Verlag, Berlin, 1936.

[Ne2] Newcomb, R. W., "Linear Multiport Synthesis." McGraw-Hill, New York, 1966.

[Ne3] Newcomb, R. W., and B. D. O. Anderson, Functional analysis of linear passive networks, to appear.

[Ne4] Newcomb, R. W., and D. A. Spaulding, The time-variable scattering matrix, *IEEE Proc.* **53**, No. 6, 651–652 (1965);

[Po1] Pontryagin, L. S., V. G. Boltyanskii, R. V. Gamkrelidze, and E. F. Mishchenko, "The Mathematical Theory of Optimal Processes." Wiley (Interscience), New York, 1963.

[Ri1] Richards, P. I., Resistor transmission-line circuits, *Proc. IRE* **37**, No. 2, 217–219 (1948).

[Ro1] Rohrer, R. A., The scattering matrix: normalized to complex *n*-port load networks, *IEEE Trans. Circuit Theory* **CT–12**, 223–230 (1965).

[Sc1] Schoeffler, J., Impedance transformations using lossless networks, *IRE Trans. Circuit Theory* **CT–8**, 131–137 (1961).

[Sc2] Schwartz, L., Theorie des Noyaux, in *Proc. Intern. Congr. Mathematicians, Cambridge, Massachusetts*, 1950, pp. 220–230.

[Sc3] Schwartz, L., "Theorie des Distributions," Vols. I and II. Hermann, Paris, 1957–59.

[Sm1] Smilen, L. I., Interpolation on the real frequency axis, 1965 *IEEE Intern. Conv. Rec.* **13**, Pt. 7, 42–50 (1965).

[St1] Streater, R. F., and A. S. Wightman, "PCT, Spin and Statistics, and All That." Benjamin, New York, 1964.

[Sz1] Szegö, G., "Orthogonal Polynomials," Amer. Math. Soc. Coll. Pub., XXIII (rev. ed.). Amer. Math. Soc.,'Providence, Rhode Island, 1959.

[Ti1] Titchmarsh, E. C., "Introduction to the Theory of Fourier Integrals," 2nd ed. Oxford Univ. Press, London and New York, 1948.

[Wo1] Wohlers, M. R., Interpolation on the real-frequency axis with positive-real or bounded-real functions, *IEEE Trans. Circuit Theory* **CT–11**, 498–499 (1964).

[Wo2] Wohlers, M. R., A distributional study of the real frequency behavior of passive systems, Proc. Allerton Conference, Univ. Illinois (1965).

[Wo3] Wohlers, M. R., Complex normalization of scattering matrices and the problem of compatible impedances, *IEEE Trans. Circuit Theory* **CT–12**, 528–535 (1965).

[Wo4] Wohlers, M. R., On gain-bandwidth limitations for physically realizable systems, *IEEE Trans. Circuit Theory* **CT–12**, 329–333 (1965).

[Wo5] Wohlers, M. R., Propagation of guided waves in linear time-invariant media: a network approach, 1965 *IEEE Intern. Conv. Rec.* **13**, Pt. 7, 30–41 (1965).

[Wo6] Wohlers, M. R., A realizability theory for smooth lossless transmission lines, *IEEE Trans. Circuit Theory* **CT–13**, 356–363 (1966).

[Wo7] Wohlers, M. R., On electromagnetic gain mechanisms in solid state plasmas, *IEEE Proc.* **55**, No. 7, 1230–1231 (1967).

[Wo8] Wohlers, M. R., A realizability theory for smooth lossless transmission lines—less transmission lines—Part II, *IEEE Trans. Circuit Theory* **CT–14**, 442–444 (1967).

[Wo9] Wohlers, M. R., Scattering matrices normalized to active *n*-ports at real frequencies, to appear.

[Wo10] Wohlers, M. R., and E. J. Beltrami, Distribution theory as the basis of generalized passive-network analysis, *IEEE Trans. Circuit Theory* **CT–12**, 164–169 (1965).

[Yo1] Youla, D. C., Representation theory of linear passive networks, MRI Rept. No. R-655-58, Polytechnic Institute of Brooklyn, 1958.

[Yo2] Youla, D. C., An extension of the concept of scattering matrix, *IEEE Trans. Circuit Theory* **CT–11**, 310–312 (1964).

[Yo3] Youla, D. C., Cascade synthesis of passive *n*-ports, Polytechnic Inst. of Brooklyn, Tech. Rept. RADC-TDR-64-332, August 1964.

[Yo4] Youla, D. C., A new theory of cascade synthesis, *IRE Trans. Circuit Theory* **CT–9**, 244–260 (1961).

[Yo5] Youla, D. C., A new theory of broadband matching, *IEEE Trans. Circuit Theory* **CT-11**, 30–50 (1964).

[Yo6] Youla, D. C., The synthesis of networks containing lumped and distributed elements—Part I, in *Proc. Polytechnic Inst. of Brooklyn Symp. on Generalized Networks*. Polytechnic Press, Brooklyn, 1966.

[Yo7] Youla, D. C., L. J. Castriota, and H. J. Carlin, Bounded real scattering matrices and the foundations of linear passive network theory, *IRE Trans. Circuit Theory* **CT-6**, 102–124 (1959).

[Yo8] Youla, D. C., and M. Saito, Interpolation with Positive-Real Functions, *J. Franklin Inst.* **284**, 77–108 (1967).

[Yo9] Youla, D. C., and L. I. Smilen, Optimum negative resistance amplifiers, in *Proc. Polytechnic Inst. of Brooklyn Symp. on Active Networks and Feedback Systems*. Polytechnic Press, Brooklyn, 1960.

[Ze1] Zemanian, A., "Distribution Theory and Transform Analysis." McGraw-Hill, New York, 1965.

[Ze2] Zemanian, A., An *n*-port realizability theory based on the theory of distributions, *IEEE Trans. Circuit Theory* **CT-10**, 265–274 (1963).

Index